The Computer Videomaker Handbook

The Computer Videomaker Handbook

A Comprehensive Guide to Making Video

Second Edition

From the Editors of Computer Videomaker Magazine

Introduction by Matt York, Publisher/Editor

Section Introductions
by Stephen Muratore, Editor in Chief

Edited by Chuck Peters, Sandi Kilcollins, Wendy Nichols

Focal Press

Boston   Oxford   Auckland   Johannesburg   Melbourne   New Delhi

Focal Press is an imprint of Elsevier.

 This book is printed on acid-free paper.

Library of Congress Cataloging-in-Publication Data
The computer videomaker handbook; a comprehensive guide to making video/from the editors of Computer videomaker magazine; introduction by Matt York.-2nd ed.
    p.   cm
Includes index.
ISBN 0-240-80435-X (pbk.: alk. paper)
   1.   Video recording-Handbooks, manuals, etc. 2. Digital video-
        Handbooks, manuals, etc. 3. Television-Production and direction-
        Handbooks, manuals, etc. I. Computer videomaker.
TR840.C66 2001
778.59-dc21   00-054350

British Library Cataloguing-in-Publication Data
A catalogue record for this book is available from the British Library.

The publisher offers special discounts on bulk orders of this book.
For information, please contact:
Manager of Special Sales
Elsevier Science
200 Wheeler Road
Burlington, MA 01803
Tel: 781-313-4700
Fax: 781-313-4802

For information on all Focal Press publications available, contact our World Wide Web homepage at http://www.focalpress.com

10 9 8 7 6
Printed in the United States of America

This book is dedicated to Thomas Jefferson for his commitment to pluralism, diversity and community. He would be happy to see readers of this book exercising their freedom of the electronic press.

Civil liberty functions today in a changing technological context. For five hundred years a struggle was fought, and in a few countries won, for the right of people to speak and print freely, unlicensed, uncensored and uncontrolled. But new technologies of electronic communication may now relegate old and freed media such as pamphlets, platforms and periodicals to a corner of the public forum. Electronic modes of communication that enjoy lesser rights are moving to centerstage.

Ithiel de Sola Pool
Technologies of Freedom
(Harvard University Press, 1983)

Contents

Preface

The editors of *Computer Videomaker Magazine* are proud to present the second edition of *The Computer Videomaker Handbook*. This edition of the *Handbook* still takes you from the planning of videos all the way through the final phases of their production, but now it takes you a couple of steps further than that. Not only have we updated and improved all the original sections, sprinkling new chapters and illustrations throughout; we have also added a whole new section.

Part VI of the 2nd Edition gives you everything you need to know to distribute your finished videos on CD and DVD discs as well as on the Internet. Discs and the 'Net promise to be the video distribution media of the future. This book will help you get started using these media now.

Finally, we have supplemented this book with a companion Web page all its own. The Web page contains links to streaming video clips that illustrate various techniques discussed in the *Handbook*. If the site holds a streaming video relevant to a given chapter, you will find this icon

near the title on the first page of that chapter. To access the page of clips, go to http://www.videomaker.com/handbook. When asked for a password, type "effects" without quotation marks.

The *Handbook* can be read cover to cover as a comprehensive course on the art and science of videography. It can be used also as a handy field reference: the simple organization and index should make it easy for you to find answers to your questions, and the glossary will define the arcane terms of the art.

Whether you are a weekend hobbyist or a practicing professional, the *Computer Videomaker Handbook* will help you use your talent and your equipment to its fullest, bringing inspiration, excitement and wonder to your audience. Which of course will bring applause, satisfaction and pride to you.

Introduction: Playing the Game

Matt York
Publisher of Computer Videomaker Magazine

Many people spend 10% of their waking day watching TV. During that period they receive a whole lot of information. Their opinions are shaped, their biases are changed or affirmed and their perceptions of the world (and how they fit into it) are altered. I believe that all of this happens even during the process of laughing at a sitcom. Television has a profound effect on people and, consequently, on the world at large.

It is exciting to know that we can all be players in the "game" of TV production. The video equipment that you can buy at retail stores today enables you to produce a TV "show" in less time and of higher quality than ever before. Just a few short years ago, this level of video production wasn't available even to a "real" TV station in many small towns. Back then, digital video camcorders and computer editing were unknown.

When you consider that many people spend 10% of their waking day in front of a TV that *could* present a TV show that *you* produced, the ramifications are fascinating. When millions of people pay attention to just one person we call it a monarchy. All of the people in the land pay attention to their king. When millions of people pay attention to each other, we call that a democracy. In fact, our society could be a democracy beyond the wildest dreams of the founding fathers of the United States. Jefferson and Franklin never imagined instant text, audio and video communication amongst the citizenry. Their expectation was government-subsidized delivery of messages sent via paper.

We have the potential to live within a democracy beyond the wildest dreams of the founding fathers, but we aren't living there yet. In our world today, billions of people are paying attention to hundreds of people. By this I mean that *most* TV programs have been produced by relatively few large corporations. The power to decide which programs get produced is concentrated within a small group of TV industry executives. The TV shows chosen, made and broadcast by these corporations are viewed, not only by American citizens, but by people all over the world, billions of them. This isn't exactly what Jefferson had in mind. But all this is changing quickly.

The Internet is enabling people to communicate quickly, inexpensively and easily-one by one-by way of e-mail and chat. The 'Net is allowing thousands of groups to create Web sites for hundreds of thousands of people to see. On these Web sites, text and graphics are now accompanied by video. The video available via the Internet is not quite as good as that typically seen on a TV, but it is getting better every year. These Web sites also allow video producers to accept orders for VHS videocassettes that can be mailed anywhere in the world so that "average quality video" can be viewed by anyone with a VCR.

That's an exciting reality, when thousands of people pay attention to thousands of people and their ideas. In the years to come politicians will have to change tactics and so will public relations firms, to attempt to garner as much "media power" as they held in the past. During the next hundred years, the power to persuade will be distributed away from large power businesses and into the hands of anyone that can hold a camcorder, speak into a microphone or write a web page. The most exciting thing of all is that *you* are one of these people.

PART I
A Peek Under the Hood

See video clips at www.videomaker.com/handbook.

Take a Look at Your Tools

Making a video is taking a journey. It requires a vehicle (electronic equipment) that is well tuned and fueled. It requires maps and compasses (pre-production planning documents). Production techniques are the "driving skills" that will determine, not only whether and how fast you will arrive at a finished video, but whether you will arrive in style. A journey is not truly complete until it is remembered and recounted. In video, we re-member, or reassemble the experience of the journey in a process called "editing" or "post-production." The recounting is done when our videos meet the eyes of their intended audiences, through various means of distribution.

The *Computer Videomaker Handbook* can guide you from the beginning to the end of this journey. You will notice that it is divided into six parts, each of which marks one of the major phases needed for a complete trip. You can read it from

beginning to end to get an overview of the trip, or open to the section you find most helpful. We recommend you keep it with you, in your camera bag and edit bay, throughout the process. It will be there for you when you need a quick definition of a bit of jargon, or a refresher in some technique. Make it *your* handbook.

When looking for additional material on any of the subjects covered herein, don't forget to visit www.videomaker .com. The search engine for the site is bound to deliver helpful information on most of the keywords you might enter.

In the first part of the *Handbook*, we will take a technological look at the video production tools themselves, to gain a basic understanding of how they work and what they can do for you. *Wait! Don't flip ahead just yet.* Do not skip this part if you don't consider yourself "technical." Don't excuse yourself by saying, "I don't need to know how the tools work; I just need to know how to use them."

In fact, if you are *not* "technical," this section is written especially for you. Would you set out on a journey without any idea of how your car works or what it can do? You don't need to be a mechanic to drive, but it helps to know what the steering wheel and brakes do. When you want to buy a new car, doesn't it help to know which features are worth looking for and which are hype? Take a peek under the hood of your video gear with us. We promise to speak plain English about things really helpful to know. You'll quickly become a smarter shopper and a producer of better moving pictures.

1
All About Lenses

Jim Stinson

Without passing through a lens, the light falling on your camcorder's CCD would be as empty of information as a flashlight beam. The camcorder's lens converts incoming light from a gaggle of unreadable rays to an ordered arrangement of visual information—that is, a picture. It's the lens, then, that makes video imaging possible. Without it, your camcorder would record an image of blank white light.

All videos are successions of individual images, each made by forcing light to form a recognizable picture on a flat surface. You can do it with just a tiny hole in the wall of a darkened room, but it's easier to use a lens.

A lens does far more than just render light into coherent images; it also determines the visual characteristics of those images. For this reason, every serious videographer should know how lenses work and how to use them to best advantage.

A Little Background

As long ago as ancient Greece, people noticed that when they put a straight pole into clear water, the part of the pole below the water line seemed to bend. The mathematician Euclid described this effect in 300 BC. But it wasn't until 1621 that the scientist Willebrord Snell developed the mathematics of diffraction. Diffraction is the principle stating the following: when light passes from one medium to another—say from water to air or air to glass—it changes speed. And when light hits a junction between two media at an angle, the change in speed causes a change in direction.

Lenses, which refract light in an orderly way, were perhaps unintended side effects of glass blowing: if you drop a globule of molten glass onto a smooth, plane surface it will naturally cool into a circle that's flat on the bottom and slightly convex on top—an accidental lens. Look through this piece of junk glass and behold: things appear larger.

Now, hold the glass between the sun and a piece of paper and you can set the sheet on fire—but only if the glass-to-paper distance is such that all the sun's rays come together at a single point on the paper.

At some unknown moment somebody thought, "Hmmn, if it works with the sun, maybe it'll work with other light sources, too." In a darkened room, this someone held the glass between a piece of paper and an open window. Sure enough, at a certain lens-to-paper distance, a pinpoint of light appeared.

But then a bizarre thing happened. When the experimenter slowly increased the glass-to-paper distance, an actual picture of the window appeared, small, to be sure and upside down, but so detailed that they could see that tree outside, framed in the opening. (You can try this yourself with a magnifying glass.)

Back to the Present

If you've ever seen a cutaway diagram of a modern zoom lens, you have a grasp on how far we've come from that first accidentally dropped blob of glass.

The camcorder zoom may contain a dozen pieces of glass or more. Some of these permit the lens to zoom, some make the lens more compact by "folding" the light rays inside it, and some correct inescapable imperfections called lens aberrations.

But since you didn't sign up for an advanced physics seminar here, we'll pretend that the camcorder zoom is a simple, one-element lens. We can do this because the basic idea is exactly the same: when a convex lens refracts light, the light's rays converge at a certain distance behind the lens, forming a coherent image on a plane still farther back.

The plane on which the focused image appears is the *focal plane*; the place where the light rays converge is the *focal point*, and the distance from the focal point to the axis of the lens is the *focal length*. Note: contrary to common belief, the focal length is *not* the distance from the lens to the focal plane.

Your camcorder's image-sensing chip sits at the focal plane of the system, behind the actual lens.

Notice also that Figure 1.1 shows an additional measurement: *maximum aperture*, or, in plain language, the lens' ability to collect light. Get comfortable with lens *aperture*, *focus* and *focal length*, and you've got everything you need to know about camcorder lenses. So let's run through 'em.

Open Wide

The *aperture* of a camera controls how much light enters the lens. In one way, a lens is just like a window: the bigger it is, the more light it admits. But a lens isn't quite as simple as a window, because the

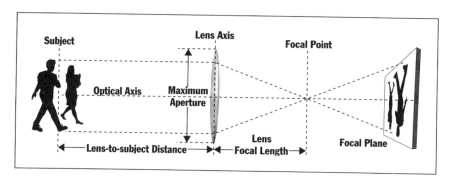

Figure 1.1 *The geometry of a simple lens.*

amount of light that gets in is also governed by its focal length (the distance from the lens to the focal point).

For this reason, you can easily determine maximum aperture—the ability of a lens to collect light. Use this a simple formula: *aperture = focal length divided by lens diameter.*

For example: if a 100 mm lens has a diameter of 50 mm, then 100 divided by 50 is 2. The lens' maximum aperture is 2, expressed as "f/2." Lens apertures are "f stops."

Since the amount of shooting light varies from dimly lit rooms to bright sunshine, all lenses have mechanical *iris diaphragms* that progressively reduce the aperture in brighter light. Your camcorder's auto exposure system works by using this diaphragm to change the lens' working aperture. In other words, the iris is changing the effective diameter of the lens.

These changes occur in regular increments called "stops," as noted. Each one-stop reduction in aperture size cuts the light intake in half. Most consumer camcorders fail to indicate these f stops. But some units—as well as most familiar single-lens reflex film cameras—indicate f stops by a string of cryptic digits: 1.4, 2, 2.8, 4, 5.6, 8, 11, 16, 22.

Why use these peculiar numbers to label f stops? Simple: long ago, lenses with maximum apertures of f/2 were very common, so f/2 became the starting point. F/1.4 is the square root of f/2; and if you look at the other f stop numbers you'll see that each is a multiple and/or root of another. (Some figures are rounded off: f/11 is not precisely a multiple of f/5.6.)

Just as confusing, these strange numbers appear to work backward. As the f stop number gets bigger, the aperture gets *smaller.* F/22 is the smallest common aperture and f/1.4 (or even 1.2) is the largest.

Why should you care how big the hole is in your camcorder lens? Because the working aperture has important effects on image quality and depth of focus. For critical applications, lenses create better images in the middle of their range of apertures. But for videographers, the crucial concern is the effect of aperture on focus.

Lookin' Sharp!

Before we can explain how aperture affects focus, we need to see what focus is and how the lens does it.

To start with, remember that the focal plane is the *one and only* plane on which the light rays create a sharp (focused) image. If you look at Figure 1.1 again, you'll see that the subject, the lens axis, the focal point and the focal plane are all in a fixed geometrical relationship. That is, you can't change one without affecting the others. You can't move the lens closer to the subject without changing the path of the light rays. And if you do that, you change the position of the focal plane.

In Figure 1.2A, the subject is a long distance from the lens, and its image appears sharply on the focal plane. Since the camcorder's CCD is on that plane, the recorded image is in focus.

Figure 1.2B shows what happens when you move closer to the subject. The geometry of the light rays moves the focal plane forward *away from the CCD.* The result? When the rays do hit the CCD they no longer form a sharp image. You're out of focus.

The solution: change the position of the lens to compensate for the shift in subject distance. As you can see from Figure 1.2C, doing this returns the focal plane to the CCD's position and the image is back in focus again.

This is exactly what happens in your camcorder lens. Lens elements move forward and backward to focus the incoming light on the CCD. Most camcorder zoom lenses feature *internal* focusing: the lenses move inside a fixed-length lens barrel. Most still cameras use *external*

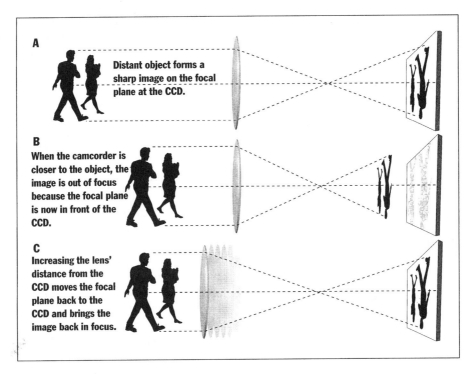

Figure 1.2 *How (and why) a lens' focus is changed.*

focusing: you can actually see the lens grow longer as its front element moves forward for closer focusing.

What's In Focus?

If you adjust the lens to focus on a subject near the camera, then the distant background will often go soft. That's because every lens at every aperture and focusing distance has what's called a certain *depth of field*. Here's how it works. Strictly speaking, the lens focuses perfectly only on one plane at a certain distance from it. Objects receding from that plane—or advancing from it toward the lens—are all technically out of focus.

But in reality, objects up to a certain distance behind or in front of this imaginary plane still appear sharp to the human eye. This sharp territory from in front of the focal distance to behind it is depth of field.

Two factors govern the extent of the depth of field: 1) the focal length of the lens and 2) the working aperture. Since we've already covered aperture, let's see how it affects depth of field.

Each drawing of Figure 1.3 represents a picture made with the same lens, at the same distance from the subjects, and focused on the same person, the woman. The only variable is the aperture. As you can see, the higher the f stop, the greater the depth of field.

In Figure 1.3A, the stop is very high (f/22) and all three subjects are sharp. In Figure 1.3B, the aperture widens to the middle of its range (f/5.6). Now the depth of field is more shallow and the man and the tree are at its front and back boundaries. They're starting to lose sharpness.

Open the aperture all the way to f/1.4 (Figure 1.3C) and the depth of field is quite narrow. Though the woman remains sharp, the man and the tree are

The Long and Short of It

The focal length of a lens affects three important aspects of the image: *angle of view*, *depth of field* and *perspective*.

The angle of view gives the lens its name.

In Figure 1.4, a wide-angle lens (here an angle of 85 degrees) includes a great deal of territory. A normal lens (here 55 degrees) is less inclusive; and a telephoto lens has a very narrow angle of view indeed (here 12 degrees). So, at any distance from the subject matter, the wider the lens angle, the wider the field of view.

Incidentally, the angles selected for Figure 1.4 are only typical examples. Each category—wide, normal and narrow (telephoto)—includes a range of angles. So while 12 degrees is a narrow angle, 9 degrees is also a narrow angle, though slightly more extreme.

As a videographer, you exploit the differences in lens angle of view all the time. For example: shooting a birthday party you may zoom out to your widest angle, to include more of the scene when the small room won't let you move the camcorder farther back from the action.

Going Soft

Earlier, we noted that lens aperture affects depth of field. Now let's see how lens *focal length* also affects depth of field.

As you can see in Figure 1.5, the wider the angle, the greater the depth of field.

In bright sunshine, a wide-angle lens will hold focus from a couple of feet to the horizon. At the other extreme, in dim light a telephoto lens may render sharp subjects through only a few inches of depth. Notice that we include the light conditions because aperture and focal length working together always govern depth of field. But the rule is, at *any* aperture, the wider the lens angle, the greater the depth of field, at any distance from the subject.

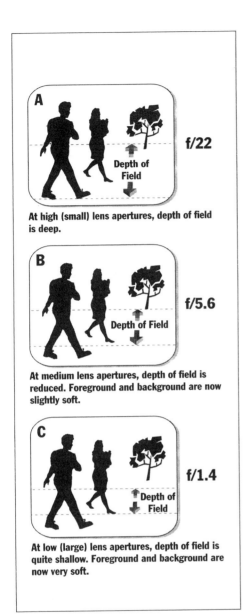

Figure 1.3 *Lens aperture affects the depth of field. (all shots made at the same focal length and focused on the woman)*

just blurs. Once again, the higher (smaller) the f stop, the greater the depth of field, and vice versa.

As noted above, depth of field is also governed by the focal length of the lens. But first, we need to see what that geometrical abstraction *focal length* really means to practical videographers.

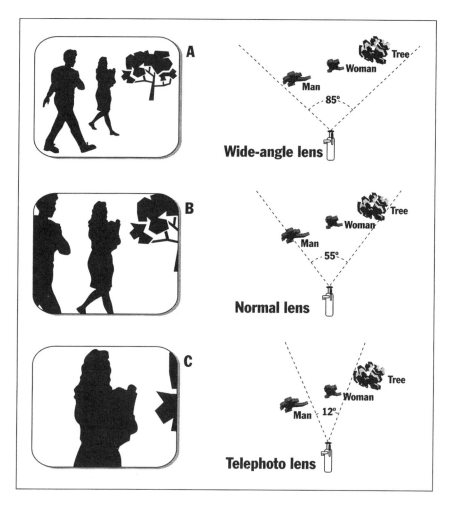

Figure 1.4 *Lens focal length affects angle view. (camera is the same distance from subjects in all shots)*

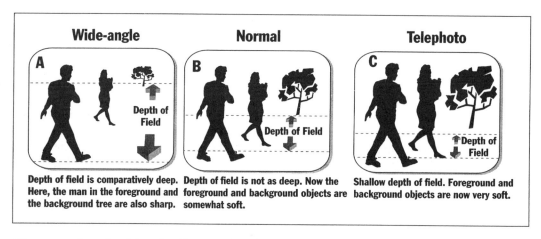

Figure 1.5 *Lens focal length affects the depth of field. (all shots made at the same aperture and focused on the woman)*

Take special note of that last phrase, *at any distance from the subject.* When some photographers can't get enough depth of field they think, "Hey, no problem: I'll increase my depth of field by going wide-angle."

Wrong! If you widen the angle you *will* increase depth of field, but you also reduce the size of the subject in the frame. To return it to its former size in the wide-angle view, you must move the camera closer. What's wrong with that? There's one last rule of focus we haven't mentioned yet: at *any* focal length (and any aperture too), the closer the lens is to the subject, the less depth of field in the image.

See the problem? Moving closer to compensate for the smaller image effectively wipes out the depth gained from going wide-angle. It's a wash.

We said that widening the angle decreases the subject size, and that leads us to the most dramatic effect that focal length has on the image: *perspective.*

Perspective and Focal Length

Perspective is the depiction of apparent depth—a phantom third dimension in a two-dimensional image.

In the real world, even people with only one functional eye can gauge distance, because the farther away objects are, the smaller they appear. Moreover, they diminish in size at a certain rate because of the geometry of the human optical system.

But other optical systems, such as camcorder lenses, may have very different geometries, and objects may shrink much faster or slower than they do in human vision. The perspectives of different lenses depend entirely on their focal lengths.

As you can see, wide-angle lenses exaggerate apparent depth.

Objects shrink quickly as they recede. Normal focal lengths imitate the moderate perspective of human vision (which, of course, is why we call them "normal").

Telephoto lenses reduce apparent depth. Background objects look much bigger and the space between them and the foreground appears compressed.

As the ground plans beside the drawings in Figure 1.6 show, you have to move the camera in order to achieve these different effects. As you change from wide-angle to telephoto, you must pull back so that the reference figure in the foreground (the man) remains the same size and in the same position in the frame. If you simply zoomed in from the first camera position, you would instead get the effect shown in Figure 1.4.

Wide-angle lenses can deliver very dramatic results. People and vehicles moving toward or away from the camera appear to hurtle past. A roundhouse punch swoops toward the lens like an incoming meteor.

But since they exaggerate depth, wide-angle lenses have drawbacks as well. Get too close to people's faces in wide angle and their noses will grow to elephant size.

On the opposite side, telephoto lenses can make great compositions on the screen by stacking up pictorial elements. For instance, if you want to dramatize congestion and pollution, get an extreme telephoto shot of a freeway at rush hour, viewed head-on. Because you're squeezing a mile's worth of cars into 100 yards of apparent depth, you make a bad problem look ten times worse.

Telephoto shots are great for suspense. Near the climax of Ferris Buehler's Day Off our hero must make it home through neighborhood backyards before his parents arrive. In one suspenseful telephoto shot, Ferris runs straight toward the camera—and runs, and runs, and runs—without seeming to make any progress. It's the telephoto focal length lens, of course, that compresses the distance he's actually covering.

What's What Here?

So far we've talked about wide-angle, normal and telephoto focal lengths without actually naming any. So what's a wide-angle

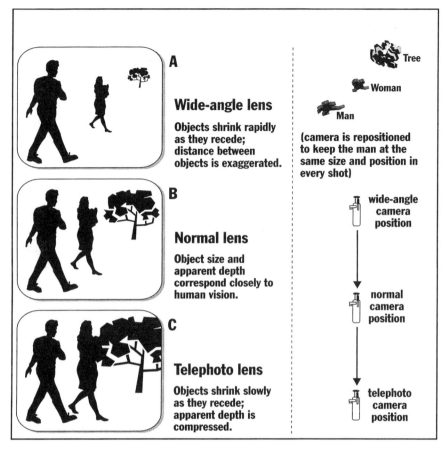

Figure 1.6 *Lens focal length affects relative object size and apparent depth. (man, woman and tree are all the same height)*

lens, anyway: 8mm, 28mm, 90mm, 200mm? The answer: *all of the above*. For a full-size VHS camcorder, wide angle is 8mm; for a 35mm still camera it's 28mm; for a 4 X 5 studio view camera it's 90mm; and for an 8 X 10 behemoth it's 200mm. In other words, the perspective delivered by a certain focal length lens depends on the size of the image it creates.

If you draw a picture of it, it looks like another dose of geometry; don't worry, it's really just common sense. The image created by a lens has to fill the camera's frame, right? But the frame is rectangular and the lens is round. That means that the lens diameter must slightly exceed the diagonal of the frame.

Conveniently, lens designers discovered long ago that for any size format, "normal" perspective is produced by a lens focal length slightly greater than the frame diagonal. That's why a 15mm lens is normal on a camcorder with a half-inch chip, but a 35mm still camera takes a 50mm lens instead. (On the larger camera a 15mm lens would be an ultra-wide.)

What does this mean to you and how do you interpret the lens markings on your camcorder? To understand the answer, you need to know what your camcorder lens is and how it works.

Zoom!

Unless you're using an older style, C-mount lens camera, or a surveillance camera discarded from a convenience store, your camcorder comes with a zoom lens. A zoom lens allows you to shift between focal lengths without changing lenses. In addition, it possesses two critical characteristics:

1. You can set the zoom lens at any and every focal length between its extremes. That means, if your camcorder lens ranges from 8 to 80mm, you could, theoretically, set it at a focal length of 43.033 or 78.25mm.

2. The zoom lens remains at the same focus throughout its zoom range. Focus on your subject at any focal length and the subject will stay in focus if you zoom in or out. *Note: some inner focus lenses do not have this capability.*

Okay, so your zoom lens is marked, say, 8–80mm. What does that mean? What's wide-angle, normal and telephoto in that range?

8mm would be wide-angle, about 15mm would be normal and 80mm would be telephoto. But regardless of what's normal for a given lens, the smaller the number (8mm in this case), the wider the angle. The larger the number (here 80mm), the tighter the angle.

Today many compact cameras use 1/3-inch CCDs, so their zoom lenses feature shorter focal ranges. In this format, a normal focal length is around 10mm, a wide-angle setting would be 5 mm, and a strong telephoto would be 50mm.

For example: the Canon XL1 Mini DV camcorder has a 16:1 zoom that ranges from 5.5-88mm. By contrast, the Fujix H128SW Hi8 camcorder's 12:1 lens ranges from 4.5-54mm. Both have 1/3-inch CCDs.

As you can see, knowing what focal lengths mean can affect your choice of camcorder. The Canon offers you a longer telephoto; the Fujix a wider wide-angle. But to interpret the numbers, you have to start with the size of the CCD. 10mm is a "normal" focal length for a 1/3-inch CCD, while 15mm is considered normal for a 1/2-inch CCD. Once you figure out your normal focal length, you can roughly calculate wide-angle and telephoto lengths as percentages of normal:

- 35 percent of normal: extreme wide-angle.
- 50 percent of normal: wide-angle.
- 70 percent of normal: mild wide-angle.
- 200 percent of normal: mild telephoto
- 400 percent of normal: telephoto
- 500 percent of normal: long telephoto.

As you can see, even the simplest lens on the simplest camcorder is a miracle of modern optical technology. A long, long way from that accidental glop of molten glass.

2
Filter Features: Camcorder Filters and How to Use Them

Michael Rabiger

Although widely available, filters aren't used much on camcorders—probably because they are not fully appreciated. Consider this: what other camcorder accessory helps you to soften picture contrast, reduce depth of field, change color intensity, shoot day-for-night, cut through haze, create star or flare effects, control reflections from glass or water, darken the sky, compose vignettes—even create fog where none exists?

Filters allow you to do all of this—and at very little expense.

Filter Principles

Filters for videography operate on two different principles. One type uses subtraction, permitting some colors of light to pass through while absorbing others. To determine what color a filter passes, hold it up to white light: its color is the color of light it conducts. What isn't conducted is absorbed, along with a certain amount of heat traveling with the light.

The second type of filter uses the principle of diffusion, allowing all light to pass but intentionally modifying how it emerges. The low contrast filter, for example, takes bright illumination from image highlights, dispersing it into shadow areas without significantly changing image resolution.

Colored filters were in use even before color film. In black-and-white photography a red or orange filter blocking blue light darkens a blue sky, rendering a cloud formation into a dramatic white sculpture against an inky heaven. On sunny days, when indirect or "fill" light contains blue sky light, an orange or yellow filter can darken shadow areas by discriminating against their blue content.

If you are shooting video for eventual black-and-white, or taping a nightmarish color sequence, consider color filtering. You can use a color monitor to test your ideas before you shoot.

Kelvin Conversion

Most color film emulsions render colors accurately under one of two main sources of light. One emulsion balances to daylight, assumed to be 5,400 degrees on the Kelvin scale; the second balances for studio lights at 3,200 degrees Kelvin.

Daylight color temperature fluctuates depending on the time of day and location on the globe. It rises to 10,000 degrees Kelvin on a mountaintop, and plummets to a ruddy 1,800 at sunrise or sunset. You can check source-light color temperature with a color temperature meter.

The Kelvin Scale involves the concept of heating a black object. As its temperature rises it emits the progression of colors—red, orange, yellow, green, blue, indigo, and violet—that emerge from a prism used to split white light into its component colors. You can memorize the "roygbiv" sequence by remembering that colorful character Roy G. Biv.

Light also exists above and below the visible spectrum. Infrared lies below the threshold of visibility, while ultraviolet shines above it. Unfortunately, both photo emulsion and video image chips pick up ultraviolet light, requiring a UV blocking filter at all times.

In white balancing a camcorder, you adjust its electronics to treat a particular light source as if it were white. This makes human flesh come out true to life. Film emulsions don't offer a white balance control, so an image shot under nonstandard light must be color corrected to render colors truly. Professional video cameras feature built-in color correction filters to maintain white balance under different lighting conditions. Colors remain consistent from location to location.

On consumer camcorders, filters can be attached to the lens itself. The most common filter is #85 orange; it converts daylight, with its heavy blue content, to match the high orange content of tungsten. Like most filters, the #85 is usually lens-mounted, converting all light entering the camera. It allows shooting in daylight with a tungsten white setting.

The fluorescent conversion filter will correct an image shot under fluorescent light to, say, 3200 degrees Kelvin. It's of limited use unless you know what type of tube is in use. Even then, the gas discharge tube's broken spectrum can make for a sickly greenish cast. The videographer should white balance under fluorescent light, but is helpless if the space contains an assortment of tubes.

Light Mixing

In both film and video, adding 3,200 degrees Kelvin movie lighting to an interior scene already lit by daylight means trouble. You are shooting under mixed color temperatures; they'll probably appear as strange orange-covered shadow areas on a face otherwise nicely color-balanced. Or you'll display a normal foreground with a lurid blue world out the window.

When you need to boost light levels in a daylit room, filter the window so incoming daylight matches the 3,200 degrees Kelvin of the supplementary lighting. Location film units generally tape large sheets of #85 gel to the windows, usually on the inside so air movement doesn't cause the gel to waver.

Another approach to color correction uses blue #80A heatproof filters over tungsten light to produce 5,400 degrees Kelvin light matching incoming daylight. This is less practical because of "filter factor" loss—too much precious light gets lost in the filter itself. The #80A's filter factor is two stops, meaning only one fourth of the light's output gets through. You need to open your camera's iris two additional stops to achieve the same exposure.

The #80A and #85 are just two of a number of color conversion filters allowing savvy videographers to unify diverse lighting sources.

Color, Contrast and Fog

If you want to shoot video with subtle and consistent scene coloration, your best bet is to balance the camera, then use a weak blue or yellow lens filter. Remember to set your white balance control to manual, or the auto white circuitry will sabotage your efforts.

Sunrise filters can enhance nature; a blue filter can simulate nighttime shooting, a process called day for night. Shoot either late or early in the day when there's sunlight and long shadows, and underexpose.

Remember to turn on the car headlights and streetlights and put lights in windows. Or just shoot at night.

UV filters screen out ultraviolet tight, invisible to the human eye but recorded on video as haze. UV filters also serve as good protection for the front of your lens. It's cheap and easy to replace a UV filter compared to lens repair.

A neutral density (ND) filter, like gray sunglasses, reduces all colors of light equally. A filter with a factor of .3 reduces light transmission by one stop; .6 reduces transmission two stops. ND filters usefully cut a len's light intake when the scene is very bright or you want to force the lens to work at a wider aperture to produce a restricted depth of field. (See Figure 2.1.)

Low contrast filters use very fine etchings on the glass to create light dispersal within the filter itself. White light redistributed from highlights is scattered throughout the image. This raises light levels in shadows and lowers the overall contrast between highlight and shadow, at small cost to picture resolution.

The low contrast filter reduces the characteristic look of video—hard contrast and saturated colors—and produces a softer "film" look with de-saturated pastel colors.

A diffusion filter softens the image, giving it a soft, dreamlike look to your scene. Fog filters are strong diffusion filters. They make the image look as though shot through mist or fog. (See Figure 2.2.)

Figure 2.1 *By reducing the amount of light that reaches your lens, a neutral density filter can change the depth of field in your shot.*

However, when something moves nearer the camera in genuine fog, the image clears—not so when using a filter.

Nylon Glass

An inexpensive and extremely reliable diffusion filter is a nylon stocking. Just be sure to empty the leg out first. My father, a makeup man often hard-pressed to generate glamour in superannuated actresses, used to speak dryly of close-ups "shot through a sock."

Another easily produced filter is a sheet of thin optical glass smeared with petroleum jelly. This produces a misty image with flares around highlights known as halations. You can limit the effect by keeping the center clear and lightly treating the edges of the frame only.

A softnet filter—fine netting laminated between clear glass—creates soft diffusion and lowered resolution without

Figure 2.3 A star filter puts stars into your shot wherever there's a point of light.

Graduated, Spot and Split

Graduated filters are half clear, with a soft transition between. A graduated neutral density filter lined up on the horizon can cool a hot sky. The clear lower half leaves the land unfiltered, so the filter functions like a tinted-top car windshield. A graduated color filter used on a static shot can make the sky a rich violet. Of course, you can't tilt the shot up or down without giving the game away.

Graduated filters can also operate vertically. You might line the filter so its dark half reduces the light entering through a doorway, creating consistent lighting throughout the whole scene.

Split field filters are those lenses that divide the field of view into two separate focal lengths, like bifocal glasses. This enables deep focus shots by dividing and thus extending the effective depth of field. You can use the fields horizontally or vertically by rotating the lens. Disguise the telltale dividing lines with a horizon, doorjamb or other eye-distracting compositional factor.

The polarizing filter (Figure 2.4) is another axis-sensitive filter particularly useful for landscape shots. It can reduce the light-polarized glare thrown off by water, plastic and glass surfaces—but not metal. It consists of a light-polarizing material that rotates until its polarity opposes the incoming reflection.

Figure 2.2 Diffusion filters can add a "dreamy" look to your shot.

highlight halations or lightened shadows. Softnets come in black, red, and skintone for enhanced effects. A white softnet acts much like a low contrast filter.

Star filters, which are pronounced diffusion filters, produce the four or six-point highlight star effects so dear to glass and jewelry advertisers. (See Figure 2.3.) Star filters are ineffective in panning shots unless you're a fan of alarming psychedelic effects. And you can't rotate the filter unless prepared not only for stars, but rotating stars.

Figure 2.4 *A polarizing filter allows you to control bright reflections.*

The polarizing filter can also effectively darken a blue sky by tuning out much of the blue light refracted from skylight moisture droplets. It works best when the lens-to-subject axis is about ninety degrees to the sun.

Mounting Filters

We arrive now at a major problem in consumer video—the mounting of filters.

Professional cameras use a matte box, an adjustable filter holder with an extendible lens hood bellows. The device holds square or round filters securely in front of the lens. No matter what lens you use or how much it rotates or extends, the standard filter adjusts and remains solidly in position. This allows rotation of axis-determined filters like stars.

The dedicated do-it-yourself type can of course improvise something. Tiffen makes square filters that fit the Cokin "P Series" filter holder, but this range of pro filters can cost the proverbial arm and a leg.

A holder can also grab custom vignette slides, such as keyhole or binocular shapes for what-the-butler-saw movies, and gunsight or periscope masks for those with warfare in mind.

In most stores, only circular screw-in filters are available to videographers, and must be ordered for a specific lens diameter. Changing is slow and fiddly. If you acquire a wide-angle lens adapter, you'll require a whole new set of larger filters.

You'll handle your filters quite a lot, so consider durability. Gelatin is optically the best material—thin and inexpensive—but it scratches and buckles easily, ruined by a single fingerprint.

Be aware that if you sandwich filters together, you tend to produce rainbow refraction circles called Newton's Rings. Manufacturers make the most common combinations. Gel laminated inside glass is durable and easy to clean, but can be susceptible to moisture. Dyed-in-the-mass glass filters vary in consistency and are expensive; semi-rigid thermosetting resin is a light and scratch-resistant material and optically as good as glass.

In Conclusion

Filters are an easy and inexpensive way to improve your camcorder footage. As such, they're the perfect upgrade for camcorder owners who want to improve their work without emptying their bank account.

One final note of caution: filters come in a variety of price ranges; don't assume that a cheaper one is necessarily a bargain. As with most videography gear, you generally get what you pay for. Caveat emptor!

3
The Magnificent Seven: Choosing a Video Format

Robert Nulph

See video clips at www.videomaker.com/handbook.

In 1954, Akira Kurosawa's Seven Samurai filled the silver screen with swords flashing and heroes saving the less fortunate. In1960 the seven came to Hollywood in true western form and became The Magnificent Seven.

Now in the new millennium the Magnificent Seven ride again, this time into our VCRs and camcorders in the form of the seven consumer video formats. These formats are available to videographers for the creation of their own star-filled productions. VHS, VHS-C, S-VHS, 8mm, Hi8, Mini DV and Digital8. This alphabet soup of formats can get rather confusing. The Magnificent Seven are here to help us, but as video victims, we are often puzzled as to which member of the troupe to call on in our hour of need.

If you are buying a new camcorder, and are considering changing from the format you are using now, this chapter will take you through the flying bullets to reach the ultimate goal of finding the format that best meets your needs. We will look at the seven formats in terms of the size and length of the tape, differences in resolution and overall video and audio quality, audio flexibility and the inherent strengths and weaknesses of each format.

Setting the Stage

The quality of video and audio is an important point to ponder when choosing a format. Video resolution and generation loss are important aspects for consideration. Resolution is the amount of detail possible with the particular format. The higher the number, the better the picture quality.

Generation loss refers to the inherent loss in signal quality that occurs when you make analog copies. You can define a generation as a copy of a copy. If you don't edit tape-to-tape, or extensively copy your videos, a format with a lower resolution may work fine for you. If however, you plan on editing your footage in a linear fashion and/or making multiple copies, resolution and the ability to withstand generation loss are important things to look for in a format.

Another area to look at concerns audio, namely the format's ability to dub sound onto a tape without damaging prerecorded video. This is very important if you will be editing in the camera, not so much if you'll be digitizing your footage and editing everything on your computer.

Editing protocols are concerns for those of you who use linear editing systems. You'll have to match the protocol used by your camcorder to that of your edit controller and record decks so that your equipment can talk to each other. Not all systems are equal and outside of the DV format, the manufacturers of equipment rarely agree on anything. Choose a format that best matches your current set-up.

It is important to note that nonlinear editing, which involves digitizing or capturing video to a hard drive for random access editing, has made terms like generation loss, audio insert and edit protocols nearly extinct. Nonlinear editing allows you to rearrange audio and video segments electronically without spending multiple generations. As a result, videographers are no longer as bound by the limitations of the format on which they shoot. In the end, edited footage is nearly identical to camera quality. In any case, both linear and nonlinear editors want to begin with the best quality footage possible.

Price may be a determining factor when choosing a format. It used to seem that the higher the video and audio quality, the higher the price. However, with the advent of digital equipment this is no longer as true. As time goes by, the prices of digital formats will continue to drop. In our last all-camcorder buyer's guide (December 1999) we saw digital camcorders, both Mini DV and Digital8 models, with MSRPs less than $1,000, significantly cheaper than some Hi8 and S-VHS models. Now that we have prepared you for the various ways to look at these seven formats, lets saddle up and take a closer look at video's Magnificent Seven.

VHS (Big Hoss)

VHS is the time-tested hero of old who is as reliable today as he was 20 years ago, when the other formats were just twinkles in the eyes of distant Japanese video engineers. Though he's not as pretty to look at as some of the newer, stronger formats, he remains a favorite of American consumers.

Most homes have at least one VHS VCR, typically used for recording TV shows or viewing rented movies. VHS is the most used yet lowest quality format available. The format's low resolution makes it susceptible to generation loss, and a bad choice if you are planning to edit anything for copying or broadcast. But, because VHS VCRs are so common, the VHS format is ideal for making distribution copies of your video productions.

You can find VHS tapes in a variety of lengths, providing up to two hours of record time in standard (SP) record mode. Blank VHS tapes are the cheapest, and most available of all formats. A two-hour tape can cost as little as $2 to $4, and you can buy them virtually everywhere. At just $1 to $2 per hour of footage, the cost of blank VHS tape is quite a value.

The VHS format records two tracks of linear audio (see Figure 3.1). VHS records these audio tracks on a separate portion of the tape than the video information, and you can edit them without affecting the

Figure 3.1   *VHS Format*

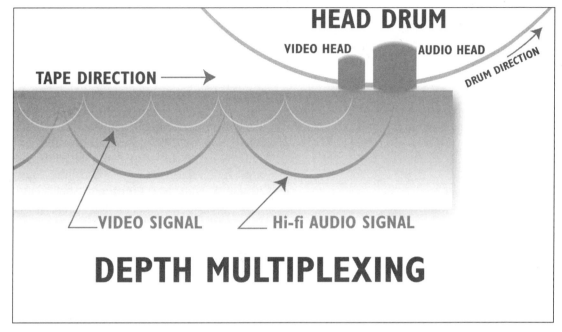

HEAD DRUM

VIDEO HEAD   AUDIO HEAD

DRUM DIRECTION

TAPE DIRECTION ⟶

VIDEO SIGNAL   Hi-fi AUDIO SIGNAL

DEPTH MULTIPLEXING

Figure 3.2 *Video and hi-fi audio are embedded in different layers of the tape's magnetic particles.*

visual image. The VHS format also includes a stereo hi-fi audio track, but this high quality, stereo audio track is recorded on the same part of the tape where the video is recorded, embedded deeper into the tape's magnetic particles (see Figure 3.2). While this track holds a solid, high-quality audio signal, you cannot edit the hi-fi audio track without also replacing the video on that portion of the tape.

VHS-C (Little Joe Video)

VHS-C, or compact VHS, as he's often called, is the good-looking younger brother in the VHS family. While he's from the same gene pool as his big brother, he comes in a smaller, sleeker package.

Engineers created VHS-C to meet the demand for smaller camcorders. This format is also inexpensive, but you can expect to pay more for VHS-C tapes than their full-sized VHS counterparts. A 30-minute tape will cost you around $4. Not too expensive, but adding up to around $8

per hour of footage. One limitation of VHS-C tape is its limited length-40 minutes max in standard mode. With an adapter, VHS-C tapes will play in your standard VCR.

S-VHS (The Muscle-bound Maverick)

Super VHS is the strongest member of the VHS family. He has all the experience of his brothers, and super strength to boot. He is a favorite of linear editors and holds his own as a legend of consumer video. Its video quality is very good, with a fair generation loss rating. The hi-fi tracks it uses for audio provide quality audio reproduction. Being in the VHS family, it allows the user to dub both audio (linear tracks) and video, a variety of editing protocols and the ability to record time code information. In the linear world of consumer video, it has become the ideal editing format. This format is a workhorse in corporate and educational video production.

However, its size and price may be its ultimate demise. Like VHS, S-VHS is a

full-sized format, making S-VHS camcorders large and heavy. While the tapes
look just like standard VHS, and will fit in
a standard VHS machine, they will only
play back in S-VHS VCRs. S-VHS tapes
are harder to find than VHS or VHS-C
tapes, and you can expect to pay $10 to
$15 dollars per two hour tape.

8mm (Junior)

Standard 8mm is the wiry little guy who
came on the scene as the new hero several
years ago, delivering both the extended
record time of full size VHS and the small
size of VHC-C. With his comfortable and
easy personality he has won the hearts of
casual shooters for nearly a decade.

The first difference of note between the
VHS and 8mmformats is the lack of a linear audio track on 8mm tape. The camcorder records the audio as an audio frequency modulation (AFM) signal on the
same track as the video. The quality of the
AFM audio is similar to the VHS hi-fi.

While this provides great audio, the
inability to dub audio makes this format a
poor choice if you are going to edit your
productions in the camera. The video signal, while a little better than VHS does
not hold up as well and has even worse
generation loss problems.

The small size and low price of standard 8mm camcorders make it ideal for
casual shooters. These camcorders fit into
tight spots and are light, for carrying
around all day at the zoo. You need to
keep in mind that 8mm will not play in
VHS VCRs, and 8mm VCRs are extremely
rare. You can count on using your camcorder as a VCR whenever you want to
play a tape, and you'll need to cable your
camera to your TV to play tapes for an
audience. If you want to edit your footage,
you'll want to do so digitally, or master to
another format.

Like full size VHS, 8mm tapes come
in lengths up to two hours. They are available in most stores for $8-$10 per
videotape.

Hi8 (Little Big Man)

Hi8 is a tough little character. He has all
the charm of his standard 8mm brother,
and though you wouldn't know it to look
at him, he has the heart of a wild stallion.
This Hi8 format is the big brother of 8mm.
Industrial and educational facilities have
used Hi8 as an acquisition format for
many years. While the cost of Hi8 tapes is
a bit high, about $15 for a two hour tape,
its small size, low equipment price and
high quality picture make it ideal for
intermediate and advanced shooters.

Its video quality is a little better than S-
VHS but it has problems holding on to that
quality as it faces generation loss. This format also has a limited number of editing
protocols and there are no Hi8 consumer
camcorders that record time code.

Like standard 8mm, Hi8 records AFM
audio under the video on the tape. Some
models include an additional audio track
called PCM Audio (see Figure 3.3). The
camcorder records this information as
stereo right and left on its own audio
tracks. The true benefit of PCM audio is
that it allows for the editing of audio
without destroying the video. Since PCM
audio has its own separate tracks, you get
a high-quality audio signal without the
limitations of hi-fi VHS or 8mm AFM.
The one drawback with this audio is that

Figure 3.3 *The Hi8 format often includes
PCM Audio.*

as you edit, you cannot listen to the audio in slow speed to find your edit points. The digitizing of the audio signal makes it very hard to hear the beginning and end of words as you slowly scroll through the tape.

Mini DV (The Sharpshooter)

Mini DV may sound cute, but don't let his name fool you; he's one mean cowboy. He can out gun any of the other consumer formats and has even begun to challenge some of the pros. He has single-handedly changed the face of consumer video, and the Magnificent Seven may never be the same.

For the first time in the history of the video industry, a consumer format, DV, has begun to trickle up to the professional ranks of video producers. Why? It could be the nearly loss-less broadcast-quality video picture, the digital dual mode audio capabilities (recording 12- or 16-bit digital audio). Because the tapes are so small, just 3x2 inches, manufacturers have been able to produce camcorders that are literally pocket-sized. Its ability to export loss-less digital images through FireWire to your computer make it a favorite of advanced videographers. This format records DV time code, supports a number of editing protocols, and is the only format that all video equipment manufacturers support.

The format also includes additional data tracks called insert and track information (ITI) tracks. These tracks, found on certain models, contain date and time information as well as pilot tone information which control playback (see Figure 3.4).

Mini DV has two audio recording modes: you can record a single stereo track that does not permit dubbing, or you may record two stereo tracks independently from the video. Both of these methods produce near CD-quality audio. DV tapes come in lengths up to 60 minutes and sell for around $20 per one-hour tape.

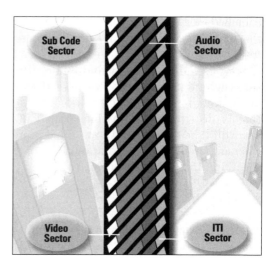

Figure 3.4 *The DV format includes extra data tracks.*

If you are a serious videographer who will be editing the footage you shoot and making copies to broadcast or exhibit, this is the format to use. While this format is more expensive than others, the quality is far superior, and prices have been steadily falling in the market place.

Digital8 (The Young Gun)

Digital8 is the newest member of the posse. The offspring of a marriage between 8mm and DV, he arrived with loud whoops and guns ablaze. And since day one, people haven't stopped talking about him. While many aren't sure what to think of this untested rookie, none can doubt that he has taken his place in history as a wild child to keep your eye on.

This format records a Digital Video (DV) signal on a Hi8 tape. Because Digital8 records digitally, the quality is extremely good and copies are virtually loss-less when you use the IEEE 1394 (FireWire) port to make copies to another digital camcorder or VCR. It records DV time code, making it easy to find shots when editing, and some models will take digital stills. The audio is CD-quality. The major drawbacks of Digital8 are the lack of playback decks (you'll have to use the camcorder), the inability to dub audio in the camera, and the uncertainty of its future.

Sony is the only company to manufacture Digital8 camcorders. However, if you are looking for a digital acquisition format that is downward compatible with Hi8, and you'll be transferring your video from the camcorder to your computer through FireWire, this may be the way to go.

Moving On

Whatever your video plight; there is a format out there for you. Carefully evaluate your needs, then pick the format (or formats) that will give you the best results. Do you shoot long programs that need 120 minutes of tape? Do you use linear or non-linear editing equipment? Do you want to do in-camera dubbing? Do you already have some equipment and have to choose a format that will be compatible with it? All of these questions and more should run through your mind as you decide on a video format. While DV and Digital8 are the hot new heroes, all seven formats have some fight left in them.

4
Solar Panel Imaging: Secrets of the CCD

Loren Alldrin

Buried deep within your camcorder lies a fabulous image sensor that sets it apart from most other image-capturing devices. This image sensor is called a charge-coupled device—that's CCD to you and me.

If you're like most videographers, you probably don't know much about this hidden treasure. And that's a shame. Knowing the hows and whys of CCDs can help make your videography more effective. It can help you differentiate one model from another and decide which camcorder to buy. Moreover, CCD sensors benefit from some of the fastest-advancing technology in camcorders; know the future of sensors and you can peek into the very future of camcorders.

The Short Explanation

Defining the CCD is, uh, simple: a CCD, or interline transfer charge-coupled device, is a tightly-packed array of tiny photodiodes consisting of silicon oxide and alternating P and photosensitive N semi-conductor regions on an N-type substrate. Every 1/60th of a second, a transfer pulse triggers a vertical transfer CCD lying between pixel rows to sweep accumulated charges out to the horizontal transfer register (H-CCD) and output amplifier. Newer designs employ an additional P+ embedded photodiode to improve signal-to-noise ratio by controlling irregular dark currents (see Figure 4.1).

Don't worry—that definition went over my head, too. Try this: imagine a huge grid made up of rows of solar panels. Each square-foot panel sits atop a small battery. Only a few inches separate each panel from those to its north and south. About a foot of space lies between one column of panels and the next; within that space you'll find a small pathway running next to each column, as well as along the bottom of the grid. The grid encompasses hundreds of panels in each direction, stretching for about 1/4 of a mile on a side.

When light strikes the panels, they charge their individual batteries. Panels

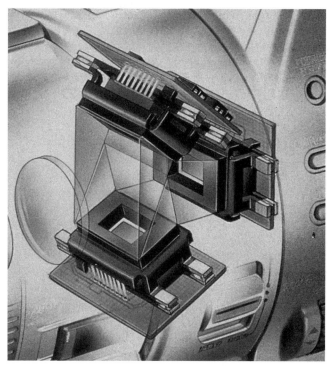

Figure 4.1 *Three-chip designs from Sony and Panasonic use three CCD sensors arrayed around a light-splitting prism.*

exposed to more light charge faster, while those in the dark build up little or no charge. After a given period of time, tiny trucks drive down between the vertical rows of panels to measure each battery's charge. The trucks then quickly discharge the battery they're measuring and move on to the next panel. When the trucks reach the end of the row, they dump their information onto a conveyor belt. This belt carries the data from the panels back to a central station. It's here that high-paid managers combine the individual measurements, evaluating the electrical output of the grid. Their final report looks a whole lot like a video image.

To relate this rather loose analogy to an actual CCD, first we need to reduce the size of the solar grid by a factor of about 75,000. Most of today's CCDs measure a mere 1/3-inch from corner to corner, and even smaller 1/4-inch designs are on the horizon.

In our little analogy, solar panels serve as the individual pixels. Today's sensors actually boast hundreds of thousands of these pixels, etched onto the top of a silicon wafer by chemical and photographic processes. The machine tolerances and cleanliness required for making sensors is truly superhuman; most CCDs come from completely automated factories where humans play minor supporting roles. A tiny speck of dust, harmless enough to us, can actually shut down the CCD manufacturing process.

The batteries represent the buildup of charges in the pixel. Since CCD pixels are photosensitive, they create a charge in proportion to the light striking them. Lots of light makes for a greater charge, while darkness leaves them with little more than the small random charges we call noise. Smaller pixels gather less light and generate weaker charges; a principle manufacturers must address to produce smaller chips and pixels. More on that later.

The trucks mimic the action of the vertical transfer registers, electronic roadways that carry charges out of the active sensing area of the CCD. These registers are necessary because the record electronics do not read charges directly from individual pixels. Instead, charges move en masse down the vertical transfer registers until they reach the edge of the chip.

The conveyor belt is like the horizontal transfer register, which unloads the charges from its vertical counterpart. The horizontal transfer register carries charges off the CCD along the edge of the sensor. Their destination: the amplifiers and specialized circuits that process the signal before recording (see Figure 4.2).

The high-paid managers represent the camcorder's record electronics, processing and modifying signals for recording on magnetic tape. Specialized chips combine color and brightness information into one signal, boost its level and then send it on to the record heads.

Generating a final report on the status of the solar grid could take hours—depending on the speed of the trucks and whether or not those high-paid managers get stuck in an important meeting. In a camcorder, however, videotape records a "final report" from the CCD sensor sixty times per second. If only our government worked that fast.

Sensor Overload

When a solar panel receives too much light, it overcharges its battery. The truck tries to read this abnormally high value, only to cook its tiny charge meter in the process. When the truck gets to the bottom of the row, it picks up a new charge meter, but until then severe damage can occur. All the readings it currently holds, as well as all subsequent measurements, are wrong. They all read maximum on the charge meter.

Something similar occurs when a given area of a CCD receives too much light. The

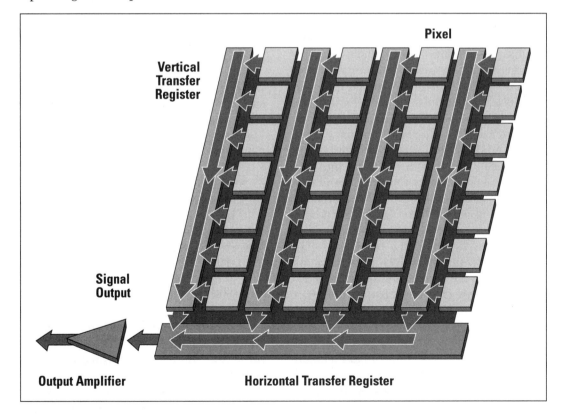

Figure 4.2 *In a standard CCD sensor, pixels feed charges to vertical transfer registers. These in turn feed the horizontal transfer register. From there, signals move through an amplifier to the record electronics.*

vertical transfer register overloads, muddling all the charges for that row. This creates a bright, vertical smear in the image, extending out above and below the offending spot. You've probably seen this before, especially when shooting a bright spot of light against a dark background.

This type of image smear is unique to CCD sensors. Metal oxide semiconductor (MOS) sensors read each pixel directly, doing away with the need for vertical transfer registers and their associated image smearing. Regardless, MOS sensors have fallen out of favor with manufacturers, probably due to higher manufacturing costs. New CCD designs address the bleed problem, resulting in chips less prone to streaking.

Shutter Shenanigans

Though you've heard the term high-speed shutter tossed about; there's actually no such component in the lens/sensor assembly. This term comes from the sensor's ability to mimic the effects of a film camera's fast shutter speed. In a film camera, opening the shutter's blades for a very short period of time exposes the film to a brief snippet of light. Thus a film camera can freeze even the fastest motion.

To understand how this works in a video camera, let's go back to the solar panel scenario. In just minutes, it will be time to measure the grid. But instead of letting the panels finish gathering a complete charge, the trucks sweep through the grid to discharge all the batteries. When the trucks return to collect measurements, the panels have been charging for just a few minutes. Output is lower, but management can still get a picture of the grid's status.

This is what happens with a camcorder's high-speed shutter. No mechanical blade assembly snaps open and shut; instead, the camcorder gives the pixels less time to charge before whisking their signals off to the recorder. If you select an extremely fast shutter speed, say

1/10,000th of a second, the pixels charge up as usual during the first 99 percent of the record cycle. Then, just 1/10,000 of a second before recording, the sensor discharges the pixels. What's recorded on tape is a brief slice of time, representing only the last tiny bit of the record cycle. Since most subjects don't move very far in 1/10,000th of a second, a high-speed shutter freezes the action. Whereas a single frame of an airplane propeller made at 1/60th of a second might show a blur, a single frame of the propeller made at 1/10,000th could show it standing still, each blade distinct.

Two matters to keep in mind when shooting with a high-speed shutter:

1. *Lighting.* Since pixels have so little time to charge, the intensity of the light must be greater to produce a usable image. The higher the shutter speed, the brighter the light required. Shooting at 1/10,000th of a second requires strong daylight. A more conservative setting of 1/2,000th of a second still requires sunlight or strong indoor lighting.

2. *Depth of field.* Because it needs more light, high-speed shutter forces the camcorder's iris to open up. This in turn reduces depth of field, a boon to creative videographers whose camcorders lack manual iris control.

If you want to soften the background behind your subject, reduce your depth of field by increasing shutter speed.

Some camcorders offer a slow-speed shutter, which has the exact opposite properties of high-speed. Slow-speed shutter delivers an image in less light, though much more image smear results. If you're shooting a stationary subject in extremely low light, slow-speed shutter may deliver an improved image. At the very least, you can use it as a unique special effect.

Here's how it works: The trucks servicing our solar array still sweep through the panels to gather readings; they simply don't discharge the batteries completely before reading the charge. This allows the panels to build up a greater charge, boosting the resulting values. Since the batteries retain some residual charge, each reading includes some values from the previous cycles. In the same way, a camcorder in low-speed shutter mode allows the pixels' charge to build up for longer than just one record cycle. It effectively "averages" the light, making fast-moving subjects smear and bleed.

Shrinking CCD Panels

Let's say that the owner of the field that contains the solar array wishes to sell off some of his land, leaving the panels with about 40 percent less area. We can't reduce the number of panels, so there's only one solution—make them smaller. To pack the same number of panels on our now-shrunken plot of land, we must cut them down to just over 7 inches per side. We buy new, smaller trucks and a shorter conveyor belt and fire up the new array. The managers are not happy.

Seems the scaled-down array now puts out about 40 percent less energy. These are lean times, and a cut in output simply won't do. The high-paid managers hire a few high-paid engineering consultants to increase the panels' sensitivity.

There you have it: the plight of the shrinking CCD. Like a tiny solar panel, a pixel's output is a function of its surface area. Shrink the pixel, and its sensitivity suffers. When sensitivity falls, so does the camcorder's low-light performance and resistance to video noise. But manufacturers can't ignore the benefits of smaller sensors—they achieve the same depth of field with smaller lenses. Smaller lenses in turn make for smaller camcorders, and smaller camcorders seem to sell better.

The solution: the microlens. Basically a tiny, translucent bubble formed over each pixel, the microlens gathers incident light that would have otherwise missed the pixel's active sensing area. CCD makers form microlenses into the CCD itself,

increasing the effective area of the pixel without actually making it any larger. Thanks to the microlens, 1/3-inch CCDs are now a reality. This microlens technology is so effective, in fact, that a 1/3-inch sensor with microlenses may outperform the larger 1/2-inch designs—like realizing even more output from our solar array after placing a glass canopy over each panel.

Another way to offset the effect of smaller pixels is through better amplification. Noise is an enemy to any kind of electrical signal, and smaller signals are the most prone to it. Amplifying a signal just as it leaves the pixel reduces noise and strengthens output. At the time of this writing, manufacturers are experimenting with a new type of CMOS (complementary metal oxide semiconductor) sensor invented by NASA/JPL called the APS CMOS (Active Pixel Sensor) that places an amplifier at each photosite.

These technologies have led to sensors a scant 1/4-inch across—a big step toward the next generation of ultra-compact camcorders.

Smaller or Better

The same technology that allows sensors to shrink allows advances in the other direction as well. If pixels offer improved sensitivity at a smaller size, then CCD makers can pack more pixels on the same size chip. Once manufacturers increase the pixel count of a given sensor, they face a tough decision. They can use the additional pixels for a higher resolution video image, or they can employ them for special image effects at the standard resolution.

Electronic Image Stabilization (EIS) is a good example of such an effect. An increase in pixel count from 410,000 to 470,000 allows the camcorder to use just the central 90 percent of the chip for imaging without resolution loss. Move this region in opposition to the camcorder's movement, and you reduce shake on handheld shots. Whereas previous EIS schemes resulted in an inevitable loss of resolution, this system shows no noticeable softness of the image.

Color Blind

Some of you may remember from your high school science class that solar panels respond only to the amount of light striking them, not the color of the light. In the same way, CCD pixels are colorblind. So how does a camcorder record a color image?

Camcorders with a single CCD sensor use a mosaic color filter placed over the pixels . Imagine a huge stained-glass window lying over our solar panel array. This window alternates panes of color—either red, green and blue or their complements, yellow, magenta and cyan. Each solar panel sits directly under a colored pane, and responds only to that color of light. When the managers tally up the charges, they make note of each panel's color.

By tracking which pixels see which color, a camcorder extracts both a luminance (brightness and detail) and chrominance (color) signal from a monochrome sensor. Color filters are relatively easy to add to a CCD, though they compromise both color and brightness portions of the video signal. Because there are a limited number of pixels responding to a given color, chrominance has only about one quarter the resolution of the luminance signal. Placing a colored filter over the pixels also reduces their sensitivity and low-light performance.

There are better, albeit more expensive, ways to coax color information out of monochrome sensors. The best system is the one professional cameras have used for years—three sensors, or chips, with one sensor devoted to each of the three primary colors. Just behind the lens, a precision-made prism splits the incoming light into its red, green and blue components. Some manufacturers use an array of dichroic mirrors to sift the light; these coated mirrors reflect only a certain color, letting the rest pass.

Because there's no color filter clouding the sensors in three-chip designs, resolution does not drop. With a chip "specializing" in each primary color, hues are very accurate and natural. The result: a better picture than a single CCD can deliver.

Future CCDs

The trend toward smaller CCDs will most likely die with current 1/4-inch designs. Sensors of this size will work with incredibly small lenses, making the transport and tape medium itself the biggest obstacles to further camcorder downsizing.

Manufacturers will undoubtedly continue in the other direction, toward larger sensors with increased resolution. Chips with pixel counts approaching one million allow for special effects and electronic stabilization without resolution loss. The advent of HDTV and the growing popularity of today's 16:9 formats will drive the market toward ultra-high resolution sensors. HDTV cameras already have 2/3-inch sensors with over two million pixels.

Advances in sensors drive other areas of the video market as well. Camera resolutions are already much greater than those of camcorder transports. As sensors evolve far beyond the recording ability of camcorders, consumers will push for new video signal formats. Sensor evolution shows no sign of slowing. As long as there's a sun in the sky sending light to CCD pixels and solar panels alike, better sensors will be here to capture it. The future of sensors is bright indeed.

5

Image Stabilizers: The Technology that Steadies Your Shots

Robert J. Kerr

If you want steady pictures, use a heavy camera. Unfortunately, today's VHS-C and 8mm camcorders fly in the face of this general rule of thumb; they're so light, the slightest external vibration can affect the quality of their images.

Enter the image stabilizer. Developed specifically to address this problem, these nifty gadgets now grace many small, lightweight camcorders. In this chapter, we'll examine the types of image stabilization systems and how they work.

A Short Stability History

I suppose that even the cave artists back at the dawn of pre-history had trouble freezing images of fast-moving antelope in their minds before they attempted to draw the beasts accurately on their cave walls. Portrait painters, too, have dealt with the problem of fidgety subjects; perhaps even famed Civil War photographer Matthew Brady cursed the artillery shells shooting past him during his long exposures.

Getting a steady image has been a problem for artists, photographers, cinematographers and videographers for as long as these arts have been practiced.

In the earliest days of photography, the size and weight of the camera and the long exposure time made the tripod *de rigeur*; it was the only way to achieve steady images. As film speed increased, so did portability; cameras such as the hand-held Kodak Brownie brought portable photography to every family. The relatively wide-angle lens further reduced the sensitivity to small camera movements. Still, the motion to depress the shutter trip lever required a steady hand for a steady picture.

That was then, this is now. Today's very fast film and electronic flash make steady still photographs the rule.

On to the silver screen. The first cinematographers also used heavy cameras mounted on tripods. Later, the shoulder supported 16mm cameras also proved heavy enough to provide steady images—if the cinematographer stood still.

Filmmakers got around this limitation by mounting cameras on automobiles and airplanes to capture moving shots. They laid down dolly track, much like train rails, to allow smooth camera movement in action scenes. In the 1970s, Garret Brown invented the Steadicam™, a stabilization device cinematographers on the move used to keep shots steady via an elaborate counter balance. Again, the success of the Steadicam™ depended on its significant mass.

The story was much the same for early television cameras, whose weight and size also required tripod mounting. Resourceful engineers developed massive camera mounts that panned and tilted effortlessly and glided smoothly across studio floors.

With the late 1970's came the introduction of lightweight battery operated videotape recorders. Getting steady video pictures with these early models—shoulder-mounted or hand-held—was a problem. A problem aggravated by addition of the zoom lens with its telephoto capability some time later. Various "body brace" mounts appeared on the market for those who wanted to improve the steadiness of their videos, but the light weight of the cameras made them very sensitive to body or other external motion.

The next major coup: the appearance of the solid state electronics color camera for broadcast news gathering. This 25-pound shoulder-mounted camera could be held reasonably steady at moderate focal lengths of the zoom, but required great stamina on the part of the cameraperson. Not to mention that long focal length shots were a real problem.

The Steadicam™ mount developed for the movies could be used for video cameras, but television applications proved infrequent. Advances in solid state electronics made studio cameras smaller and lighter; the addition of housings and heavy zoom lenses kept them heavy enough to provide smooth operation. Enter the 1980s and the rapid growth of the portable color camera/recorder and camcorder industries. Light weight was a priority; the introduction of the CCD camera chip made VHS-C and 8mm video cameras smaller than ever. The palm-corder overtook the steadier, larger shoulder-mounted models; its small size—coupled with a zoom lens-made steady pictures a problem. Fortunately, the same technology also supplied the microcircuit advances needed to provide the solution.

About this time, Garret Brown invented a smaller, lighter version of his Steadicam' stabilization device. The ever-popular Steadicam JR™ helps to stabilize the images of camcorders weighing less than 5 pounds. It completely isolates the camera from rotational body movements, thanks to a delicate balancing system featuring a low-friction gimbal between the camera and the support handle.

This device, although useful, is not one you'd carry with you on vacation. It will fold to a shoulder mount configuration, but is best applied to specific shooting problems you can plan in advance. When properly deployed, the Steadicam JR™ can yield steady video pictures.

Not too long ago, numerous manufacturers including Panasonic, JVC, Hitachi, Sony and Canon introduced different systems that reduce image jitter problems. Unlike the Steadicam JR™, these systems are integrated into the camcorder itself and do not employ any external hardware.

There are two main image stabilization systems: optical stabilization and electronic stabilization.

The All Electronic System

The electronic system operates by first reducing the area of the CCD chip from which the video image is read (see Figure 5.1). This smaller image then increases in size to fill the whole screen. The exact area scanned then shifts electronically to compensate for unwanted external movement of the camera. Since this system does not actually sense the movement of

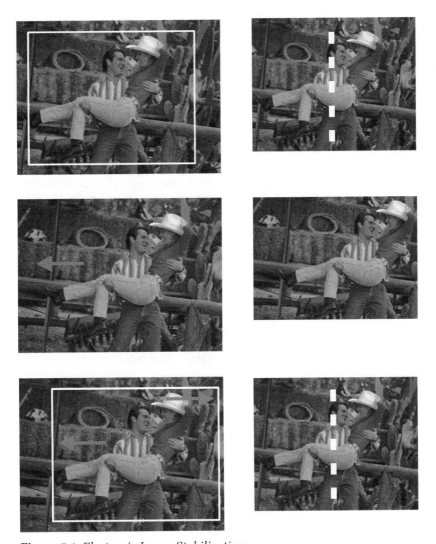

Figure 5.1 *Electronic Image Stabilization*

the camera it must sense camera shake from the image only. The trick is to tell camera movement from movement of the subject.

Some manufacturers use a motion detection method based on fuzzy logic. How much to compensate for movement is a decision based on comparing the two images. An image freezes in computer memory and divides into numerous quadrants. A processor compares the differences between the individual quadrants of the frozen image and the current image. If all quadrants change in the same direction, the processor deduces that camera movement caused the differences between the current and stored images. The area of the CCD being scanned then shifts in the opposite direction to cancel the movement.

Changes in fewer than all quadrants indicate subject rather than camera movement and no compensating action occurs. If quadrant analysis indicates that both the subject and the camera moved, fuzzy logic calculates the image shift needed just to compensate for the camera movement.

One criticism of this system: the loss in image quality brought about by reducing

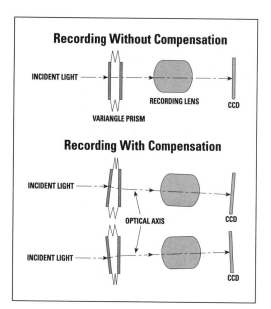

Figure 5.2 *Optical Stabilization System*

the number of pixels used to create the picture. This loss is noticeable to varying degrees on most camcorders, virtually invisible on others. By the time the video signal goes to the tape and back, especially on standard 8mm and VHS models, image loss is negligible. Most videographers will find the added stability to be worth the tradeoff.

Optical Image Stabilization

Optical stabilization operates very differently from the electronic system. Instead of sliding an undersized image around the CCD camera chip, the optical system corrects for camera movement before the image reaches the chip. This way, the full resolution of the CCD occurs at all times. The result: no image degradation.

The key optical component is a variable bend prism. As light passes through a prism, it bends in the direction of travel. The amount of bending—known as *refraction*—is a function of the angle at which the light strikes the "in" side of the prism, the relative angle of the "exit" side and the refractive properties of the prism material.

Refraction is what you see when you look at an object at the bottom of a pool or stream. If you look straight down on the object, the light reflected from it passes straight through the surface of the water. No bending or refraction occurs, and you see the object in its actual position. If you view the object in the pool or creek from an angle, thanks to refraction you'll think the object is considerably higher than it really is. If you've ever been spear fishing you'll know what I mean. Try to spear fish from above the water, and you'll have to aim the spear below the spot where you "see" the fish.

Back to the prism. When you think prism, you probably think about how it breaks up light into the color spectrum. Like how raindrops make a rainbow. You see the rainbow because different colors of light bend by differing amounts as they pass through the prism.

Rainbows may be pretty, but they're not so desirable in image stabilization systems; to eliminate any potential rainbow problems manufacturers choose prism materials carefully and restrict the angle of refraction to no more than 1.5 degrees.

The prism used in the optical stabilization system is a unique variable angle design consisting of two glass plates joined at the circumference by a flexible bellows. A silicone fluid with controlled refractive properties fills the space between the lenses (see Figure 5.2).

When the two plates are parallel, light passes through undisturbed. If, however, the plates contract at any point on the perimeter, the light path bends away from the compressed area. Thus the system can actually *steer* the optical image by manipulating the prism.

The next step: how the system can tell when to perform such steering.

The optical system requires two motion sensors, one for pitch (tilting up and down) and the other for yaw (panning side to side). The sensors amplify and process the motion signals to determine where and how to move the image. The results convert to electric current applied to two drive actuators, one for pitch and one for yaw. These actuators adjust one of the glass plates in the prism relative to the other, directing the image back to its proper position on the CCD sensor.

Field Testing the Two Systems

The easiest way to show how effective these two image stabilization systems are is to test them under adverse conditions. In this case, the test consisted of video recorded on a road of moderate roughness by one camera equipped with electronic image stabilization and then by a second, fitted with optical stabilization.

I did the videotaping from the passenger side of a car, while my business partner drove. We completed one complete trip over the test course for each camcorder. We used a medium telephoto zoom setting to exaggerate the effects of camera motion. Then we brought the tapes back to the studio and compared results.

Both systems provided extraordinary improvement in the stability of the image. With either type of camera, shots of the cars ahead of us stayed steady in the picture—even as the dashboard of the test vehicle shifted up and down at the lower part of the picture. Certainly video from moving vehicles proves much more usable when you engage the image stabilizers.

The optical system makes no use of video information, so it cannot wrongly interpret moving objects as camcorder motion. The same is not true, of course for the digital system. The question was, how well would its fuzzy logic compensate for an actual moving subject combined with camera pitch and yaw? The answer: fuzzy logic did an excellent job; road images remained steady—even when the car dashboard bounced up and down in the lower part of the picture.

Being almost completely mechanical, the optical system experiences some inevitable delay as its components move and adjust. This makes it somewhat slower to respond to quick jolts, but this does not adversely affect normal operation. A happy by-product of the mechanical system: its remarkable smoothness.

The electronic system is very fast, and tried to compensate for even the most instantaneous bumps in the road. Some image jump occurred, as though the electronics eventually gave up on fixing the jump and instead started fresh with a new image.

An interesting test result with both systems pan or tilt actions, with the stabilizers engaged, the movement of the image in the viewfinder lagged behind the camera movement, or "floated." The effect is only noticeable when moving the camera while looking in the viewfinder; I didn't notice it when viewing the recorded videotape. This simply tends to demonstrate that two dramatically different approaches provided almost exactly the same satisfactory result.

Cynic that I am, I tend to view a lot of "features" on the higher priced camcorders simply as opportunities for the video department salespersons to move customers to higher priced models.

Not so, however, for the two image stabilizers described here. They work, and prove very useful in many situations, particularly hand-held "shots of opportunity."

And So

Image stabilization is a feature well worth having, particularly on today's small, lightweight camcorders. Use it, and your images will be easier to watch; shoot hand-held telephoto shots without it, and—well, just try it. You'll see what we mean.

6
The Viewfinder:
Window to Your Video World

Loren Alldrin

When you think of the interface between you and your camcorder, what comes to mind? The buttons? The hand grip? The zoom rocker?

While all of these are a part of the unit's "user interface," the most important link between you and your camcorder is the viewfinder. With it, you decide on composition, monitor focus, check the status of your camcorder, even watch what you've already shot.

Yet this vastly important part of the camcorder rarely gets a second thought from most videographers. One came on your camcorder, and that's all there is to it.

Actually, that's not all there is to it. Understand your viewfinder, and you'll be a better videographer. With a thorough grasp of viewfinder technology, you'll use your camcorder more effectively. And when it's time to buy a new one, you'll make a more educated purchase.

How It Works

Viewfinders boast tremendous diversity in design. Units twist, swivel, extend and disconnect—and some of today's latest designs are even more versatile. All these adjustments have a two-fold purpose: to make the viewfinder comfortable for the user, and make the camcorder itself smaller and more adaptable. Use a camcorder whose makers ignored viewfinder comfort, and you quickly learn to appreciate the engineering time spent on better designs.

The working principle of the viewfinder is quite simple: a tiny display sits a scant one-inch from your eye, its light passing through a diopter lens. The lens changes the effective distance between the eye and the display. This makes the viewfinder seem larger and farther away than it really is. If you've ever tried to focus your eye on an object this close, you understand why the diopter is necessary.

You can adjust the diopter lens on most camcorders. This compensates for vision problems, making eyeglasses unnecessary for most videographers. Some viewfinders collapse, placing the diopter lens extremely close to the display. This effectively disables the diopter, making the viewfinder visible from a few inches or more. Magnification also decreases, so that the viewfinder appears closer to its actual size.

The ability to see the viewfinder from a distance comes in handy when shooting at extreme camera angles. Most viewfinders tilt for this very purpose, allowing you to frame your subject properly at all times. Tilting the viewfinder up 90 degrees and collapsing the diopter makes extreme low angles a breeze.

Most viewfinder designs dictate that you hold the camcorder in your right hand while viewing the image with your right eye. A few models address this right-hand bias with a rotating viewfinder for both left and right eye shooting. When you flip the viewfinder from one side to the other, a switch automatically inverts the image.

Tiny B&W TVs

If you've never pulled the lens assembly off a viewfinder to check out the actual display, you probably don't realize just how tiny it is. Most viewfinder displays are a scant 1/2-inch from corner to corner, no bigger than the average person's smallest fingernail. The technology that packs a simple TV into such a small space is truly amazing.

Today's viewfinders use one of two methods for displaying images, depending on whether they are black and white or color. Black and white—or monochrome—viewfinders function like tiny televisions, while color liquid crystal display (LCD) viewfinders work much more like a common digital wristwatch.

Whether a color or monochrome viewfinder proves most comfortable for you depends mostly on your shooting style. Both have good points and bad points—one offers lifelike color at the expense of resolution, the other delivers crisp, color-free images.

Monochrome viewfinders use technology that dates back to the black and white origins of television. Like a standard cathode-ray tube (CRT) TV set, the viewfinder has a cathode emitter that fires electrons toward the face of the viewfinder. This surface has a coating of phosphors, which glow when struck by the electron beam.

This electron beam passes through a changing magnetic field controlled by the incoming video signal. This magnetic field actually bends the beam between the emitter and screen. Precisely controlling the beam's direction and intensity allows the viewfinder to draw individual lines of the image, one after another. Two full sets of these lines, offset and intermeshed, create what we see as a complete picture.

Like a monochrome television, most viewfinders adjust for brightness and contrast. Brightness sets the overall intensity of the viewfinder image. Boosting this control while shooting in a sunny locale can help you see the image despite competing light and reflections. When shooting in a dark location, reducing brightness will cause less eyestrain. You'll also experience less of a shock when you look away from the bright viewfinder.

Contrast adjusts the difference between the brightest and darkest parts of the image. You should set this control so that white areas of your image are bright without blooming. Dark areas should be black and not grey. Brightness and contrast controls interact—you'll probably need to adjust them both to achieve the proper look. When using manual exposure, correctly setting brightness and contrast becomes critical.

On consumer camcorders, these controls usually fall on the bottom of the viewfinder. Often recessed, adjustments require a small blade or Phillips screwdriver. Most people find factory settings adequate in diverse shooting situations. Nevertheless, keep such a screwdriver in your camcorder bag for those rare adjustments.

On professional camcorders, it's very easy to reach the knobs that control color and brightness. News reporters and documentary videographers don't have time to fumble with screwdrivers in the heat of a shoot.

LCD Technology

Black and white viewfinders have been the industry norm for many years. Manufacturers, however, couldn't help wondering if a color viewfinder would sell more units. Yet making a 1/2-inch color CRT was impractical—there's no affordable way to cram three electron guns and a color phosphor mask into such a tiny space. There had to be another way.

The answer turned out to be the color liquid crystal display. Once manufacturers succeeded in shrinking LCDs for other applications, the next step was to reduce them to the diminutive proportions camcorders need. Japanese engineers are the undisputed kings of miniaturization, and they accepted the challenge with aplomb.

Their labors resulted in the world's first one-inch color viewfinder, appearing here in the U.S. in 1988. First-generation designs were not that popular, but recent advances greatly improved the LCD's performance— and further reduced its size.

LCDs operate on a principle entirely different from the CRT viewfinder's. Instead of firing an electron beam at phosphors, LCD viewfinders use a special liquid crystal fluid that changes properties in response to voltage. When paired up with polarized glass, this fluid either allows light to pass, or blocks it. When just certain areas of the fluid receive a charge, only those spots change opacity. Control the voltage to enough tiny zones, and you can create an image.

Normal LCD displays, such as those found on wristwatches or camcorder status panels, use reflected light to generate images. Light passes through clear portions of the liquid crystal and reflects off a shiny backing beneath the display. Those areas of the crystal that remain dark block the light. If there's no light shining on this type of display, there's no image.

A camcorder's LCD viewfinder differs from this example on a few points. First, a color viewfinder must generate its own light. It can't reflect light from the viewing environment; there isn't any with your eye pressed against the rubber cup. Instead, a tiny light source sits behind the LCD panel. Thus the liquid crystal in a viewfinder or other backlit display blocks direct light, not reflected light.

Second, a camcorder's LCD viewfinder must generate a color image. Liquid crystal, by its nature, does not generate different colors. It can only block or pass whatever light strikes it.

The key to getting color images from an LCD is the use of a color mask, much like the one found on a single-chip camcorder's CCD image sensor. This mask consists of alternating squares of red, green and blue material. Each lies above its own discrete pixel of liquid crystal. As each pixel lets light pass it sends its particular color to your eye. You see the blend as a full-color picture.

Most LCD displays have color and brightness adjustments; manufacturers permanently set contrast to maximum. The brightness control has the same effect as that of the CRT viewfinder, allowing you to compensate for especially bright or dim shooting locales. On an LCD, this control actually adjusts the brightness of the backlighting element. The color adjustment works just like the color level control on a normal television set. Increasing this control intensifies your image's colors, while decreasing it all the way produces an almost monochrome image.

LCD displays offer one more control that CRTs don't: hue or tint. This adjusts the color balance of the viewfinder without actually affecting what's recorded on tape. If you're relying on your color viewfinder to verify white balance settings, this control can cause problems. More on this later.

Eye to Eye

When comparing CRT viewfinders to LCDs, two major differences become apparent.

Resolution. CRTs offer higher resolution. Most 1/2-inch or 2/3- inch CRTs offer resolution approaching that of a high-priced monitor. It's not uncommon for viewfinders in high-band camcorders to offer 350 or 400 lines of resolution. Professional cameras often have 1.5-inch viewfinders resolving over 500 lines.

LCD viewfinders offer limited resolution—the standard 96K-pixel LCD resolves around 200 lines. This limited amount of detail can make manual focus difficult, obscuring the telltale sharp edges of a crisp shot.

LCD developers are pushing hard to overcome the LCD's resolution problem. Newer designs promise much higher resolutions, and future LCD displays will approach the detail of CRTs. Current designs far surpass first generation LCDs in resolution and sharpness, in spite of their smaller size.

Contrast ratio. This is the difference between the brightest and darkest areas of the image. Contrast ratio is superior on CRT viewfinders. Some color LCD viewfinders look "washed out," rendering brightness extremes as shades of pastels instead of true dark and light. This is another area seeing rapid improvement.

Liquid crystal displays often show higher levels of image smear and lag than CRTs. Due to the amount of time required for the liquid crystal fluid to change states, fast moving portions of the image leave a blurred trail behind them. Again, this drawback of LCDs is far less pronounced in newer designs.

With all these disadvantages, why would anyone want to use an LCD viewfinder? The answer is simple: color. Our eyes expect to see our world rendered in a full palette of different hues, not just a range of grays. Camcorder users have made the adjustment to monochrome viewfinders out of necessity. But when a grizzled camcorder veteran first uses a color viewfinder, he might find the difference almost breathtaking.

Color viewfinders do more than just render our world in pretty, lifelike color. Sometimes color is the only thing that distinguishes a subject from its background. Trying to find a colorful bird against a busy backdrop of leaves, for example, is often a lesson in frustration with a black and white viewfinder. It's hard to find the bird looking for shape alone. With a color viewfinder, however, the bird's bright colors stand out like a beacon.

Second, a color viewfinder allows you to watch for proper white balance while shooting. It's easy to forget to switch white balance settings when moving from one lighting condition to the next. But with an LCD viewfinder, you'll notice inaccurate colors before you start recording. With a monochrome viewfinder, there's no way to know if your white balance is correct unless you attach an external color monitor.

There is one *caveat* involved in relying on a color viewfinder for white balance: you must set the hue control correctly, or colors will look wrong even when they're not. Thankfully, the default setting is accurate on most camcorders. If it's not, you can easily adjust your LCD for true colors.

Here's how. Put your camcorder in a location with a known, single color temperature light source. This might be strong sunlight, or a room well-lit with only tungsten bulbs. Set your camcorder's white balance to the appropriate color temperature, and point the unit at a piece of white paper, color bars or some other multi-colored object. If the colors in the viewfinder match those seen with your naked eye, your hue or tint control is correct.

If the colors don't match, adjust your hue control to match them as closely as possible. Remember this setting, as you may need to return it to this position each time the camcorder powers up. This will not guarantee correct colors in all lighting conditions—that's still a function of your camcorder's white balance. But with your color LCD adjusted properly, you'll know when your camcorder fails to deliver accurate color. You can then make corrections before you shoot, instead of after the fact.

Menu Madness

In a camcorder, the viewfinder does more than just show the recorded image. It can also display a wealth of data about the camcorder's operation mode.

On most units, everything from battery status to microphone setting shows in the viewfinder. This is both helpful and distracting, depending on how much space remains for the actual image. In some viewfinders, having numerous special modes engaged floods the display with words and numbers.

Many camcorders allow you to access advanced features through menus instead of discrete buttons. This represents an effort to simplify the physical design of the unit, as well as streamline operation.

Unfortunately, in-viewfinder menus rarely make the videographer's life any simpler.

Why not? Part of the reason involves the lack of any kind of standard between manufacturers. Each camcorder maker devises a unique way of stepping through the menu and selecting functions; some are more intuitive than others. On most camcorders, a function that would otherwise be a single button-press becomes a multi-step operation. What's worse, you must keep an eye pressed to the viewfinder during the operation.

Industry insiders propose a universal viewfinder interface using a small joystick or trackball, similar to the graphic user interface growing more common on home computers. With this scheme, the videographer would move a pointing device to select menu options.

At this point, however, manufacturers have little motivation for developing such an instrument.

The situation with in-viewfinder menus can only get worse, especially as camcorders continue to shrink in size. Space considerations will eliminate almost all buttons, making the camcorder more reliant on menus and software controls. Hopefully, manufacturers will begin to see the user interface as worthy of research and development dollars.

Bright Future

There's no question that viewfinder advancements in versatility and image quality are coming on fast. For the first time in the history of consumer camcorders, the viewfinder is actually the most distinct characteristic of certain models. With color LCDs showing up on camcorders of virtually every format, viewfinder type has in fact become a significant buying point.

And the trend shows no sign of stopping. Future viewfinders will deliver even better image quality to keep up with high-resolution formats. Unique designs are on the drawing board, many of which remove the viewfinder from the camcorder itself. Where will all this lead us? Out from behind the camcorder, that's where. After all, we've been there long enough.

7

What's Under the Hood: Inside Your VCR or Camcorder

Robert Nulph

Did you ever notice that sometimes the movies that you rent from your local video store don't play very well on your VCR? Perhaps that ball game that you taped last week has a white fuzzy streak running across the middle of it, or your camcorder munches your tape as you record that once-in-a-lifetime event.

These are symptoms of your VCR or camcorder being out of whack. This is either because the tape you are watching is bad, or the natural wear and tear from hours of use. You can fix most of these problems. All you have to do is look under your VCR or camcorder's hood, recognize what you are looking at and care of the problem. For serious problems, take the time to talk to a reputable VCR technician.

To help you understand the gadgets, wheels, buttons and pegs under your VCR's hood; we will look at the tape transport system. We'll also look at the various kinds of VCR head drums. We'll look at the differences between formats

and the advantages and disadvantages each format has with tape transport and recording styles. Finally, we will discuss some troubleshooting techniques that could save you money on expensive repair bills, and explain why some tapes just don't want to play on your machine.

The Parts Department

Before you can fully understand a system, it is a good idea to know the parts. In a VCR, it's important to understand the functions of a number of parts. The largest part of the VCR's tape path mechanism is the head drum. (See Figure 7.1.) You will find this large, precisely tilted cylinder at the center of the VCR. The tape travels across it diagonally in a method called a helical scan. As you look at the drum, you will notice small objects called "heads."

The heads of a VCR are small electromagnets and either erase the signal from a tape, read the signal that is already on the

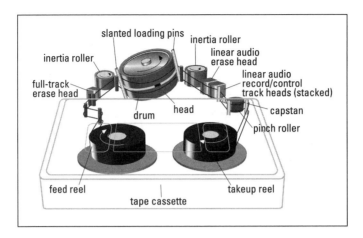

Figure 7.1 *The tape path and its parts.*

tape or record a signal to the tape. They do this by reading or changing the polarity of the oxides found on the surface of the tape.

In a VCR, there are eight different types of heads. The full erase head, found on VHS machines, is a stationary head located just before the head drum in the tape path. It erases everything that is on the portion of the tape that passes over the head. This head erases the video track, all audio tracks and the control track. It does this so the new information that you record onto the tape will have no trace of the old information to interfere with it. (See Figure 7.2.)

You will find a flying erase head, located on the main drum assembly, on some VHS and virtually all S-VHS, 8mm and Hi8 format VCRs. This head erases along the same path as the record heads, and gives the VCR the ability to handle precise, glitch-free editing. (See Figure 7.3.)

The video playback heads, located on the drum assembly, convert the signal recorded on the tape back into an electrical signal and feeds it to the outputs of your VCR.

You will find the audio playback heads in different locations for different formats. The VHS hi-fi, S-VHS hi-fi and all 8mm audio playback heads are found on the drum assembly. If your camcorder or VCR is a VHS unit, you will find the standard analog audio playback head on a post located next to the drum assembly.

Each video record head records one field of a video signal. There are usually two video record heads on the drum. Each head records a video field, and the two fields combine to make up the video frame. Some VCRs or camcorders have four heads. Increasing the number of heads by two gives the recorder the ability to record video using a smaller drum. It also wraps the tape three-quarters of the way around the drum instead of the usual one-half. Since there are four heads, the recorder can switch between the working heads to ensure proper placement of each video track. On some VCRs the four heads split, so that two heads record video and the other two heads provide clearer slow motion and pause playback.

There are a couple of different types of audio record heads. The hi-fi stereo audio head, located on the drum assembly, records high-quality stereo in the same track as the video. The audio signal records deep within the oxides of the tape, and the video signal records in a layer closer to the surface of the tape. This is referred to as "depth multiplexing" as shown in Figure 7.4. Although this provides for great audio, it makes it impossible to dub new audio to the tape without destroying the video on top of the audio. Hi-fi VHS, S-VHS and AFM 8mm all have this type of audio record head.

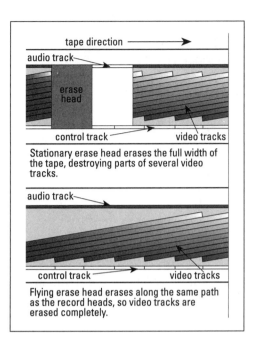

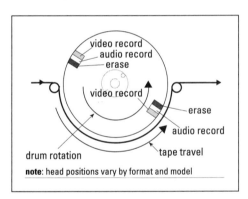

Figure 7.2 *Stationary erase head vs. flying erase head.*

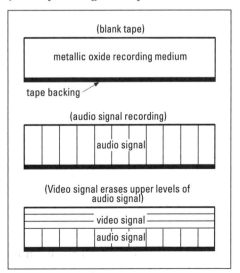

Figure 7.3 *Overview of various heads on the spinning drum.*

Another audio record head is located after the drum assembly for linear VHS and S-VHS audio. It records a linear analog audio track. This track is often mono, but some models split the audio into low-fidelity stereo. To accompany the linear audio record head, decks that are capable of audio dubbing will also have an audio erase head. This head, located immediately before the linear audio head, erases anything already on the audio tracks.

The final type of head is the control track head. This head records a sync sig-

nal onto the tape to enable the VCR to play back the tape at the same speed every time. This recorded signal also controls the counter on the front of your VCR. Located on the same post as the linear audio record head on VHS family models, this head lays down a continuous linear control track. On 8mm and Hi8 recorders, the control/tracking head, located on the drum assembly, lays down a signal on the same track as the video signal.

The Tale of the Tape

When you slide your favorite movie or a blank cassette into the VCR, a number of things happen. The protective flap on the video cassette opens, and when you hit play or record, two slim, smooth pins called loading pins slide up behind the tape and pull it into the VCR's mechanism.

Rollers and guide pins gently press the tape against a series of heads located on the VCR's head drum. These various heads alternately erase, record or play back the audio and video information depending on what buttons you press on the front of the deck.

Let's look at what the tape does on its journey through the system. In a standard

Figure 7.4 *Cross-section of videotape showing depth multiplexing of video "over" audio signal.*

VHS VCR or camcorder, the tape first encounters the full erase head. When you press the record button, this head erases the tape as it passes across it. The tape is now ready to record a new signal.

After erasing the tape, it then wraps part way around the spinning drum found at the center of the VCR. This drum contains all or some of the following: video record and playback heads, stereo audio heads and a flying erase head (see Figure 7.5). Remember helical scan? As mentioned before, the drum is spinning at a high rate of speed at a very precise angle. The tape passes across this slanted drum creating a pattern called a helix, as shown in Figure 7.6.

This helical scan is the secret to how VCRs and camcorders can record so much information on a relatively short tape. Because the tape runs across the drum at an angle, the recorder can lay down more and longer parallel tracks. This gives the record and playback heads the chance to read or record onto the slim diagonal strips provided. As the drum spins and the tape feeds over it, it records or plays back the video, and in some formats, the audio information.

In the VHS family, the tape then passes across two fixed heads, stacked one on top of the other. These two heads are the linear audio record head and the control track head. If your VCR is capable of dubbing linear audio, the tape will pass a

linear audio erase head before it gets to the audio record head.

The final major part of the tape transport system is the capstan. This slender metal post rotates at a very consistent and precise speed. A rubber roller presses the tape against the capstan so that when it spins, it pulls the ape along through the mechanism at a constant rate.

From there, the tape then goes to its take-up reel.

Different Formats—Different Tracks

As we described earlier, the VHS tape has one or two linear analog audio tracks along the top edge of the tape. There are video tracks in the center of the tape, and if your equipment permits, hi-fi stereo audio tracks. Along the bottom of the tape is a control track. However, not all formats are created equal.

Unlike the VHF family, the 8mm family of tape formats has no linear tracks and there is no need for a stationary full-track erase head. The drum assembly contains all the heads; audio, video, control and flying erase heads. The tape is 8mm wide, two-thirds as wide as VHS tape.

The tracks on the 8mm tape are all diagonal, including the tracking section that is equivalent to the VHS linear control track. Sectors divide each diagonal track. This format combines video, AFM audio and the tracking signal all within the same sector. It also includes a digital stereo PCM audio sector and a time code sector. Like the hi-fi tracks in VHS tape, you cannot change the AFM audio without changing the video. You can edit the PCM audio sector however, without changing your original video.

DV formats have a similar transport system as the 8mm format. There are no stationary heads that could become misaligned. The drum contains all the necessary heads. Like the 8mm format, all the tracks are diagonal, and there are no linear tracks. The big difference in this

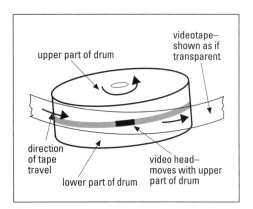

Figure 7.5 *Videotape moving past spinning heads on the drum.*

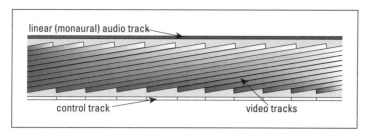

linear (monaural) audio track

control track   video tracks

Figure 7.6 *Helical and linear tracks of VHS tape.*

format is the way it records all the different information. The DV format divides each track into four sectors: sub code, video, audio, and Insert Track Information (ITI). Each sector contains digital information and has a specific layout and function.

The sub code sector contains the time code information and index identification that you can use to locate specific shots on the tape or identify specific still pictures.

The video sector is unique among all the formats. Each video sector contains one-tenth of a frame of digital video information. This is possible because the DV format head is revolving at 9000 Revolutions Per Minute (rpm) compared to 2700rpm for camcorders and 1800rpm for standard VCRs. The video sector also contains useful information such as the recording time and date, the input source and the video processing data needed for recording in the 16x9 aspect ratio.

The audio sector contains stereo digital PCM audio. You may choose between one 16-bit, CD-quality stereo signal, or two 12-bit, near-CD-quality stereo signals. You can edit either one of these choices.

The ITI sector contains the signal needed to maintain tracking, and allows video and audio insert editing. While the DV format is relatively new, its high-quality and versatility, as well as simple tape transport mechanism, should make it a star in the video production field.

Cleaning and Troubleshooting

Now that you know the parts and function of the tape transport system, lets look at some typical problems related to the system. If you insert a tape in your VCR and get nothing but snow, one or two simple things may be wrong, or a few serious problems may have developed.

The first and most obvious answer is that the tape is blank. The second and more troublesome answer is that the playback head is dirty. You can use a good cleaning tape quite successfully to take care of this problem. Just follow directions. There are two types of cleaning tapes available, a wet type cleaning tape that uses a cleaning fluid on the tape and a dry abrasive type cleaning tape. The wet type is better for your deck. Read your owner's manual and follow the suggestions given. Snow may also be an indication of more serious problems. If cleaning the heads does not help, see your local repair shop.

If you'd like to get into your machine to clean it manually, take the time to buy "head cleaning sticks" with chamois tips and head cleaning fluid. Do not use cotton swabs on video heads! The cotton fibers can stick in the heads and cause more problems. Always clean across the heads in the direction of tape travel. Don't clean up and down or perpendicular to the tape travel. Remember to be careful in there, bumping tape guides or other parts can throw things out of alignment. Always remember to disconnect the VCR or camcorder from power before opening it. Unless you have problems with video snow, intermittent color or loss of hi-fi audio, leave the video head drum alone!

If the tape you are watching is suddenly "eaten", it is probably due to a dirty or worn idler tire (which controls the take-up reel), preventing the take-up reel from turning. If this seems to be the problem, consult your dealer or VCR repair service.

If the top of the video is constantly shifting, or it rolls inconsistently, the tape may need to be "packed," which is accomplished by fast-forwarding to the end of the tape and then rewinding back to the beginning. If the tape is in the cassette unevenly, it can pull the tape mechanisms out of sync and create problems. By fast-forwarding and rewinding the tape, you distribute the tape evenly in the cassette and alleviate this problem.

The most common problem you may come across is bad "tracking." If you recorded the tape you are playing in a different VCR or camcorder, its alignment might be slightly off. If this is the case, the picture will have a fuzzy, noisy area in one section of the screen. You can fix this problem by turning the tracking control knob on your VCR. This knob moves the guide pins in your VCR to duplicate the alignment of the machine that recorded the tape. The change in alignment may be very slight, but in the videotape transport system, being a little out of whack can cause a whole lot of problems.

Shutting the Hood

Taking the mystery out of the tape transport system should give you a better understanding of how the various format systems work. Keep this in mind when you go shopping for a new VCR or camcorder. If you have specific editing and audio needs, you want to make sure you buy a format and VCR that will enable you to perform the functions you need.

8
Tape Truths:
All Exciting Overview of the Making of Videotape

Loren Alldrin

Ask any videographer about the craft of video, and you'll likely get an earful. Try it sometime—grab any videographer and ask about the topic of your choice. Be it lighting scenes with glowbugs, shooting from atop a moving train or even audio production with 8-track cartridges, chances are you'll find you have an expert on your hands.

Then, slyly, slip in a question about videotape. Don't make it too hard. Start with something simple like, "What is the difference between grades of videotape?" or "How is videotape made?" You'll probably get a different response altogether.

You'll probably get silence.

Surprisingly few videographers know what separates one brand or grade of tape from another. This is largely the fault of tape manufacturers, who've introduced a slew of confusing buzzwords and acronyms in an attempt to create some distinction for themselves within the market. The result is that videotape, the very medium of our visual expression, has for many years been the victim of numerous half-truths and marketing ploys.

Fortunately, the essential nature of that thin black ribbon is not hard to comprehend. The more videographers know about the materials and manufacturing of videotape, the better informed their buying decisions will be.

Advances and Variations

On a basic level, all magnetic tape is the same. Whether audio, video or computer data tape, there's still that thin layer of magnetic particles covering a flexible mylar backing. By passing this thin ribbon over an electromagnet, information is stored and retrieved.

A tape's magnetic particles number billions per square inch and function like tiny bar magnets. Though each particle is physically anchored in the tape's coating, its magnetic polarity is free to change and swivel when a magnetic force is applied.

Before recording, the particles are oriented randomly. During recording, the video heads arrange the particles into patterns dictated by the changing voltage of the video signal. These patterns are then picked up by a playback head; amplified and processed to become the video image.

Improvements in videotape over the years have been dramatic, keeping pace with advances in hardware. Today's tape offers frequency response and noise levels that match or exceed the decks and camcorders they enter.

This doesn't mean further advances in tape manufacturing are impossible. On the contrary. The closer tape manufacturers come to the perfect tape, the closer we get to realizing the full potential of video gear.

There are many variations in the tape manufacturing process, because there are many different manufacturers. Each puts a personal twist on methods and materials, hoping to achieve an edge in the market. Considerable research and development is invested in tape manufacturing, with special attention directed toward high performance—and high priced—formulations. Capturing the dollar of the uncompromising video purist is quite the competitive industry.

Secret Formulas

While videotape may appear simple, it's actually the culmination of years of audio and video research.

From the early days of magnetic audio recording on thin metal wire to the first rust-covered tape with paper backing, magnetic recording technology has steadily advanced. Today's manufacturing techniques benefit from the latest research in physics, chemistry and electronics, yet the basic methods resemble those of yesteryear.

The actual magnetic medium in videotape starts out suspended in a liquid known as a binder. Where binders were once simple glue holding the magnetic particles in suspension, today they've become a complex molten brew of adhesives, lubricants, cleaners, solvents, dispersion agents and static-controlling compounds. Each manufacturer has its own blend of binder ingredients, and jealously guards the details of its exclusive mixture.

To this hi-tech soup are added the actual magnetic particles. Mixed in liquid form in large sealed vats, the binder and magnetic particles are computer-monitored for temperature, humidity, pressure and time. When conditions are perfect, the binder is applied to the tape's base film, bonded by chemical action. The tape is then passed through large ovens where the binder is dried and hardened, the magnetic particles suspended and dispersed evenly on the surface of the tape.

Initially, the magnetic particles are oriented randomly in the binder, scattered through the liquid like pigment in paint. Yet the physical alignment of the particles is crucial to efficient magnetic recording. To orient the particles in the same direction, the tape is passed through strong magnetic fields as the binder hardens. The more uniform the dispersion and orientation of the particles, the better the tape performance. Early techniques used a single magnetic field; today manufacturers achieve improved uniformity by passing the tape through two or more fields.

A smooth finish on both surfaces of the tape is crucial. This is accomplished by compressing the tape through large, polished rollers under extreme pressure. Called "calendaring," this deceptively simple process affects the noise level, friction and overall stability of the tape.

At this point the tape is still in large rolls, each many feet wide and weighing thousands of pounds. Before being loaded into cassettes, it must be slit to the desired width and wound onto large "pancake rolls." These are then placed in automatic tape loaders, which add a small section of leader and wind the desired amount of tape into the cassette. Pancakes are also sold to cassette duplicators, who load the precise length needed into shells before duplication.

A high degree of precision and cleanliness is necessary throughout the entire manufacturing process. The tape must be slit within microns of the desired width to insure smooth operation in VCR or camcorder. It must be properly loaded into a well-designed cassette, or jamming and breakage will occur. Specks of dust or backing material picked up at this stage will manifest as dropouts or clogged video heads.

Making the Grade

The star of every tape is the tiny magnetic particle, solely responsible for picking up and carrying the video signal. Particle size, composition, density and distribution play a large part in determining a tape's performance, and these are the areas where tape manufacturers concentrate most of their efforts.

Early videotapes used magnetic particles that were relatively large. They were easier to formulate, disperse in the binder and distribute evenly along the tape. But the size and relatively sparse distribution delivered limited frequency response and high noise levels.

In the old days, ferric oxide was the most common magnetic material used; cobalt was added to the particles to stabilize and improve their magnetic properties. Chromium dioxide—chrome—was employed by some manufacturers.

Early research was devoted to reducing size and increasing particle density. Particle size decreased steadily, but manufacturers soon discovered smaller particles were more difficult to disperse evenly in the binder. New binder formulations and application techniques were then developed in response, causing videotape performance to improve dramatically. Longer, elliptical particles were created for even greater magnetic densities.

Today's formulations benefit from smaller, needle-shaped magnetic particles that can be packed incredibly tight on the surface of the tape. New production technologies allow particle orientation to be controlled with a high degree of precision.

Greater magnetic density is one of the major differences between the different grades of videotape. High-grade tapes use smaller particles in a greater concentration than normal-grade cassettes. This results in improved S/N ratios, better frequency response and a greater amount of magnetic retention. High-grade tapes cost more, as they require more expensive materials and stricter manufacturing methods. Extremely high-grade cassettes are usually manufactured in much smaller quantity, a factor contributing to their cost.

The difference in performance between normal and high-grade tapes is often dramatic, with the latter delivering greater detail, truer colors and less noise. Some high-grade tapes, when recorded in the slower EP speed, will outperform normal tapes recorded in SP. Multiple generations hold up better on high-grade tapes: third-generation high-grade tapes may look better than first-generation normal-grade.

Super-VHS formulations take high-grade even higher, using even smaller and more densely packed particles. More magnetic energy can be stored at a higher frequency, as required by the format. Manufacturing tolerances are even more stringent than for high-grade cassettes, with quantities considerably lower. S-VHS tapes command the highest price of any half-inch format; they also deliver the best performance.

Some videotapes are labeled "hi-fi," promising increased audio performance. In reality, most manufacturers feel the relatively few hi-fi tapes purchased don't justify producing a special hi-fi formulation. Instead, manufacturers may use a different sort of selection process for determining tapes with optimum noise and dropout figures. Other manufacturers simply adorn a high-grade tape with the "hi-fi" logo to increase sales. According to tape manufacturer Scotch 3M, virtually all high-grade tapes deliver excellent hi-fi performance.

Heavy Metal

In the '90s, a slightly different particle formulation appeared under the aegis of VHS tape manufacturers JVC and Maxell. Called magnetite, this cobalt-doped material offers a 20 percent increase in magnetic potential over standard cobalt ferric oxide. Magnetite was first researched over a hundred years ago, but found to be too unstable for magnetic recording. JVC and Maxell have succeeded in encasing volatile magnetite in a sheath of stabilizing material, and have begun using it in all grades of VHS videotape.

Metal particle tape (MP), common in consumer-level 8mm formats, uses normal manufacturing methods with magnetic particles of a different composition. Based on an iron molecule, MP formulations deliver dramatically higher performance than standard cobalt ferric oxide tapes. This allows high quality pictures to be recorded on tape with significantly less surface area than half-inch formats. While MP formulations could conceivably be used in VHS and S-VHS formats, the magnetic powders involved cost quite a bit more than cobalt ferric oxide. Whether the resulting difference in quality would justify the cost to consumers is questionable. Many professional formats, including digital video systems, use MP formulations.

A more recent advance in videotape technology, metal evaporated tape (ME), uses a different manufacturing method to deposit magnetic particles. Instead of being carried in a binder and painted onto the tape, ME magnetic particles are vaporized from a solid and deposited onto base film. Inside a vacuum chamber, an electron beam heats metal to thousands of degrees. Inside the chamber the metal vaporizes, adhering in an extremely fine layer to the specially prepared base film.

A protective coating is then applied to the magnetic layer. The result is a smooth, thin, densely-packed film of pure magnetic particles.

Due to the extreme purity of the magnetic layer, ME tapes deliver performance many times that of standard VHS formulations. No binder is used, allowing the individual particles to mesh with a density approaching solid metal. While early ME tapes were known to suffer from dropout problems, newer methods have reportedly reduced dropouts to levels comparable to other formats. ME tapes therefore represent the pinnacle of consumer tape manufacturing, and it is the technology used by the various DV formats.

Advances in tape manufacturing have affected all grades of tape, with today's least expensive name-brand VHS tapes offering much higher performance than normal-grade tapes of the past. Ultrahigh-grade tapes available today deliver video quality unobtainable even a few years ago.

Frequency Response

As with most forms of manufacturing, making videotape involves tradeoffs and compromises. Advances in one area create challenges in another. But tape manufacturers continue to believe videographers will choose whoever promises a better videotape.

One area of tape performance caught in compromise is frequency response—the range of frequencies a tape is capable of capturing and reproducing at the same signal level. Wide frequency response guarantees the tape will accurately reproduce the entire spectrum of video information. This characteristic relies in large part on the size of the magnetic particles, with smaller particles generating better luminance detail through extended high-frequency response.

Unfortunately, such small particles don't do as well with the lower frequency color information. The result is a crisp, detailed image with compromised color accuracy.

Using larger particles extends low frequency response and delivers better color reproduction, but high-frequency detail suffers. One solution involves using two different magnetic layers, the lower using larger particles for optimum low-frequency response, the upper incorporating smaller particles for optimum luminance detail. This method, used in Fuji's Double Coating videotape line, is said to offer broad frequency response and reduced noise levels.

An ultra-smooth magnetic surface provides better signal-to-noise ratio by improving the contact between video heads and magnetic media. At the same time, a smoother surface also increases friction between tape and head/transport assembly. The solution lies in integrating advanced lubricants into the binder to reduce friction, insuring the heads enjoy an easy trip over the tape's surface.

A completely different challenge arises at the back of the tape. Here, friction is actually an ally to smooth, consistent movement of the tape through the transport. Coating the backing with layers of carbon increases friction and makes it more uniform, insuring the transport drive surfaces get a good bite on the tape. The carbon also serves to control static buildup, an enemy of any kind of recording.

Tape Strength

Binder composition presents tradeoffs as well. It needs to be solid enough to hold the magnetic particles in proper alignment but flexible enough to not impede tape travel or shed oxide. Early binders were deficient in preventing magnetic particles from flaking off in the transport, which reduced the life of the tape as well as the hardware. Modern binders are considerably more durable and flexible, some nearly eliminating oxide shedding completely. A number of manufacturers use a multi-layer binder, with a slightly different composition on each layer. This allows a more rigid binder to be used on one level, with a more flexible composition above or below it.

One of the most obvious tradeoffs in videotape manufacturing is the relationship between tape thickness and durability. Longer record/playback time is a significant purchase point for many videographers; manufacturers have scrambled to be accommodating by fitting more tape into cassettes. The convenience, however, comes at a price.

A videotape's tensile strength is determined by the thickness and composition of the base film (see Figure 8.1). Longer tapes require a thinner backing, which can compromise tape resistance to breaking and stretching. Temperature extremes, poorly designed or maintained tape

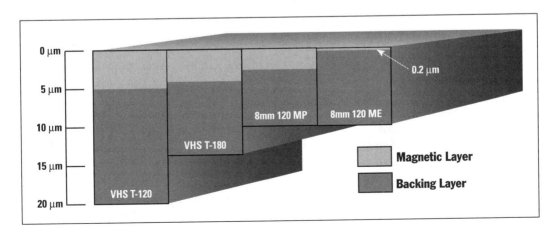

Figure 8.1 Videotape Thickness

transports, even frequent play/rewind cycles can put significant mechanical stress on a videotape. Even though some manufacturers claim to have developed stronger, thinner backing layers, shorter tapes still have a better chance of surviving such rigors.

In addition to snapping or stretching the tape, mechanical stresses can also cause small sections of magnetic material to shed off the backing. Visible as fleeting white lines knifing through the video image, dropouts spell disaster—or at least frustration-for the video producer. An otherwise perfect shot can be rendered useless by just one poorly—timed dropout.

Selection Confusion

The actual performance difference between brands of tape within the same grade may be hard to measure and even harder to see. All tape manufacturers face the same trade-offs and compromises, and all have experienced advances and setbacks at roughly the same pace. Unless scrutinized in a side-by-side comparison, with all other variables eliminated, it's hard to determine one tape's superiority over another.

Other factors add to the confusion. There is no industry-wide standard for judging tape performance; instead, each manufacturer measures videotape against its own "reference" tape. Most manufacturers use a high-grade tape, but each reference is a little bit different.

The bottom line is this: when comparing tapes between manufacturers, the numbers mean very little.

Comparing different grades of videotape requires hardware up to the task. If you expect to see a dramatic difference between normal and extra-high-grade tape on your $189 VCR and twenty-year-old TV set, you're dreaming. The hardware is simply not capable of tapping the increased potential of the higher-grade tape. Perform the same comparison on a high-end VCR and monitor and the differences will become obvious.

Sometimes, differences between batches from the same manufacturer are greater than those between brands. Computer-assisted quality control has helped eliminate inconsistencies between manufacturing runs, but there's still some variation. Batches of binder and magnetic material may differ slightly; the large sheets of base film may possess slight surface deviations. Fortunately finding yourself the victim of a "bad batch" is becoming a rare phenomenon.

The key to avoiding substandard videotape is to find a brand that offers consistent results and stick with it. Many videographers swear by a given brand of tape, touting its merits with almost religious fervor. The more you understand about the materials and techniques that go into tape manufacturing, the less you'll feel at the mercy of chance when selecting videotape stock. For every pound of advertising jargon, it's the ounce of fact that should guide your decision-making.

If a manufacturer comes out with a promising new development, don't be afraid to give it a try. Remember that most manufacturers offer a free replacement policy if you're not completely satisfied with the videotape. Don't be afraid to return a tape riddled with dropouts, offering poor video performance or emitting questionable noises.

Lastly, use your new knowledge of videotape to turn a critical eye to manufacturer claims. Analyze cassette boxes, brochures and advertisements with an eye to gleaning fact from fiction. Beware of "breakthroughs" that are nothing more than marketing hype. Yet be ready for legitimate advances that could have a real impact on the quality of your video productions. They do happen.

9
Degeneration: Limiting Signal Loss

Jim Stinson

Your Hi8 camera original looks so great that you think it might pass for DVC. The dub to S-VHS for editing still looks good, if not quite so crisp. But in the edited master tape the lines are fuzzier, the colors blearier and the bride's face looks like it was caught on a surveillance camera.

Face it: every serious videographer lives with tape generation quality loss.

The decay goes on. The release prints you make from the edit master are foggier, grainier, smearier still. And Uncle Ezra goes around bad-mouthing your videography skills because he's so cheap he duped his copy for free from the one his son bought. Of course it's crummy looking, Ezra. You've got a fifth generation copy, for crying out loud!

With all-digital cameras and decks, the generational quality loss problem vanishes. But until that happy day when all of us use all-digital gear, we must take every possible precaution to minimize quality loss from dub to dub.

So let's review what quality loss is, what causes it and how to avoid as much of it as possible.

First, a quick run-through on the symptoms of video degeneration.

How Pictures Break Down

When you review a copied tape, the first thing you notice is a vague, general loss of sharpness. Details are harder to make out and faces in long shots become less recognizable.

Edges suffer too, growing softer; you see an increase in video graininess and snow.

By the time you reach the third or fourth generation, colors start to shift out of alignment both horizontally and vertically.

In a fifth generation copy, all these defects worsen. And you may see yet another: vertical straight lines bulging into curves.

If you go down yet more generations, the image decay grows so pronounced that the video becomes unwatchable, and eventually, indecipherable. (See Figure 9.1.)

Figure 9.1 *8mm video copy degeneration: 1st generation to 4th generation to 7th generation*

As you dub from generation unto generation, the sound quality suffers too, especially the high frequencies. The decay from copy to copy is less objectionable, however, for two reasons. First, the audio signal is less complex than the video, so small losses affect it less. Secondly, our ears tend to be more tolerant than our eyes.

Who's the Culprit?

Several factors cause generation loss, each affecting different aspects of the picture.

Overall fuzziness can be the result of signal loss. Signal loss starts in the camcorder, which may not capture 100 percent of the signal sent from the CCD. It continues with the tape, where not every single microscopic rod of oxide is magnetized to record the signal. The strength also weakens when the signal travels over cables that are too long, too thin, or both. And the process of signal loss continues through every tape generation.

Signal loss also causes the decay of edge sharpness. How? Failure to retain the high frequencies that create the sharp transitions in screen brightness.

Another major villain is noise. Noise is any element of the current signal that was not part of the original video signal. In other words, noise is anything added to the signal courtesy of electronics or environment.

Audio noise most often takes the form of hissing. Video noise appears as snow and grainy picture quality. Noise also degrades overall picture detail and edge sharpness.

Like signal loss, noise begins at the camcorder chip and builds up from component to component, on its way to the editing VCR. On top of that, each tape dub carefully copies the noise as well as the program signal—adding noise of its own, to boot. After three or four generations, the cumulative noise level is high enough to obscure the subtle changes in the program signal—changes that carry the details of picture information.

Color shift is that annoying effect that makes the video image look like a badly printed newspaper photo; you know, where the movie star's lipstick appears a quarter-inch to the right of her lips. All the colors are slightly awry. The multiple edges make people on the screen look like cartoon characters rattled by an earthquake.

Color shift happens because the luminance (brightness) and chrominance (hue and intensity) signals take separate paths through cameras, VCRs and monitors. They usually come back together for composite transmission from one piece of hardware to another. These often divided signals fall slightly out of sync each time they recombine; the cumulative effect can create a serious mismatch.

Time-base error causes wavering vertical lines known as "flag waving" or "flagging." How these video scan lines appear depends upon a sophisticated timing process driven by the timing information on the tape. Compromise this crucial timing info, and all the scan lines will not start at the same point on the screen. Some will start further to the left or right than others; vertical lines will no longer be vertical.

Time-base error is the result of tape stretching, variance in head speed caused by excessive camera movement during recording or incompatibilities between recorder and playback device. Whatever the cause, it accumulates from dub to dub; wavy lines or even double images worsen with every generation.

If the time-base error grows too great or the timing signal becomes too weak, the image may not appear at all. My video students love to make low-rent special effects by manipulating the shuttle/jog controls on their editing source decks for stills and slow or reverse motion. But when they copy their programs to play at home on consumer VCRs, they sometimes lose the picture completely during their special effects.

Why? Because the timing signals sent during special playback modes are weaker than normal. Though the powerful cir-

cuitry of the editing decks picks up the weak timing signals, the home playback decks can't find them.

So there you have the causes of copying quality loss: signal loss, noise, color shift and time-base error. But that's like saying that a virus causes the flu. The important question is, what can you do about it?

Like the flu, copy degeneration can't be cured. But you can practice healthy video habits to minimize it and take steps to combat its symptoms. This means shooting to avoid it and editing to reduce it.

Videotaping for Maximum Quality

For obvious reasons, the better your initial tape signal, the more quality loss it can absorb. You can maximize original tape quality if you:

1. Use the best camera you can afford. Three-chip cameras produce cleaner signals with less noise and better high-frequency response than single-chip models. Even in single-chip cameras, high-end models boast sophisticated circuitry that enhances signal strength and reduces noise. The newest technology incorporates signal amplifiers at the chip level—or even at the individual light sensors.

2. Use the very best tape you can for the camera original. The characteristics you're looking for: high "coercivity" (the ease with which the tape can accept a signal); and high frequency response. Remember that edge sharpness in particular is a function of the high-signal frequencies.

3. Always shoot at the fastest tape speed, if your camera has more than one. Faster tape speeds lower noise and improve high frequency response.

4. Light your subject. Your camera may indeed shoot in illumination as low as three or even one lux, as advertised. But it does so by electronically amplifying

the initially weak signal it generates from low light. Remember: when you amplify the signal, you magnify the noise right along with it. The result is a screen crawling with grain. A little light on the subject will improve quality dramatically.

A/B Roll Shooting

If you edit A/B rolls (to create wipes, dissolves and other effects), make a first generation B-roll as you shoot.

If you go into post production with only your camera original, you'll have to dupe it to create a B-roll. That way, you lose a generation and produce a visible mismatch between the original footage and the dupe.

To avoid this, cable your camera to a VCR containing a second tape. When you shoot, put both camera and deck in record mode; the VCR will make a first generation copy of everything the camera shoots.

When shooting with two cameras, take the idea a step further by switching live and "on the spot." When my students tape a program in the school auditorium, we create a master assembly tape by recording main titles before the show and then cueing up the tape in the VCR just after those opening credits. During the show, the director and technical director use a digital switcher to send signals from camera A or camera B to the master tape.

After the show, we roll in end credits. Ta da! A first generation reproduction master that will create release tapes only one copy away from the original.

Whether you make a duplicate original or switch between two cameras, always do two things to minimize signal loss and color shift:

1. Keep cables between camera(s) and deck as short as possible.

2. Use Y/C connections rather than composite video. (More about this in a moment.)

A show is predictable enough to edit a first generation master on the fly. But most events are too unmanageable for that. Even then, you can create release tapes that are only third generation-as long as you plan your editing strategy to minimize duplication.

Plan for Savvy Editing

The trick is to plan your strategy from the start. This way, you can perform all editing operations on the second-generation program assembly tape.

First, include all titles as you build your assembly. Don't create a separate title tape to use as an editing element. Instead, lay your titles in directly from your titler, computer or camera-and-title cards. This is easy with plain titles, but more difficult with moving titles and those enhanced with effects.

For example: if your titles roll up the screen, you can't lay them over several live action shots unless you edit those shots beforehand and then superimpose the title crawl over the edited dupe. Oops! You've lost a generation. To avoid this, choose a single shot long enough to roll titles over and then lay titles and live action down at the same time.

You can also save a generation in A/B roll editing by creating two camera tapes, as described above.

TIP: at the very head of the tape, record something to use as a sync point; I shoot the digital face of a running stopwatch. Then use the sync reference on each tape to null the time counters of the A and B edit decks. This makes it easier to get and keep the two VCRs in sync.

Editing Audio

Since audio quality loss is less obvious than that of video, save video generations by pre-mixing audio tracks. Then lay the composite onto the second generation video.

One way to do this: mix and record audio on a duplicate of the video assembly tape. The video on this copy tape is used solely for timing the audio. When mixed to your satisfaction, dub the audio across to the original assembly video. The result: second-generation audio and video.

Depending on your equipment and experience, you can handle more than just music and narration tracks. You can loop the original stereo audio through a mixer to combine it with the music and effects. Alternatively, you can take a separate tape of ambient sound and run it through your mixer. Whatever the technique, the basic idea is the same. If you need extra generations of audio, build them separately. Then re-lay them onto the second-generation video.

More Hot Tips

Here's a grab bag of ways to reduce quality loss in the edit bay.

- Don't dub 8mm or VHS-C tapes to VHS for editing. Instead, use the 8mm camera as a source deck; or work directly with the VHS-C tape in its converter cartridge.

- Use your camera as a source deck, it will produce the most reliable sync signal. Why? Because it created that signal to begin with.

- You do want to dupe your camera originals if you have access to a fully professional format like 3/4-inch or Beta or Beta SP. Because these professional formats incur far less generation loss, you can transfer small format originals to them and do all your editing in the larger format. This way, you can still create release copies of better quality than you could have achieved otherwise.

- If your VCR has a switch marked "edit," by all means use it throughout the editing process. The edit switch disables circuitry designed to improve the picture on a TV or monitor. The circuitry actually reduces quality when dubbing.

Lose the Black Boxes

Videographers often employ stand-alone signal massagers like digital switchers, color processors and video enhancers. However useful, these black boxes all contribute to quality loss. Every time you run a signal into an input and back out again through an output, the circuitry between the two contributes to signal loss, noise and time-base error. Unless those connectors handle luminance and chrominance separately (Y/C connectors), color shift increases as well.

That's why so-called dubbing enhancers are no longer so popular. They degrade the signals in some ways even while enhancing them in other ways. And they add noise in the process.

A rule of thumb: take processors out of the loop when you don't need them. Say your next few edits are straight cuts. Remove the temporarily unnecessary digital mixer. You can always patch it back in later when you need an A/B effect.

And even when you do need a piece of hardware, be careful how you cable it into the signal path. For instance, many monitors include audio/video output jacks. It's tempting to route your signal from source VCR through the source monitor and then to the assembly VCR. But if the source VCR has two sets of output jacks, better to use one for the monitor and the other for the assembly VCR. That gets the source monitor out of the recording signal path, and minimizes the chance of signal loss.

Making Connections

In a way, the humble cables between editing components are as important as the hardware they connect. Not even copper wire is a perfect conductor; cables mean resistance and capacitance. Resistance spells overall signal loss; capacitance reduces high frequencies.

The thinner and/or longer the wire, the greater resistance it offers. The voltage changes that make up a video signal are measured in millivolts, or thousandths of a volt; even minor losses to resistance can degrade your video and audio.

To avoid the problem, buy hefty, shielded cables. Avoid the cheaper, flimsier cables sold to connect stereo components. And keep all cables as short as possible. Don't use a six or twelve-foot cable for editing. With a little planning, you can connect everything in your edit bay with three-foot cables.

Down the Highway

If you think that correcting every detail that can degrade signal quality is a pain, you're right. But for now, it's your only practical alternative to muddy copies. High-end pros don't have these hassles; they take a high-quality original—like Beta SP—and bump it up to one of several digital formats. From there, they can make as many generations as they need, without losing any quality whatsoever. And now, with the advent of the digital formats (Mini DV and Digital8), videographers of all experience levels can get their hot little hands on digital video. This is a good thing, because prosumers have long wished for a better format; the VHS and 8mm formats have been pushed and pulled and squeezed as far as they can go.

Analog vs. Digital Copying

Though desktop computers and music CDs are old hat by now, many people still don't fully understand the difference between digital and analog recording and copying. (A CD or a DVC tape is digital; a VHS videotape is analog.) Briefly, where

an analog recording imitates its source; a digital recording encodes it.

In analog audio, for example, the changing voltage of the recorded electrical signal mimics the changes in air pressure of the sound that created it. In xerography, the copy imitates the appearance of the original document. With a player piano, the holes in the piano roll tape imitate the pressures of the human fingers that originally played the piece.

In the same way, the video signal, modulating up to six million times per second, mirrors the electrical current variations of hundreds of thousands of light sensors on the camcorder chip. Many of those variations are as tiny as 1/1,000 of a single volt.

Whatever the medium—electrical signals, paper and toner, piano roll—the copying method creates a record analogous to, but not precisely the same as, the original being duped. When the original zigs, it zigs; when the original zags, it zags. It is, literally speaking, a copy. An analog copy, imperfect, yet still very similar to, the original.

And there you have the problem with analog recording and duplicating. Sounds, documents, photographs, video images—all these originals are full of very subtle details. But each copy loses some of those subtleties. Why? Because (unless we invent matter duplication in some future era) no copy can be an absolutely perfect replication of its parent. There are inevitable changes introduced in the copying process.

In video copying, the imperfections creep in from progressive signal loss (especially the high frequencies) and from the buildup of video noise. Noise accumulates from copy to copy, degrading more and more of the program signal.

In digital audio and video, the recording process does not imitate the original. Instead, the digital recorder analyzes the electrical signal repeatedly, and assigns each sample a number. It then converts this number to binary—a long string of ones and zeros. Instead of a complex and subtly changing signal that mimics the

Figure 9.2 *Digital video copy degeneration: 1st generation to 4th generation to 7th generation*

This means that a digital signal does not have the minute fluctuations that plague the analog signal. If the recording system has a one-volt threshold, for example, the signal strength is either one volt (encoding "1") or zero volts, (encoding "0"). No subtlety here—the recorded signal is either high or low. And since the equipment reads anything under one volt as a "0", it's blind to noise, which typically begins at about a millivolt.

In making an analog copy, you must copy the entire signal, noise and all. Because the copying process adds yet more noise, the overall noise level builds from generation to generation.

In digital copying, by contrast, you're not really copying anything. The playback unit sends a "1" or a "0" down the wire; the recorder writes a new "1" or "0" to tape or disk. Any noise is left behind. The new noise that you add is too far below the threshold of readability to interfere with the digital pulses. The result: digital video post-production that may involve many generations with no quality loss whatsoever. (See Figure 9.2.)

For certain esoteric reasons, digital video is not capable of an infinite number of perfect generations, as people who create highly sophisticated digital animation know. But for regular production needs, it's a gift from heaven.

Is It Really Worth It?

Minimizing inter-generation quality loss is a highly incremental process. No single tactic will produce a dramatic improvement in your release tapes.

So start with high-quality tape and hardware, and carry quality control all the way through the editing process to the final release tapes. You'll gain improvement that anyone can see at a glance.

original sound or image, the result is a series of strong, simple, discrete pulses.

Why does digital encoding permit perfect copies? Because the digital signal is so simple that it's: 1) not subject to signal loss; and 2) not affected by noise.

10
Resolution Lines

Bill Rood

With video equipment manufacturers increasingly engaged in spec wars over lines of resolution it seems appropriate to investigate those figures, what they mean and why they're so often misleading. Knowing how to measure resolution will help you make a smart purchase the next time you look for a camcorder, VCR or monitor.

Much of the confusion centers around the use of the term "lines." Lines of horizontal resolution should not be confused with scan lines. In America, the National Television Standards Committee (or NTSC) television system mandates that the television picture will consist of 525 vertical scan lines, each scanning from left to right on the screen. This fact does not change, no matter how sophisticated the video gear.

So when a manufacturer boasts that a device features "400 lines of resolution," the reference is not to vertical resolution, or the number of scan lines. What's under discussion is horizontal resolution, or, more specifically, horizontal luminance

resolution. The chroma resolution in the NTSC system is as little as one tenth that of the luminance, depending on the particular hue. So for our purposes, I'll discuss only luminance resolution.

When a Line Is Not a Line

While the number of scan lines is fixed and can be counted, the number of "lines" in the term "lines of horizontal resolution" is in fact strictly a unit of measurement. There are no actual lines you can count, except with a special test chart (see Figure 10.1). You can put your face right up to the picture tube and see the scan lines, but you can't see lines of resolution.

We should actually refer to horizontal luminance resolution as "video frequency response." It's expressed in megahertz (MHz), usually with a tolerance, just as with audio equipment. Unfortunately, consumer video equipment manufacturers apparently believe this too complicated for the average consumer to understand, so they use the questionable lines method instead. Measurements stated in lines also sound more impressive than those expressed in megahertz. Three hundred lines sounds better than three-and-a-half megahertz.

So how does this frequency response differ from vertical resolution? You can measure video frequency response by examining just one of the 525 scan lines, provided you are displaying a test signal of vertical bars. The frequency response of the entire picture should be the same on every line.

As an example, let's examine one scan line of a black-and-white picture. As the scan line traces from left to right, we'd see that the brightness of the line at any given point is a function of the picture content at that point. If the picture consists of a white picket fence against a dark background, the line would start off dim, then brighten as it reproduced one of the pickets. It would go dark as it passed between pickets, then brighten again when it hit the next one. And so on. This sequence would continue until the end of the line.

What happens if we make the pickets on the fence closer together? The line still must switch between light and dark, but faster. In effect, we've upped the frequency of the input signal we're trying to reproduce. At some point, a given piece of gear cannot make the changes fast enough; thus, we arrive at the limits of its resolution. As the pickets got closer, they would begin to appear less bright. Finally, you would no longer distinguish a picket from its neighbor. The black and white pattern will melt into a neutral gray.

Specs Game

So how do we measure video frequency response, or obtain horizontal resolution specifications?

For equipment handling video signals in a purely electronic form, use a test signal known as "multiburst". This signal contains a series of bursts, nothing more than white/black/white transitions. Tiny picket fences, if you will, each higher in frequency than the one previous. When

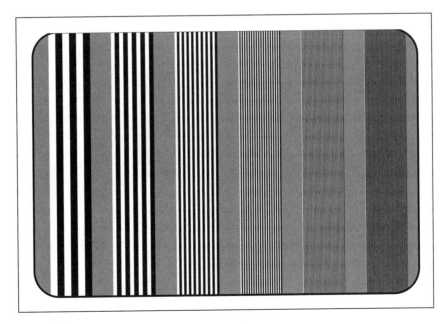

Figure 10.1 *Horizontal Resolution Test Chart*
Credit: Thomas Fjallstam

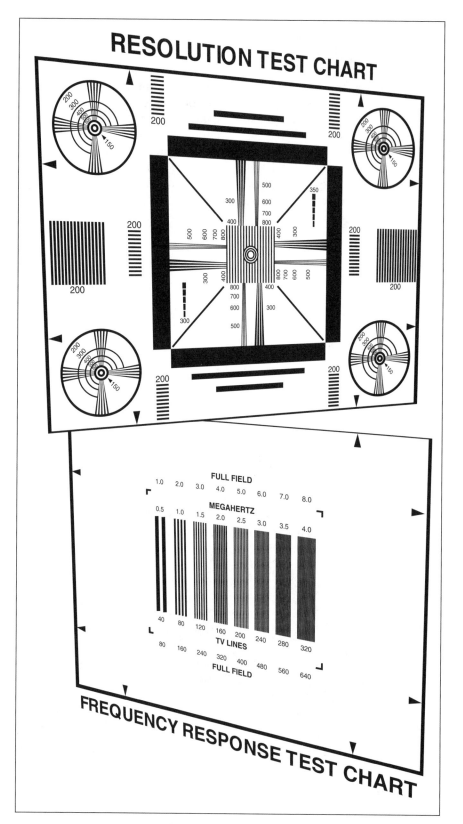

Figure 10.2 *Resolution and Frequency Response Test Charts* Credit: Earl Talken

viewed on an oscilloscope, each burst should offer the same amplitude.

If the signal drops off as the frequency rises, you know you're seeing high frequency response roll-off. The frequency of each burst is fairly standard; usually 0.5MHz, 1MHz, 2MHz, 3MHz, 3.58MHz and 4.2MHz. The beauty of this method is its extreme precision, with no sloppy guesswork.

For devices which pick up or display images, like cameras and monitors, virtually the only way to determine horizontal resolution is to display a special resolution chart (see Figure 10.2). This chart features little wedges: a series of converging black lines on a white background. As the lines come closer together, you again begin to encounter a point where the change from black to white to black occurs so quickly the lines become indistinct.

Along the outer edge of the wedge at regular intervals are line numbers, reading 200, 250, 300, 350, 400 and so on. This is where "lines of resolution" comes from. As you might guess, it's a rather inexact method at best; the scale is quite coarse, and open to interpretation as to where exactly the lines blur together. It also doesn't take into account the fact that while black/white/black transitions may be visible, they may appear at a substantially reduced level, indicating high-frequency roll-off.

Unfortunately, this method is used to spec virtually every piece of consumer video gear, be it camera, VCR, disc player or whatever. This means you should take any stated resolution spec with a grain of salt the size of Nebraska. There's much room for error, interpretation and fudging.

Lines to Megahertz

Approximately 80 TV lines equals 1MHz of video bandwidth. So a piece of equipment rated at 300 lines would feature a video frequency response of roughly 3.75MHz. This is all well and good, except the 300 lines figure most likely came from observing the wedge pattern, and may be less than accurate. Also, it's impossible to tell by this method if the response is flat at 3.75MHz, or even if there's any response at all.

With an off-air broadcast, video frequency response cannot rise above 4.2MHz. Using the above formula, we could then say that TV broadcasts feature 336 lines of resolution. The reason for the limit? The design of the television transmission system from way back when. It places the audio carrier at 4.5MHz; picture information must vanish by that point.

Most TV broadcasts originate from digital video, stored either on hard drives or from digital videotape, such as D-1. These machines also master virtually everything seen on pre-recorded home video, from VHS, to DVDs. These digital masters are a "full" 480 lines of resolution, or 6MHz of "flat" video frequency response. Unfortunately, the best resolution figure you can hope to see from broadcast is 360 lines—80 "lines" X 4.5MHz.

Until all manufacturers come clean and start stating horizontal resolution in terms of megahertz, and within a certain tolerance, we'll never have a reasonable standard with which to compare products.

Until then, read the specs if you will, but don't make judgments based upon them alone.

11
An Inside Look at Cables and Connectors

Joe McCleskey

Visit any place where video editors work and you'll likely find more than your share of cables: S-video cables, audio cables, power cables, composite cables, headphone cables, microphone cables and cables for cable TV reception, among others. And if the cables aren't confusing enough for you, there's a whole host of connectors to go with them, with names like BNC, DIN, RCA, phone, phono, XLR and stereo mini-plug.

In this chapter, we're going to take an in-depth look at some of the cables commonly used for video. We'll take a look inside the four main cable types that home video producers deal with: composite (RCA-style) cables, S-video cables, RF cables and DV (IEEE 1394) cables.

Composite Cables

Composite cables are perhaps the most commonly encountered cable type in consumer video. They often come in the box with home VCRs and camcorders.

Sometimes, they come in groups of three attached cables—one with yellow RCA-style connectors for video and two with red and white connectors for stereo audio. Also included in this category are those cables that have yellow, red and white plugs on one end and a mini-plug connector on the other end. This type of cable often comes packaged with miniature camcorders that make use of the single connector on the camera body. (See Figure 11.1.)

Often, video gear (especially professional video gear) will have BNC (bayonet nut coupling) connectors for composite video. The chief advantage of the BNC connector is its ability to lock in place with a push and a twist; this prevents connections from wiggling loose, a common problem with the typical RCA-style connectors.

On the video side, these cables carry composite signals, so called because the signal is a composite (or mixture) of all black-and-white and color information contained in the video signal. They

Figure 11.1 *Composite cable and composite input.*

usually consist of two wires running parallel to one another down the cable; one wire (power) corresponding to the tip of the connector and the other wire (ground) corresponding to the outside ring portion of the connector.

Why aren't these cables just called video cables? Because they actually carry two separate signals that are composited into one. Inside your camcorder and/or VCR, the black and white (luminance) information is dealt with separately from the color (chrominance) information. Back when television technology was black and white only, the only type of video signal a television had to cope with was a black and white signal. When color came along later, it was dealt with as another layer of information on top of the black and white signal, in order to make color TV signals compatible with existing black and white televisions.

To send both chrominance and luminance information together on the same cable, the two signals must be mixed together in a process known as modulation. The modulation of the signals—and subsequent de-modulation at the other end—is what makes the composite video signal highly susceptible to generation loss (the tendency of a signal to degrade whenever you make copies of it). The only type of video cables that are more susceptible to generation loss are the RF cables used for cable television and antenna hook-up.

RF Cables

RF cables gave cable television its name. The ubiquitous RF cable can be found mostly on television sets and VCRs, as well as a few old-school camcorders. RF stands for radio frequency. The name is appropriate because this is the cable most commonly used to transfer radio-frequency signals from an antenna to the VCR or TV set—or even from the cable TV station to your house.

RF cables are generally made of coaxial cable, a cable that carries two metal leads, one inside the other. The two wires do not carry two different signals; instead, they carry the power and ground for a single signal, with the ground (outside) cable providing limited shielding from radio interference. (See Figure 11.2.)

The RF cable (sometimes referred to as an F cable) provides the best means of long-distance signal transfer. This is why it's most commonly used to connect cable TV stations to clients; with a heavily shielded cable and amplifiers in far-away neighborhoods, it's possible to send an RF signal miles and miles without serious degradation. However, it is the worst solution for making copies or editing. Here's why: remember how we told you that composite video signals mix the color and black-and-white signals into one? Well, the RF cable does this one better: it carries audio along with the mixed color/black-and-white

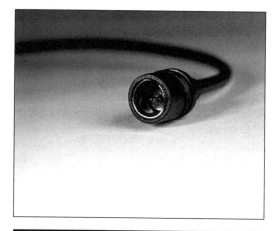

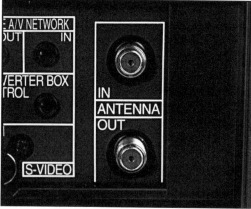

Figure 11.2 RF cable and RF input.

video signal. This means that problems arising from modulation/ demodulation of the signal are multiplied in RF cables. Generation loss accrues at a much faster rate and even second-generation copies tend to have an abundance of video noise, audio noise and bleeding colors.

Camcorder enthusiasts who have television sets without composite video inputs (yes, they do still exist) must find some way to connect the video outputs of their camcorder to the RF-style connectors of their TV. VCRs usually provide the needed connections, but for some who still have older VCRs without video connectors, the RF-to-composite connection may require a separate inexpensive device called a modulator, readily available at consumer electronics outlets. Unfortunately, you can't just whip up an adapter to connect composite cables

directly to RF cables without a modulator, because RF cables transmit signals differently than composite cables.

In short, you should only use RF cables for viewing video and use some other type of cable for copying and editing.

S-Video

One obvious way to solve the modulation problem that composite video and RF cables present is to leave the color and black-and-white information separate and send them down a pair of wires in the cable. That's what S-video or Y/C, cables accomplish. (In video technical parlance, Y is the symbol for luminance and C is the symbol for chrominance.) Because S-video cables do not carry two signals modulated into one, they provide a very robust means of editing and copying video.

Looking at the end of an S-video connector (see Figure 11.3), it's easy to see that it consists of two pairs of wires instead of a single pair. One pair of wires (power and ground) carries the color information, while another carries the black-and-white information.

S-video connectors are found only on high-bandwidth cameras and VCRs. You'll find them on S-VHS, Hi8 and DV equipment, but not on 8mm or VHS gear. More and more televisions are coming equipped with S-video connectors, but they are far from universal.

S-video cables do have some minor limitations. The most prominent is perhaps the length issue: when the cable goes beyond twenty-five or thirty feet, you notice significant degradation of the signal. For this reason, they are not the best solution for long-distance cable throws. There are devices (called line amplifiers) that can solve this problem, but they usually add a significant amount of noise to the signal.

There are adapters that convert Y/C signals to composite, but beware: because composite signals are modulated, the use

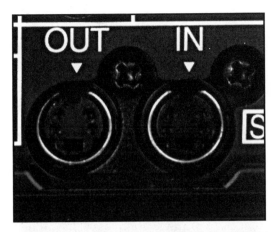

Figure 11.3 S-Video cable and S-Video input.

of such an adapter negates the benefits of the S-video connection. In fact, it may make the problem worse. Inside these adapters is a small ceramic filter that demodulates the composite signal before sending it down the S-video cable. This demodulation is the same sort of thing that adds noise to a composite signal. The use of an S-video-to-composite adapter just adds one more place where the signal can degrade when it's copied.

S-video connections work best when used throughout your entire system. In other words, you won't get all the benefits of S-video technology if you use composite video cables between, say, your special effects generator and your record VCR, with S-video cables everywhere else. It is possible to safely connect a monitor to the output of your record VCR with composite or even RF cables. Because they sit

outside the main signal stream, these cables won't affect the quality of your final product one bit.

FireWire

The cable types we've dealt with so far are all similar in that they're designed to carry analog signals. IEEE 1394 or FireWire or i.LINK, is special because its main purpose is to carry digital signals. Thus, you'll find FireWire connectors only on equipment that's designed to handle digital signals. In the consumer video market, this equates to Digital8 and Mini DV camcorders and VCRs.

Briefly, the difference between analog and digital signals is this: analog signals carry a continuously varying voltage, which corresponds directly to the type of signal that analog VCRs and televisions were designed to interpret. Digital signals, on the other hand, consist of long strings of numbers in binary notation (zeros and ones). Because digital signals only consist of two values—zero and one—they're much more resistant to noise and other forms of signal degradation. This means that it's possible to copy, say, twenty generations of DV or Digital8 footage without noticing the slightest loss of picture quality.

It's important to remember that although the DV and Digital8 formats are the most prominent applications of FireWire technology to date, in essence FireWire technology has nothing to do with digital video. It's just a way of transferring digital data from one location to another at a high rate of speed. FireWire is a serial data protocol, number 1394, approved by the IEEE (Institute of Electricians and Electronics Engineers).

FireWire connectors come in two basic types: the six-pin connector, agreed upon as a standard in the mid '90s, and the smaller four-pin connector, which has become most prominent on digital camcorders and VCRs. While one connector type will not fit the other, both are interchangeable for all other purposes, which

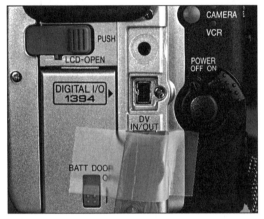

Figure 11.4 *FireWire cable and FireWire input.*

means it's possible to have a FireWire cable with both types of connectors, one at either end. (See Figure 11.4.)

Like S-video, FireWire connections must remain throughout the signal stream for users to reap all its benefits. In order to get copies or edits without a hint of generation loss, you must use FireWire connections throughout your system. Unfortunately, there are currently no titlers with FireWire connections, and only one special effects generator (Videonics' MXPro DV). FireWire connectors are becoming quite popular on home computers, which have reached a point where they have plenty of muscle for video editing.

FireWire's primary drawback: cable lengths are limited to 15 meters (about 45 feet).

Making the Connection

One final note about cables: in general, it's best to buy the highest quality you can afford, to insure the stability of your system and keep generation loss at bay. When purchasing cables, video editors are faced with a wide array of options. Even if you just want a simple yellow-tipped composite video cable, you still find yourself faced with a plethora of types from which to choose. Some boast greater shielding; some even come with built-in hook-and-loop cable ties to help you organize your workspace; some add gold tips. While gold certainly is an excellent conductor, it's usually the case that a gold-tipped connector is an expensive and unnecessary addition to a quality cable. More often than not, an ordinary nickel-plated connector will perform identically to its gold-plated brethren and gold plating has a tendency to flake off in time.

To summarize: whenever you're making analog video connections, use S-video cables if at all possible to minimize generation loss. If you can't use S-video cables, use high-quality composite cables. Avoid using RF cables for anything but viewing video on a monitor. And if you have two or more pieces of video equipment with FireWire connectors, then by all means use FireWire and avoid the ill effects of generation loss entirely. With all this in mind, go forth and cable your system with confidence.

12
Try a Tripod: Some Valuable Features in the Three-Legged Race

William Ronat

Few pieces of video support equipment are as useful as the tripod. Simple in concept, elegant in function, the tripod has a long history of bringing needed stability to the world of photography.

In pre-video days, tripods served still and motion picture cameras. As far back as the 1860s, people like Matthew Brady were lugging tripods onto battlefields to help steady huge still camera equipment.

And before tripods became popular for image gathering, they supported the surveyors' levels used to map out the countryside.

Why, you may wonder, a *tri*pod? Why not a monopod, or a quadrapod, or an octopod? This question doesn't require an Einstein to answer. One leg: camera falls down. Two legs: camera falls down. Three legs: camera stands up. Four legs: one more leg than you need.

The triangle is one of the most stable configurations for a support device. Ask a karate expert, or an offensive lineman. Or, for that matter, a tree.

Why a Tripod?

I know what you're thinking. "What do I need with a tripod? It's just one more thing to tote around."

Maybe you're right. If you have a steady hand, and/or a lens stabilizer, you may never encounter a situation requiring a tripod. But if you shoot professional video and work with heavy equipment, you know that working hand-held for any length of time can get darned uncomfortable. Just try holding your hand on top of your head for a couple of hours to simulate the experience.

Aside from avoiding pain, tripods are handy if you want to be in your own shot.

Say you're shooting a news story for a cable access show and you want to do a stand up. This is the shot where you appear on camera, looking solemn, finishing up with, "for Cable Access, this is John Smith, reporting from Bosnia."

Using a tripod, you can stop somebody passing by, make him or her stand in the spot where you'll be standing when you talk to the camera. You can then compose the shot using this surrogate John Smith. Lock down the tripod. Take the passerby's place and say your piece.

Three S Theory

A tripod's purpose can be described in the famous Three S Theory, which I just made up. Tripods keep it *Steady*, keep it *Straight* and keep it *Smooth*.

Put a camera on a tripod and it will be steady. It won't bob, wave or float, assuming it's locked down. It will sit there like a rock until you get ready to move it. That's steady.

Keeping your shot straight is a little trickier. Let's say you've set your camera on the tripod so your shot is looking out across a flat desert. The horizon is that line where the sky meets the earth. In a standard shot, the horizon should be kept parallel to the top and bottom of your frame. Cancel this if you're trying for the Dutch angle so popular in the old Batman TV shows, where everything is tilted. It's possible for a shot to start out looking straight, horizon parallel to the top of the frame. But when you pan, the horizon will start to go downhill.

This happens because your tripod legs stand in such a way the camera isn't level to the horizon. Your tripod is sitting with two legs on either side of the front of the camera; the third leg points behind the camera, and is shorter than the front two legs. Even though your shot looks level when the camera's pointed straight ahead, when you pan, the camera begins to lean in the direction of the third leg.

Or, as I like to put it: look out, the world's tilting.

On the Level

Some tripods come with a leveling bubble, a handy gizmo that is nothing but a bubble floating in liquid.

You position the bubble either inside a circle or between two lines on a tube. By moving the bubble to its correct position your camera becomes perpendicular, relative, I think, to the gravitational pull of the Earth (but don't hold me to this). The result: you can pan your camera 360 degrees, the horizon staying straight in the frame.

You can position the bubble by raising or lowering the tripod legs or by adjusting the tripod's head—if the head attaches to the tripod with a claw ball. The latter allows you to loosen the head and position the leveling bubble without touching the legs. A nice feature.

The last S in the Three S Theory is keep it Smooth. The part of the tripod responsible for this action is the head. Some tripods don't have heads: cameras attach directly to the tripod. But on more sophisticated tripods the camera attaches to a plate, the plate attaches to a head and the head attaches to the tripod.

Using smooth resistance, a head helps make camera movement smoother. This resistance, known as drag, is usually adjustable. With a small amount of drag the camera pans or tilts easily. Add more drag and moving the camera becomes more difficult.

If you don't want the camera to move at all, you engage the locks. There are separate drags and locks for both the pan and tilt functions of the head. If you want a pan but no tilt, you can lock the tilt control and the camera will only pan. And vice versa.

Heads and Legs

Heads come in two flavors: fluid head and friction head.

A friction head creates resistance by pushing metal against metal. A fluid head floats on a bed of oil or some other viscous fluid. Friction heads aren't as smooth as fluid heads, but they're also cheaper, which is the way things usually work in this world.

Tripod legs generally extend by telescoping. This is necessary to position a tripod level on a hill or stairs. With tripods that extend you can get your camera high up in the air, useful when you must ascend to eye-level with NBA players. Some tripods have a center column that cranks even higher.

A word of caution here, if you get up too high, your camera, tripod and everything else can tip over. So put a sandbag in the center of the tripod to make it more stable.

Some tripods allow the legs to straighten out until the head is resting almost on the ground. Good for low shots. Good angle for your remake of *Attack of the Fifty Foot Female Mud Wrestler*, featuring a point of view shot from the terrified town's perspective. Coming soon to a theater near you.

Wheels

With a nice smooth floor you may be interested in tripod dolly wheels.

What's a dolly? That's a movement of the camera and tripod. These moves can take the camera around the subject, or the camera can follow people at the same speed as they move. They require your tripod to have wheels or they require you to place your tripod on a wheeled device. These shots are very pretty, but they're also very difficult. If you don't have a smooth even surface every little dolly bump will translate into a very big video bump.

Wheels are handy, however, as transportation. Just leave your camera, extension cords, a grip bag and a light attached to the tripod and roll on to the next location. Sure beats carrying them.

Another feature you may want is quick release; a plate or shoe attached to the bottom of the camera. The plate fits into the head to secure the camera. But if you want to go hand-held in a hurry, you flip a switch or push a button to immediately release the camera.

If the head screws into the bottom of the camera, it will obviously take a lot longer to turn, turn, turn the knob to get it off again. Most professional model tripods feature quick release.

The Envelope, Please

Before leaving the subject of tripods, we should explore the Steadicam™. You may say, "Say, that's not a tripod!" And you'd be right. But it performs some of the same jobs, so we'll give it a glance.

The Steadicam JR™ is a system that balances your camera so completely the image seems to float on air. It eliminates shakiness, allowing a camera operator to walk up stairs or run along the ground without applying objectionable jiggle to the image. It's a slick little system, creating videos that look like feature film.

But a Steadicam™ is not the same as a tripod. Although it has a stand, you can't lock it down on a shot. Also, some people think a Steadicam™ is like a gyroscope, forcing your shot to remain horizontal. Wrong. There's a bubble level on the monitor to show the operators when the shot is level, but it's up to the operator to keep it there. My conclusion: Steadicams™ are great tools, but should supplement a tripod, not replace it.

If you're in the market for a tripod, shop around. Try the model before you buy. As you test drive the tripod think about the three Ss: keep it Steady, keep it Straight, keep it Smooth. If you watch your Ss, you should be O.K. And if at first you don't succeed, try another tripod.

13
Blending a Sweet Sound:
Audio Mixers

Jim Stinson

You've got your camcorder; you've got your VCR. What you might want next is an audio mixer. That's because a good soundtrack is crucial to quality video, and an audio mixer gives you greater control over the soundtrack. The trick is to get the right one.

In this chapter, we'll review what a mixer does and how to choose the best mixer for your particular needs. You'll learn how to find the unit that delivers the features you want at a price you can afford.

Audio Mixer Basics

Though a big professional unit comes with more intimidating knobs, sliders, buttons and displays than the cockpit of a 747, audio mixers are really straightforward critters that perform two basic tasks:

They balance the volume of sound elements—camera sound, music and narration—coming in from several different sources and blend them into a single audio program.

They process sound elements by adjusting their volume, timbre and perceived location.

Why bother to balance and process sound? First, because effective sound editing contributes powerfully to any video program. (Hollywood's known that for over 60 years; that's why they give Oscars for sound.) Audio mixing improves sound clarity by balancing elements so that the important sounds dominate the track. The process can intensify the drama of your images with realistic effects, and enhance the mood of your images with music.

Another reason for building creative audio is that the process itself is so satisfying. Enhance your video with a sophisticated audio track, and you'll fall in love with the creative challenge that is audio editing. In short, audio editing is fun.

Audio mixers come in a wide variety of flavors and prices. You can work with two mono channels or with 48 stereo

channels—or more. You can pick up a simple box for 50 bucks or shell out the price of a Mercedes for a unit as big as a pool table—or spend any amount in between. You can twiddle a couple of simple knobs or play arpeggios on an instrument with more bells and whistles than the collected works of Spike Jones. But mixers with eight channels or fewer are most likely to meet the needs (and the budgets) of prosumer videographers.

Mixers fall into four different classes: production, DJ, studio and audio/video. Each class includes a number of units with a range of features and prices.

On-Site Production

Production mixers are different from the rest in one crucial way: they're designed to mix sound elements during the actual videotaping rather than during post-production. In other words, they record mixed sound on the original camera tape, not the edited assembly tape. (All professional video crews use production mixers—usually run by the person who wields the mike boom.)

You can use all production mixers during the editing phase; just remember that many of them offer fewer features than those units made expressly for post-production use.

Disco Meets Video

A number of vendors offer audio mixers designed primarily for the disc jockeys hosting most contemporary parties, banquets and balls.

These hard-working artists have special needs. A disc jockey wants to:

- Fade from one music source to another one in real time, using a cross-fader control.

- Prepare the next piece of music privately by listening to it through headphones patched to the cue channel.

- Detonate soothing and tasteful sound effects like machine guns and sirens, either with built-in digital effects or with a pre-programmed sampler.

- Dip the levels of music and machine guns together while performing voice-over announcements via the mixer's talk-over feature.

What's any of this got to do with video post-production? Lots, when you think about it. Like running a disco, video sound editing is often a real-time operation. Since you must record your final composite audio in a single pass, you are often juggling several sound sources at once. You may want to mix from one sound source to another, so a cross-fader control would simplify the task. You may need to locate the next sound effect or piece of music while recording, so a cue channel would let you audition a source without recording it. You may want to lay down perfectly synchronized sound effects, so a sampler that stores effects and puts them at your fingertips would sure help.

And if you're adding narration, it would be great to have a talk-over function that would consistently dip both music and ambient sound to a preset level whenever you began speaking.

With all that said, however, a DJ audio mixer can work beautifully in video production—if you already have one. But if you're buying a new, full-featured mixer, a studio unit might be preferable.

In the Studio Audio

In general, studio mixers tend to offer more versatility in input/output functions and signal processing.

Those mixers tend to cost more than the others discussed in this chapter do, but as with most audio equipment, you get what you pay for.

The most basic advantage of these mixers is simply better sound. They are designed for audio alone, so they offer the most in audio quality. The manufacturers of these units tend to be more careful about such considerations as audio noise and shielding.

They also offer more in the way of versatility—stereo line inputs, mike and phono inputs, analog or digital VU meters, equalizers, cueing channels, cross-faders, and so on.

Still, you may prefer to trade some advanced features for convenience and lower cost. Maybe you should check out an audio/video mixer.

Audio/Video Processing

Audio/video mixers combine video and audio processing functions in one piece of hardware. In the video department, they mix A and B rolls and often supply special effects ranging from simple color wipes to elaborate digital processing. If you can live with their usually limited audio processing abilities, A/V mixers offer two big benefits:

• They put both audio and video functions under your fingertips in the same place and using the same style controls.

• Some compare in price to stand-alone audio mixers with similar features.

For example, some A/V mixers include two audio inputs for line and mike signals. That means you can mix original camera sound with music and narration—which is as much as many videographers need to do.

Some even offer video fades and wipes—something you can't expect from any audio mixer, no matter how expensive.

And if you want to go upscale a bit, you can get fancier audio/video features, such as stereo inputs and cross-fade controls between A and B rolls.

Two Sound Tricks

Many audio mixers have few channels and even fewer sound processing features. Here

are two simple ideas for getting around these limitations.

FIRST TRICK: Premix unsynchronized tracks. Audio tracks fall into three types:

1. Those you must perfectly synchronize, such as original camera sound with dialogue,

2. those that should be closely but not perfectly synchronized, such as narration and music and

3. those that are unsynchronized, such as atmospheric background tracks including surf, traffic and restaurant chatter.

If you have four tracks—camera, narration, music and background—but only three audio inputs, you can premix all but the camera audio to make an audio subassembly tape, and then make a final mix of this tape and the camera.

This is a practical solution, because the generation-to-generation quality loss in audio signals proves much less obvious than in video.

SECOND TRICK: Daisy chain processors. If your mixer does not have a built-in equalizer or limiter, and it also lacks a send/return capability, simply patch your signal processor between the source and the mixer.

For example, background traffic noise can often ruin your on-camera interviews. To reduce that noise without affecting your narration or music channels, connect your video source's Audio Out to a graphic equalizer and then patch the equalizer's Line Out to your mixer's Source In.

Or, if you want to equalize the entire audio program, connect the graphic equalizer between the mixer and the assembly record deck.

Decision Time

As you can see, choosing the right audio mixer for you means deciding what you want to do with it, how you want to do it and what you want to pay. To help make those decisions, here are some general suggestions.

If you have not yet performed much sound mixing, start with a simple, inexpensive unit. Then, if you enjoy the process of audio editing and feel the results are worth your effort, you can move up to more versatile equipment.

If you are an experienced sound editor and know your current and future audio needs, get the best and most versatile mixer that will meet those needs and still suit your budget. One more time: high-quality sound is well worth the investment.

And whichever route you take, remember that sound is not a post-production chore but one of the most creative aspects of video production.

Give it the care and imagination it deserves. You won't be sorry.

PART II
Map the Journey

Pre-Production Planning

People who like budgeting love pre-production planning. Trouble is, no one likes budgeting. Setting and sticking to budgets is like taking cod liver oil; we do it because it's good for us. They help us meet our goals. It may be hard to keep within the chocolate budget, but doing so allows us to buy a new camcorder. It may not be much fun in the beginning, but in the end, you're gland you did it. Pre-production planning is like that.

Creating a good production plan helps us assess our resources and our needs for resources: people, equipment, time and money. It helps us anticipate problems (there is no A/C power at the mountain location) and therefore solve them in advance (bring extra camcorder batteries and a generator for the lights). Most importantly, it helps us exploit creative opportunities to their fullest (while we're shooting the fourth panel of the storyboard on the mountain, we should take some additional panoramic shots to use

for the title in panel number one). A good production plan is a map and a set of driving instructions that will take us from our first shot to the last tear in the eye of our viewer.

Elaborate productions *require* production plans, but even your simple vacation video can benefit from one. Whether your plan fills 50 pages or one side of a cocktail napkin, it will make your production run more smoothly and will let you harvest better results.

Depending on the type and scope of your production, your plan may include one, some or all of these elements:

- Description of concept and treatment
- Description of target audience
- Plan for distribution
- Plan for promotion
- Script
- Storyboard

- Descriptions of the various roles for talent

- Descriptions of the various crew positions

- List of equipment needs

- List of supply needs

- Shooting schedule

- Editing schedule

- Budget. Yes, budget. It's good for you.

This section of the *Handbook* fills in the details of the major elements in a production plan. Make yours as detailed as your project requires. Remember, every minute spent planning can save you ten minutes (and sometimes even hours) in execution.

Once you've completed your plan, carry a copy with you through all phases of your project. While you're on the road to finish your video, your map would do you little good sitting in your desk drawer.

14
Honing Your Ideas:
From Concept to Finished Treatment

Stray Wynn Ponder

See video clips at www.videomaker.com/handbook.

If you're like most videographers you probably have more project ideas than you can shake a camcorder at. So with a little talent and the right equipment, you should be able to produce top quality video work, right?

Right. Then why do so many great ideas fizzle out somewhere between that first blinding spark of inspiration and the final credit roll?

The answer is simple: before the lights come up, before the cameras roll, even before you write the script, you must take two essential steps if your video is to find and follow its true course:

Step One: clearly define your concept.

Step Two: write a concise treatment.

A concept nails down your program's primary message, and the manner in which you will deliver it to your primary audience. Later, as you navigate the winding curves of production, you'll think of the concept as your destination. A treatment is a written summary of the video's purpose, storyline and style. It will become your road map. These tools will help you maintain solid and continuous contact with the video's intended direction every step of the way.

These are probably the most overlooked steps of pre-production, but if you conscientiously pursue them on every project—no matter how simple—you'll save time and add polish, propelling your work to new horizons of quality.

Developing the Concept

How does a concept differ from a raw idea? Let's look at a couple of ideas and watch how they change as we develop them into concepts:

The Trees of New England; and *Car Repair.*

Each of these has possibilities as a video project; but if we were to pick up a camera, or to start writing a script at this point, we'd suffer a false start. Before we can set out on our creative journey, we need a clear understanding of our destination.

Admittedly, many ideas don't deserve to survive. Who among us hasn't pulled the car off a crowded freeway to jot down a "great idea"—only to read it later and find that great idea somewhat less than overwhelming.

Take our first idea: *The Trees of New England*. This sleeper might die right on the drawing board. Why? Because, for the videographer trying to earn a buck, it lacks profitability. And for most hobbyists, it involves too much time and effort. The visual effect could no doubt be stunning, but who would purchase (or finance) a video about trees when public television carries a variety of nature shows that feature similar subjects every week?

To succeed in the marketplace, your work must effectively deliver a primary message to a primary audience. To prove worth the effort, *The Trees of New England* would have to distinguish itself from similar programming through style or content to appeal to existing markets. Another option: *The Trees of New England* could deliver its message in a way that would captivate audiences in a new market niche. *Note:* if you can see a way to make money with this tree idea, please feel free to run with it.

You may find yourself shelving many ideas that survive this kind of initial scrutiny; these ideas typically lack some element necessary to a profitable video, such as reasonable production costs or a viable market. Or through research you may discover that someone else has already produced your idea. That's okay; you can always generate more ideas. Don't get too caught up in creative decisions during these first stages of exploration. In the process of transforming a germ of an idea into a viable concept, necessity will make many decisions about a project's direction for you.

How about our second wannabe video—*Car Repair*? This one offers a mul-titude of development possibilities. But remember, you can't please all of the people all the time. Avoid the temptation to create a "do-all" video. As producers, we always want the largest audience we can get—up to a point. Create a repair program that appeals equally to master mechanics and interior designers, and you'll get a show without a specific destination. In other words, your project could end up running out of gas in the wrong town.

Your first move: define the audience. Let's find a target group who could use some information about car repair.

Brainstorm A-comin'

Here's where brainstorming becomes indispensable. There are as many ways to brainstorm an idea, as there are people, so there are no hard and fast rules. Basically, you need to distract the left (logical) side of your brain so that the right (creative) side can come out to play.

Here's what works for me: I speak my thoughts aloud, no matter how silly they sound, while bouncing a rubber ball off the concrete walls of my basement office. This technique gets the creative hemisphere of my brain churning; my subconscious coughs up ideas from a well much deeper than the one serving my logical hemisphere. I write down the more coherent mutterings on a dry erase board as they erupt. All in all, it's probably not a pretty sight, but you're welcome to adapt this method to your own brainstorming technique.

Here's a condensed version of my brainstorming session for the car repair idea. I flip the ball. It hits the floor, the wall and then slaps back into my hand.

"Repair," I say to myself, as I continue to bounce the ball. "Maintenance... mechanics... men... women... children... women... smart women... independent women... car maintenance... where's the need?... when would they have the need?... college! ... BINGO!

When young women go away to college, they no longer have Mom or Dad around to watch the oil level and check the belts. The same is surely true of young men, but I decide to target women as the larger of the two potential audiences. Should I go after both in hopes of selling more tapes? Absolutely not. Since the buying characteristics of the two groups will be different, I must tailor the style of the production to one audience or the other.

Through brainstorming, the original idea "car repair" has now become its simpler cousin, "car maintenance." Do we have a real concept now? Not yet, but we're getting there; we know our target market and our message. Still to be considered: the production's style, or the best manner in which to convey our message. This will eventually encompass shooting style, lighting style, acting, wardrobe, makeup and dozens of other factors. For now, however, we'll break style down into two parts: 1) getting the viewer's attention; and 2) keeping it.

Hook, Line and Profit

A hook is the attention-getting element that yanks viewers away from their busy day, and into our product. The need for a good hook is the same in every communication medium, whether it's an advertisement, a popular song or a training video. Human beings are frenetically busy creatures; you must seduce them into giving their attention away. After delivering this interesting hook and convincing them to look our way, we must follow through and give them a storyline that will hold their interest for the duration of the program. There are a number of ways to engage and keep the viewers' attention:

- Shock value

- Self-interest

- Visual stimulation

- Glitz and glamour

- Comedy

To decide which combination of elements will work best for our car maintenance video, we need a better understanding of our target market: 18 to 22-year-old females needing to perform simple car maintenance themselves. As with many aspects of concept development, most of our decisions are made for us as we discard what will not work—which leaves us with what will.

My gut says to skip shock value in a program that deals with cars. Self-interest is definitely an important consideration for a young lady who is both: 1) trying to assert her independence for the first time (ego self-interest); and 2) living on a budget (financial self-interest).

Visual stimulation? Our target group comes from a generation accustomed to the kaleidoscopic imagery and lightning fast cuts of beer commercials and music videos. Let's use this one.

Glitz and glamour are obvious shoo-ins for this age and gender. Comedy can be an excellent tool for communicating many subjects, as long as you execute it well. Let's keep humor in mind, too.

Simply being aware of these tools is not enough. More important is an understanding of the ways they will impact our target audience. If we can effectively use one or more of them in our production (and our marketing package), we may just have a moneymaking project on our hands.

To recap: we need an eye-catching (visually stimulating) presentation that offers college-aged females something they clearly need (self-interest) in a manner consistent with their accepted versions of self-image (glitz and glamour). If we can discover ways to enliven this delivery through the use of comedy, all the better.

Even if we are unable to meet all these criteria, we must be aware of them, so at the very least we avoid working against the psychology of our target audience.

More ball bouncing is probably called for at this point to help us predict how we'll apply these general ideas to our intended audience. But rather than put you through that again, I'll just tell you what I came up with for our sample project: *A Young Woman's Guide to Minor Car Maintenance*. The package resembles that of a concert video or a compact disc more than an instructional videotape jacket. Lots of neon colors surround a snazzily dressed college-aged woman, who leans confidently over the open hood of a small automobile. Her posture says, "I have the world by the tail, and so can you if you take a closer look at this."

The back of the jacket explains that you'll need no tools to perform most of the tasks covered in the program. These tasks are simpler than you ever thought possible, even fun once you give them a chance. Best of all, you'll feel an exhilarating new sense of independence after you master these simple skills.

Writing the Treatment

We've come a long way from the original idea. By asking the right questions, we've developed a potentially viable concept. We understand it in terms of:

- to whom the video speaks,

- how the video will speak to them and

- what the video will say.

Now we can write a treatment, which will help us pursue our project without losing sight of our concept. By clearly defining our direction in this way, we can hold true to our original vision for the project.

Depending on the complexity of a production, its treatment may be long or short. Some in-depth treatments resemble scripts; others simply document mood changes and/or visual effects, with technical annotations along the way. Regardless, the treatment should always move the reader chronologically from the beginning to the end of the program.

There's no established manuscript format for a treatment. Just try to tell a story in as readable a way as possible. The treatment for our car maintenance video might begin like this:

Project Name: A Young Women's Guide to Minor Car Maintenance

Statement of Purpose: The main goal of this project is to provide information about basic car maintenance to female college students under the age of twenty. These young women face the full responsibilities of car care for the first time in their lives.

In the interest of hooking and keeping the attention of the target audience, we'll present this information in a series of three music videos. Cuts will be as short as possible. A different actor/musician with a distinct personality will demonstrate each automotive maintenance task.

Most important, the tasks will not be overly technical in nature. Our audience needs to understand only the basics of car care: how to check belts, check the oil and other fluid levels, change a tire, fill the radiator, replace a burned-out fuse and so on. The frequent use of common-sense metaphors will remove any feelings of intimidation this subject may arouse in viewers.

The video jacket layout resembles that of an album cover rather than an instructional videotape. The songs contained in the program will be remakes of popular rock-and-roll songs, with lyrics pertinent to the mechanical tasks.

Summary

The opening credits emulate the digital-animated effects common to music video TV stations. These lively visual effects are choreographed to heavy guitar and powerful drums. The monolithic CTV (Car Television) logo vibrates in time with the music.

Cut to a perky female vee-jay who says, as if continuing a thought from before the latest station break, "We'll hear more of the latest tour information soon, but first let's take a look at this new release from Jeena and the Jalopies..."

Cut to close-up of female lead singer in the middle of a concert. We hear the giddy cheering of a large crowd as she introduces the next song. Her tormented expression prepares us for a tale of love's cruelty; but when she speaks, it's about how her car has done her wrong. The hand-held cameras circle like vultures on the fog-drenched stage. Her dead-earnest performance mocks the lyrics, which seem comically out of place.

Cut to a dressing room interview with Jeena. "Yeah," she says, "almost every song I write is taken from my own life. I hated that car. (She takes a drag from her cigarette.) "And I loved it. Know what I mean?" Music from Jeena's live performance fades up as the camera holds on her face.

(Music continues.) Cut to Jeena standing next to her car, a late model import. She wears the demeanor of a child instructed to shake hands with an enemy, but stubbornly refuses to do so. She casts occasional guilty glances at the camera, but refuses to look at the car, with which she is obviously quite angry. "My old car wasn't like this," she claims, shaking her head. "I could see the dip stick—easy. Check the oil and be done with it. So, you know, easy." Video dissolves to a memory sequence of Jeena opening the hood of an older automobile.

That gives you an idea of how the beginning of our treatment might read. It paints a much more complete picture than the words Car Repair. This video will probably be around 30 minutes in length; its treatment will run about ten pages, typewritten and double-spaced. If that sounds like a lot of writing, compare it to the amount of money and work required to reshoot even one minute of video.

More Treatment Tips & Tricks

Some productions, like our car maintenance video, will involve fairly hefty budgets financed by outside investors. The treatment then becomes a sales tool for communicating the project's value to potential investors.

Depending on the type of video you're producing, other uses for a treatment include:

Seeking client approval,

giving a "big picture" of the program to the technical and creative staffs and

making sure that you can arrive at your destination.

Perhaps the most important benefit of writing a treatment comes as a result of the writing itself. In moving from the general concept to the specific steps to develop that concept, your treatment will pass through many incarnations. Problems will crop up at this stage of the video's development; you'll solve them by revising the treatment. In overcoming each of these obstacles on paper, you will save yourself from facing them later on the shoot itself.

Production Planning Tools

Videographers have traditionally used several tools to help them navigate the circuitous pathways of production. In filmmaking, there's the storyboard, a comic book style layout of sequential drawings that tell the visual story of a movie. Some videographers use storyboards as well; but for many low-budget productions storyboards prove too expensive a luxury.

This is certainly true for our car repair video. For this production, our treatment must do the storyboard's job—by creating compelling, descriptive images with words. The treatment must clearly map out the avenues we'll travel without necessarily describing every fire hydrant and blade of grass along the way.

A general rule of thumb: gear the sophistication of your treatment to the purposes it must serve. If you need to impress the board of trustees at a major cable network and feel you are out of your league in terms of writing skills, hire a freelance writer to prepare the treatment. The earlier in the creative process you bring this person in, the more benefit you can gain from his or her experience.

Don't sell yourself short, though. If you feel reasonably sure that you can tell your video's story from the beginning to the end, in a readable way that your colleagues will understand, do it.

Planning Counts

The worst mistake: skipping these crucial planning steps altogether.

Even the simplest video can flounder if you neglect the proper planning process. The meticulous development of concept and treatment allows you to cut and polish your rough project. The goal is to move into the later phases of the work with a crisply faceted jewel that will withstand the rigors of scripting and production.

15

The Sands of Time: Scheduling Your Video for Success

William Ronat

Maybe your idea of preparing for a shoot is to confirm that you've charged your batteries and there's blank tape in your camera bag. A schedule might consist of calling your friend on the telephone and telling him to meet you at the park. If this is the case, then you probably haven't been involved with any videos that have the complexity of *Gone With the Wind*. Or if you have, you've been very lucky to pull it off.

But if you want to take as much good and bad luck out of the loop as you can, you should consider doing some work before you walk out the door. Planning and scheduling a video shoot is just as important as directing and camera work, especially for those of you who want to make money in this crazy business.

There are a million video demons out there just waiting to ambush your project. With a good plan, you can anticipate where they're most likely to strike and have your exorcism kit at the ready. With no plan at all, they can easily overwhelm you, and soon all the little problems that

arise will have your head spinning. (See Figure 15.1.)

Even if you're a video hobbyist who doesn't need all of the forms and schedules outlined in this chapter, it would do you some good to at least brainstorm a plan before you go out and shoot blindly.

So take a little time, make some plans and avoid the hassles. A well-ordered production always runs more smoothly and saves you time and tape; read on and we'll help you learn how it's done.

Paper Works

For planning purposes, it's a good idea to have a *script*, a *shot list* and a *shooting schedule*. These documents can help you work efficiently by helping you to take steps in the proper order. Not this serious about your production? Then drop the fancy names and scribble out a real simple version of these documents.

The script is the written form of the video. Usually it breaks the information

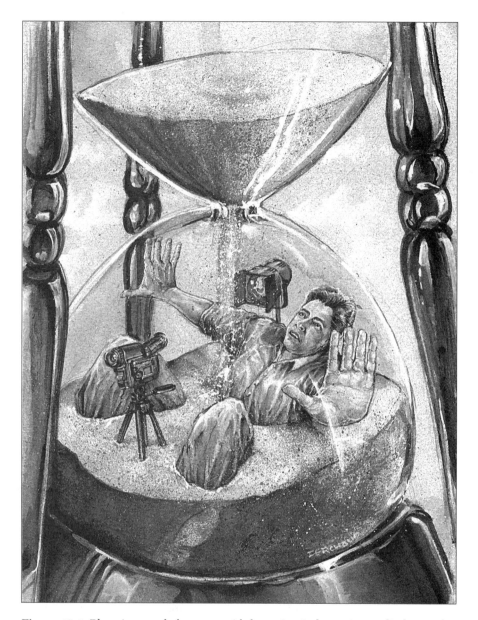

Figure 15.1 *Planning can help you avoid drowning in last minute glitches and details.*

into video (everything you see on screen) and audio (everything you hear). It includes each scene with appropriate notations: *Scene 1.* Exterior. A small house. You can also break scenes down further into shots: *1a.* Wide shot of house, *1b.* closeup of window, *1c.* closeup of porch.

By creating a script before shooting, it's possible to avoid overshooting, where you take more shots than you will actually need. However, you must sometimes shoot your video first (as in news or documentary programs) and write the script later. Even without a script, you can still create the other documents (shot list, shooting schedule) during the planning process.

When you start with a script, creating a shot list is easy. You simply transfer each shot and its corresponding number to another sheet of paper. For example:

1a.
Wide shot of house.

1b.
Closeup of window.

1c.
Closeup of porch.

At this point, the shots are in the same order they were in the script. If you aren't working from a script, you should create a shot list based on what you know you will find at each location. If you are shooting at a hospital, for example, you would want an establishing shot of the building, a closeup of the sign, the lobby, the patient's bedroom, the operating room, various closeups during the operation, etc. This list then becomes your shot sheet.

A shot list doesn't have to be anything fancy or formal. If your work is of the unprofessional, just-for-fun variety, simply jotting down a few ideas on paper before you start shooting will help you organize your ideas into a plan.

Once in the field you would merely cross off or put a check mark next to the shots as you captured them on tape. But on a complex job that requires several days of shooting at numerous locations, you will want to create a shooting schedule.

Right On Schedule

The schedule is all about time management, especially crucial if you're a professional charging your client for your time. If you need people to appear in shots (also called talent), then you'll have to pay these people for their time, unless they are friends or associates willing to help you for free. Rented props or vehicles cost money—sometimes by the hour. If you miss a shot because the sun went down before you could get to it (i.e. you ran out of time), then crew, equipment, people and props will have to return the next day—possibly at a considerable cost. Time management is money management.

Even the amateur videographer probably doesn't have all the time in the world to spend behind the viewfinder. Hobbyists, too, can benefit from a little bit of scheduling; it helps them to maximize their precious videography time, which is all too often in short supply. The key to keeping everything running smoothly is to know how long each individual shot will take to achieve. You can't know for sure, of course, but there is a way to estimate. Assuming that you have a professional crew (which is large enough and made up of individuals who are competent to cover the complexities of the job), then you can use this simple formula:

- For an elaborate shot involving actors (and microphones), complex camera moves and/or critical timing between elements (*the dancing hippo crashes through the door just after the ping pang ball bounces off the igloo for the third time*)—allow 30 minutes to get the shot.

- If the shot is a wide shot or a simple medium shot involving actors and microphones—allow 15 minutes to get the shot.

- For closeups and shots where the audio is not important—allow 10 minutes per shot.

Obviously, the dancing hippo shot would take longer than 30 minutes to perfect, but you could probably finish the closeup with no audio in 30 seconds. Eventually, the variations in time will even out.

Live Long and Prosper

Now that you know how long (approximately) each shot will take, your next step is to *group* the shots in a logical way. When you add the estimated minutes for each of the shots in a group, you will know roughly how long you will be at a location, about how much time you will require for your talent to stick around and when to rent your props.

You should always place a shot in a group based on the most important element in the shot. If you can only shoot the exterior of a house at 8 a.m. because the front of the building faces east, then *time of day* is the most important element (unless you can light the shot and/or manipulate your shot angle for the same effect). If your main actor is only in town until the 15th of the month and then leaves for Tibet, then you must schedule around this concern. Many shots don't have a critical element, so you can shoot them anytime. These shots would logically go together by *location*.

It might be easier to group shots if you first break them down as shown in Figure 15.2.

I've grouped these shots together because the interior shot is at the same house as the exterior, allowing you to move immediately from one to the other. If the interior shot was at a different house, it would not be a part of this group. You now know that you should allow 30 minutes for this location. You also know that you have to be at the location a little before 8 a.m. to get the exterior shot. Using logical groupings, the rest of your shooting schedule should fit together like a jigsaw puzzle.

Back to the Future

Once you've completed the shooting schedule, you have in your hands the most important document for putting people and equipment at the right place, at the right time. You also know how many days of shooting there will be, which will help you work out a way to finish everything before your deadline.

A note about deadlines: if your project is a professional one, your client will probably provide this for you. If not, then you'd do yourself a favor by setting one for yourself, because projects without deadlines have a way of never getting finished.

Let's say that you have to complete your video by a July 15 deadline, and it's now the beginning of May. Plenty of time, right? Well, let's look at it in terms of a professional project schedule. A useful exercise is a *back time schedule*. This is where you begin your plan with the deadline and then move backwards until you reach the present.

You don't want to wait until the last minute to deliver your video to the client, and you also want to leave some breathing room in case something goes wrong. Let's put client delivery a week before the deadline, or July 8. Since that is a Saturday, move it back a day, to July 7. We'll say you have estimated that you need six days for final post-production, special effects,

Shot #	Location	Actors/ Extras	Props	Time of Day	Duration
1 - Wide shot of house	Exterior - house	None	None	8 a.m.	10 Min.
26 - Med. shot, Nelly typing	Interior - house	Nelly	Typewriter	N.A.	10 Min.
27 - C.U. of typewriter	Interior - house	Nelly	Typewriter	N.A.	10 Min.
				Total:	30 Min.

Figure 15.2 *Breakdown of shots grouped by location.*

audio sweetening, etc. You probably don't work on weekends, so this would take you to June 30. Except that the holiday lands right in the middle of that week and you're going to be out of town. Cross out the 3rd and the 4th. Now you start final post on June 28th.

Before the final post, you're going to do a rough edit to show the client what the show will look like without special effects, etc. and where they can make final changes. They will need three days to make their decisions—move back to June 23. The actual rough cut will take five days—back to June 16. Reviewing footage after the shoot and before the rough cut will take five days—back to June 9. Three days of shooting (plus three days of rain days in case of bad weather) and you are back to June 1.

If you need a week to put together your shooting schedule (May 25) and two weeks to write the script (May 11) then you can see that you had better get started. Take the back timed dates and reverse them:

May 11-24/Scriptwriting

May 25-May 31/Scheduling

June 1-June8/Shooting

June 9-June 15/Review footage

June 16-June 22/Rough cut

June 23-June 27/Client approval

June 28-July 7/Final edit

July 15/Presentation to client

What have you achieved when you put together a realistic schedule such as this? From the client's viewpoint the schedule does several things. It lets them know that you know what you're doing. It tells them what needs to happen (they need to approve the script, for example, by May 25th or else the deadline is in jeopardy) in order for them to get their video on time. It helps them understand the necessity of keeping on schedule for approvals, etc. It gives them the secure feeling that all is right with the world. And if the client is

you, then you will be the lucky recipient of that good feeling.

Professionals can also use schedules as a handy tool for dealing with the client. "As you can see, if you don't approve of Nelly, or at least some actress by tomorrow, we'll have to delay shooting by a week. Then, to meet your deadline, I'll have to work on the Fourth of July, which means I'll have to double your bill. Sorry, company policy."

Without a schedule, it's easy to fall into the complacent attitude of, "Gee, I have until the middle of July to get this video done, why hurry?" But when you have a calendar in front of you with clear milestones to reach every week or so, you're more likely to stay on track. This applies to any video with a deadline, be it big corporate production or homespun holiday greeting video.

What's On Your Mind?

The more detailed shooting schedule is useful in telling everyone what you're thinking, and how they can help you achieve your goals. With thousands of project details swimming in their heads, directors can sometimes forget that the crew and the client have no clue as to what's going on unless the director tells them. A clear schedule can answer most of the questions that you would otherwise have to address during shooting.

Knowledge is power. This is true in scheduling as well, but only if you spread the knowledge around. If everyone knows where they should be and what they should be doing, you're more likely to create a powerful presentation.

Some people find a symbolic representation of the project easier to follow; in which case you might want to build a *Gantt Chart*. (See Figure 15.3.) These charts depict progress in relation to time and are especially helpful when major portions of the project overlap one another.

I once directed a documentary-style video for the local school board. There was a broad spectrum of information, fifteen locations (different schools), dozens

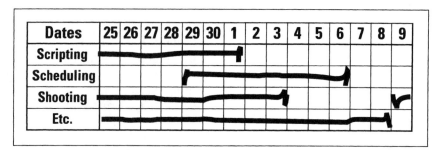

Figure 15.3 *Gantt Chart depicting shooting schedule.*

May 12

Location	Subject	Approx Length	Arrive Time
Rockledge	Drafting and Design	2 hrs.	7:00 am
	Home Economics	1.5 hrs.	9:00 am
Travel		.5 hr.	
BCC	Cosmetology	1.5 hrs.	11:00 am
Travel		1 hr.	
Cocoa High	Space Engineering	2 hrs.	1:30 pm
Travel		1 hr.	
Harris Corp-Bldg. 62	Vibe	1.5 hrs.	4:30 pm
Total		11 hrs.	

May 13

Location	Subject	Approx Length	Arrive Time
Rockledge	Technology Studies	2 hrs.	7:00 am
	Automotive Tech	2 hrs.	9:00 am
Travel		1 hr.	
Merritt Island	Photography	2 hrs.	12:00 pm
Travel		.5 hr.	
Edgewood	Technology Studies	1.5 hrs.	2:30 pm
Total		9 hrs.	

Figure 15.4 *Portion of overall shooting schedule.*

of interview subjects and thousands of potential problems. Additionally, my company was also producing a brochure that needed still images. This meant that the video crew would go in, shoot footage and capture interviews and then the still photographer would set up for his shots. Meanwhile, the video crew would move to the next location.

Figure 15.4 is a portion of the overall schedule for this project (there were several shots in each location, not noted here, but included in my detailed shot lists).

It was an ambitious schedule that had to click like a Swiss watch. Through our clients on the school board, we contacted every teacher who was to be interviewed or whose classroom was to be used. They were not only told verbally what time the crews would show up and what would be needed, the teachers were also sent the sheet of information in Figure 15.5.

We personalized these information sheets for each teacher. The sheet noted the shots we wanted to capture, the steps the teacher should take to prepare the room and the responsibilities of the students (if they wanted to be in the show, they had to have a signed talent release before shooting).

Was all this preparation necessary? I am absolutely certain that we couldn't have completed the video on time without it. My crews worked their tails off, the teachers were wonderfully prepared, and our clients on the school board smoothed the way at every step. We used every minute we had, and even challenged the space-time continuum by squeezing in a few minutes that we didn't have. We got everything we wanted. The show has gone on to win numerous awards at the state level in Florida. More importantly, it has been a success for the client.

Are planning and scheduling important? Only if:

Patient Care Assisting   **Palm Bay/ 2 hrs.**
Betty Smith   **Approx. time of arrival: 10 am**

Basic plans for your session: **Date: Friday, May 6**

Video—
1. Wide shot showing beds.
Several students going through procedures (some as patients, some caregivers—as they would in a normal class)—various shots.
2. Close-ups taking pulse, blood pressure, any applicable procedure.
3. Students working with the CPR model.
4. Students using the equipment to lift patient from bed.
5. Teacher working with student at desk (colorful textbook illustration, if possible).
6. Interview with teacher.
7. Interview with student (at Ms. Smith's discretion, not mandatory).
Photography (see above for descriptions)
1.
2. or 3.

Notes for Instructor

A. Any student not returning signed talent release prior to shooting cannot be allowed to partipate. All students under 18 require parent or guardian to sign.
B. Review appearance guidelines on take-home sheet with your students.
C. Simplify your room, by removing any superfluous boxes, papers, etc.
D. Have critical areas as you would like them to be seen, beds made, and so on.
E. CPR models need to have clean T-shirts.
F. Have patient lift out and available.
G. Identify a superior student for interview (optional, if no appropriate student).

(We will also take suggestions from the teacher. **Please have appropriate technology up and running before the crew arrives.**)

Figure 15.5 *Example of personalized information sheet.*

a. Time is important to you, and it should be—time is money,

b. you dislike confusion and chaos, and I'm not assuming anything, some people like chaos or

c. you want to look good.

When a shoot goes smoothly, you will look *good*—to your crew and talent, to your client and to yourself. Plan it. Schedule it. And you'll look *mahvelous.*

16
Script Right:
Video Screenwriting Tips

Stephen Jacobs

Even the best videographers often balk at writing scripts. Take my friend Mike Axelrod. A video computer programmer since graduating from college, Mike recently quit work to return to school.

As a film buff who's created a few videos with friends, he decided to combine his two interests and major in computer animation. He signed up for a required screenwriting course and experienced no difficulty in writing individual scenes three to five pages long.

Then he learned the second half of the course required completion of a screenplay for a fifteen-minute film, fifteen to twenty pages in length. The news left him pale.

"Fifteen to twenty pages! I'll never write something that long," he howled. "I'm not a writer!"

Well, maybe not. But good video begins with good writing. In this chapter, we'll show how even non-writers like Mike can put together a serviceable screenplay.

Videographers as Storytellers

Some people think writing is a talent reserved for those lucky enough to be born with it. This is not necessarily so. Writers who seem to have come from the womb pen in hand are the exception, not the rule. Most have to work at it. In many ways, writers are no different from videographers. Both are people trying to communicate specific information, a story, a particular point of view.

When you pick up that camcorder and press the little red button, you start to tell a story. Whether you're shooting the science fiction epic of your dreams, documenting a wedding for the 300th time or producing a thirty-second public service announcement for the local Red Cross, you're still telling a story.

So before you begin you should first consider what message you are trying to convey, and to what sort of audience. If you don't know what you're trying to say or whom you're trying to reach, you'll have a hard time getting your message across.

You may begin with the goal of writing a screenplay for an introductory videotape on washing machine maintenance; suddenly you find yourself showing the viewer how to make a washing machine from scratch. Or you write something for an experienced group of professionals, then find yourself explaining the basics.

In fiction, the routine Hollywood script features an easily identifiable protagonist, the hero; and an equally transparent antagonist, the villain. The hero confronts a problem and spends much of the film working toward a solution. The antagonist seeks to block this progress. Along the way the hero inevitably undergoes a change in character.

Some films slightly alter this formula. Director Barry Levinson has said his film *Rain Man* was a real challenge because the protagonist of the story, the autistic Raymond, by definition couldn't experience a change in character. It's his brother who must change.

Outline and Outpouring

For some people, outlining is extremely beneficial. Others find it gets in the way of the actual writing, another little task aiding and abetting procrastination. Some projects are more conducive to outlining than others. The larger the project, the more likely it will benefit from some pre-scribble structuring.

Beginners should definitely outline. It helps you understand where the story's heading. Does outlining mean the formal, Roman numeral structuring they taught you in elementary school? No. Think of it as simply jotting down a temporary table of contents. Once that's complete, you're ready to start writing.

As a videographer, you try to avoid situations where you must edit in-camera. You know the more footage you shoot, the more you repeat shots and obtain different angles, the more you'll have to work with during editing. It's not unusual to read about professional film directors with raw footage ratios of 20 to 1.

As a writer, you should give yourself the same freedom. Begin by throwing anything and everything down on paper. This isn't always easy to do. But if you criticize every sentence as you write it, you may not finish the first paragraph. Let it all out first; you can edit and revise later.

Show, Don't Tell

Writing for video is different from other types of writing in two important ways:

The lengthy description and dialogue that distinguish some writing is notably absent here. When writing for video, you need to show things, not talk about them. So skip those long passages of narration. Instead, use different camera angles and shots of the same scene to highlight the features of the landscape you wish to emphasize. Don't use long sections of dialogue to explain anything. Show it.

Screenplays happen in the here and now. They describe the action as the viewer will see it unrolling across the screen. Use active voice and present tense. For example, write, "John looks into the mirror, then does a double take. There's a second head growing out of his neck. He screams in terror and blacks out..."

Don't write, "John looked in the mirror, then did a double take. He saw a second head growing out of his neck. He screamed in terror and blacked out..."

Or, "We see our John look into the mirror and do a double take. His reflection shows a second head growing out of his neck. We hear him scream and see him slump to the floor. He's blacked out..."

NOTE: These examples are based on an actual film, *How to Get Ahead in Advertising*, in which the main character does indeed wake up one morning to find a second head growing out from the side of his neck. As they say in Hollywood, you can make a movie about darn near anything.

Relevant Questions

As your script nears completion, ask yourself the following important questions:

Does every single scene in the script a) advance the action and/or b) develop the character?

Do all scenes, dialogue and narration serve a legitimate purpose? Does the third scene featuring John arguing with his new head show us anything new about them, or did you just like the dialogue?

If you've shown the viewer how to remove the left front tire of a '58 Thunderbird, do you really need to show removal of the right front? Where else can you cut? Where should you expand?

Remember: you want your dialogue or narration to sound natural. That's a hard one. A good way to check your writing for a natural feel is to put it away for a few days, then reread it. Does it still sound right? Read it aloud. Do you stumble, fumble, fall? Then your actors or narrators will as well.

Finally, ask a few trustworthy and literate friends to read your work with a kind and constructive critical eye. You don't have to take their advice, but you should at least consider it, especially when the same problem is identified by more than one critic.

Writing is rewriting. Put your ego aside and polish your work. Once you're satisfied that it's as good as you can make it, it's probably ready to go.

Script Requirements

It's not enough to write a good script. You also have to format it according to screenwriting conventions. There are two main scriptwriting formats. Fictional scripts, or scripts with a lot of dialogue, use a format that evolved from the theater.

Documentaries, industrials and other productions which match narration to the visuals typically use a split page, two-column format.

There are a number of variations within these two categories (see Figure 16.1), but all will fall under one type or another.

In the theatrical format, scenes are numbered; it's best not to number your scenes until you complete your final draft. Manually renumbering scenes can grow tiresome.

Number the pages in the upper right hand corner. When a scene crosses over from one page to the next, place the notation "continued" at the bottom right corner of the first page. Mark the top of the second page with the scene number and a "continued" notation.

It's important that the finished screenplay follow the general visual format, allowing for simple identification of scene location, direction and dialogue. For example, the first time Dan's name appears in the script his name is set in capital letters. The second time it isn't. This makes it easy for the reader to keep track of new characters. When indicating dialogue, character names always appear in caps.

Technically, each time the camera moves it inaugurates a new scene; we have three scenes now, rather than just the one. Some screenwriters also capitalize all audio cues—TYPING, TICKING—for easy identification.

With documentary and industrial video, you are frequently more interested in matching certain visual information with spoken word audio. That's why it's easier to format this type of script in two columns. The description of the visual information goes in the left column and the audio information goes in the right.

As with the dramatic screenplay, these types of formats vary slightly from script to script. In some, a thin third column to the left indicates shot numbers. Other variations use an abbreviation for the shot as part of the description.

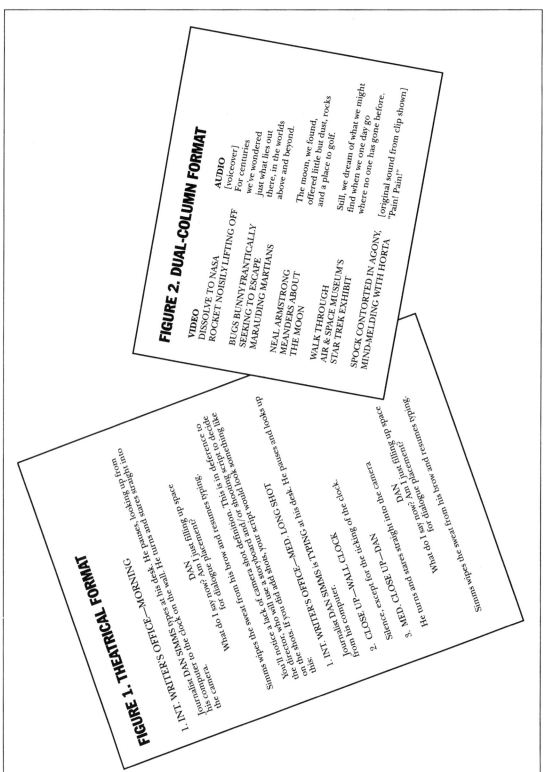

Figure 16.1 Samples of theatrical and dual-column scriptwriting formats.

It's a Wrap

Yes, learning to write can pose a challenge, but take a look at Mike. He gritted his teeth, put his stuff down on paper, and found he actually enjoyed himself.

The fifteen-minute film he was assigned to write turned into three five-minute sections for a feature film script he vows someday to complete. He scored himself an "A" in the course. Most important, he engaged in work he felt good about, and gained a new perspective on his capabilities. Now it's your turn.

Writing Tools for Videographer:

There's a wealth of help out there for script writers, from how-to books to formatting software. Below we list some of the most popular writing tools. For more, contact The Writer's Computer Store 1-800-272-8927 (www.writerscomputer.com)

How-To Books

Screenplay: The Foundations of Screenwriting by Syd Field, ISBN: 0440576474, Dell Publishing, $13.95.

The Complete Guide to Standard Script Formats: The Screenplay by Cole and Haag, ISBN: 0929583000, CMC Publishing Inc., $18.95.

The Elements of Screenwriting by Irwin R. Blacker, ISBN: 002861450X, MacMillan Distribution, $9.95.

The TV Scriptwriter's Handbook by Alfred Brenner, ISBN: 187950510X, Sicman James Press, $15.95.

Scriptwriting Tutorial

WritePro; for DOS, Windows or Mac, eight lessons, $199.

Script Formatting Programs

Final Draft, for Mac or Windows, $249

Screen Writer 2000, for Mac or Windows, $299

17

Look Who's Talking:
How to Create Effective, Believable Dialogue for Your Video Productions

John K. Waters

A script is a story told with pictures—but silent pictures they're not. Since 1927, when the debut of *The Jazz Singer* transformed "moving pictures" into "talkies," dialogue has played a crucial role in making successful films and videos.

But with everything else you worry about as an independent videographer—maintaining your equipment, getting the shot, getting paid—dialogue may be low on your list of concerns. Still, you don't want to underestimate its power.

Good dialogue works hard. It keeps things moving, ties individual segments together and unifies your piece. On the other hand, bad dialogue discredits your work, destroying your credibility and ultimately costing you your audience.

Whether you're shooting independent features, corporate image spots, local TV commercials or your sister's wedding, what your subjects say can make the difference between an amateur show and a powerful piece of professional videography.

Here's how to make sure what they say works.

What is Dialogue?

Any time you put words into the mouths of your on-camera subjects, you are writing dialogue. That definition includes hosts, commentators and spokespersons, as well as actors playing parts.

The primary purpose of dialogue is to move your story forward. It accomplishes this by revealing character, communicating information and establishing relationships between characters. It can also foreshadow events, comment on the action and connect scenes.

For example: in my script *Sleeping Dogs Lie*, the hero, Ray Sobczak, is a reporter working on a story that is annoying some very important citizens. Here, a local politico delivers a veiled warning.

JAMIE

What happened to you?

SOBOZAK

Zigged when I shoulda zagged. How's the campaign going?

JAMIE

Oh, I just love spending obscene amounts of my mother's money— and what your paper charges for ad space is truly obscene.

SOBOZAK

Something tells me your mother can afford it, Mr. Bockman.

JAMIE

My father, who could also afford it when he was alive, was Mr. Bockman. Call me Jamie.

SOBOZAK

Okay, Jamie. Think she'll win?

JAMIE

(ignoring him) How's the story coming?

SOBOZAK

Which story is that? (Jamie gazes out at the cluster of downtown buildings and the hills rising behind them.)

JAMIE

Salinas is growing, Ray. Over a hundred thousand at last count. But, underneath, it's still a small town.

This exchange probably won't go down in history as the most memorable in films, but it has many earmarks of good dialogue. It tells us something about the characters: Jamie is rich, has an ambiguous attitude toward his mother's campaign for mayor and has well-informed connections; Sobczak is tough, and won't let a bump on the noggin keep him from getting the story.

It foreshadows future events: people are watching. And it moves the story forward: this encounter gives Sobczak an idea, which leads him—and the story—in a new direction.

Keep It Lean

The above example also demonstrates another important quality of good dialogue: brevity. Good dialogue is a lean exchange between people, composed of short phrases. On paper, it looks like lots of white space; big blocks of type are definite warning sign that you are overwriting your dialogue.

"The good stuff is a dance of two and three-liners between characters," says scriptwriter Madeline DiMaggio. "It's a bouncing ball that keeps your audience riveted."

When it comes to writing dialogue, DiMaggio knows what she's talking about. She has written over 35 hours of episodic television for shows ranging from Kojak to The Bob Newhart Show to ABC's After School Special. A former staff writer for the daytime soap opera Santa Barbara, she is also a teacher and the author of How to Write for Television.

DiMaggio says the kind of close-to-the-bone dialogue you want for your videos comes only through rewriting (see Figure 17.1), "It doesn't happen on the first pass," she says. "At first your dialogue is cardboard—and that's the way it should be. It's only later, when you go back and take five lines down to two-and-a-half, and then two-and-a-half lines down to one, that you find the real gems."

DiMaggio says she plays a game with herself during her rewriting process: if she can whittle a piece of dialogue down to four lines, can she cut it to two? If she can chop it to two, how about one?

"If script writers were doctors," she says, the best ones would be surgeons. "Cut, cut, cut!"

Figure 17.1 *Rewriting and rewriting and rewriting again is key when writing dialog.*

Make It Sound Real

A tried and true technique for developing your ear for natural-sounding dialogue is to surreptitiously tape conversations and then transcribe them later.

I do this especially when I'm writing about types of people I don't know well. Ethical questions aside, this has worked well to awaken my sense of how people talk.

The first thing I noticed when I began doing this was how fragmented conversations are. The following is an example from my files.

MAN #1

Hey, what's up?

MAN #2

I dropped by the place. Thought I'd say hi, but nobody was, you know...

MAN #1

I was over there yesterday and...

MAN #2

... home. You know?

MAN #1

Nobody? Man, I...

MAN #2

That's because of the, you know, holiday 'n stuff, and her car was there and everything, but...

MAN #1

What a piece of *#@*!, man, that car...

MAN #2

Yeah, and, you know, I left like, a note.

MAN #1

No way!

On paper, this conversation looks like an exchange between two orangutans, but they sounded perfectly normal. That's why, unless you're making a documentary, you can't just transcribe tapes of real conversations and use them raw in your scripts. And even documentaries require judicious editing.

"If you went out to a coffee shop with a tape recorder," DiMaggio says, "and went home and put what you recorded into a script, it wouldn't work. Good dialogue isn't actually real. It just gives the illusion of reality."

DiMaggio says one of the best ways to learn how dialogue sounds is to record dialogue.

"That's how I got to know Santa Barbara," she says. "There were so many characters, and they all had their own voices! I would audiotape the show and then listen whenever I was driving. When you cut off the other senses, your ear becomes much stronger."

DiMaggio also recommends audiotaping shows to develop an ear for genre dialogue.

"Comedies, mysteries, dramas—it's an incredible way to learn," she says.

Good dialogue also has a spontaneous quality, as if your characters were speaking their lines for the first time.

"When the dialogue is stilted or too formal," says corporate video writer/producer Susan O'Connor Fraser, "the audience just laughs at it. When they start doing that, you've lost them."

O'Connor Fraser is the creative director for Tam Communications in San Jose, California. She's been writing and producing videos for Fortune 500 companies for the past 15 years. Her company produced a reality-based show on paramedics in San Jose, which aired on the local ABC affiliate.

"I don't think corporate video is that much different from features," O'Connor Fraser says. "Dialogue is dialogue, and every story has its own reality. Star Wars has a reality, and so does a corporate sales presentation. Everything must play and be believable within its own reality."

According to O'Connor Fraser, one of the most common dialogue errors she sees is characters addressing each other by name too often.

"I've seen it done in every passage," she says. "It's, 'Well, John . . . Well, Lisa... What do you think, John...I'm not sure, Lisa.' It's just not real."

She says reading your script aloud is one of the easiest ways to spot dialogue errors. (See Figure 17.2.)

"You're writing for the ear. So you need to find out how it sounds. You don't need actors, though they are a wonderful luxury. Just read it out loud with a friend, or by yourself while you're sitting at your computer. You'll hear many of the problems right away."

Stay in Character

When I write, I become the characters I'm writing about. This is pretty easy when I'm writing about thirtysomething white guys from the Midwest. But what about when the character I'm writing dialogue for is a New York cop, or a Southern doctor, or a black female Vietnam veteran with a Harvard MBA, two grown children and a neurotic obsession with alien abductions?

You simply cannot write dialogue that rings true unless you acquaint yourself with the kind of people appearing in your video. This is where real world research is essential. I'm talking about stuff you can't find in the library. But that doesn't mean that you have to spend a week on a fishing boat in Alaska or infiltrate the local Jaycees to get the right slang and jargon for your script, though those are tried and true approaches.

"If I don't have any personal experiences in my own life I can draw on," says O'Connor Fraser, "I track down the kind of person I'm writing about and take them to lunch." In her corporate work, she tends to deal with a limited number of "types," mostly from the high tech world; after 15 years she knows them well. But she still checks her "voice" with face-to-face interviews—especially when she's writing a script for an on-camera presentation by a company executive.

"When I go out and interview a president or vice president who will be on camera," she says, "I listen very carefully, so I'm really hearing them talk. I don't want them to sound like they're reading from the inside cover of an annual report. I want them to sound very natural and comfortable, as though they were talking across the table from someone.

"Interviewing your clients is also one of the best ways to pick up the buzz words of their professions. Listen closely and make a list of unfamiliar words or phrases. Ask for clarification so you understand them in context. When you sit down to write your script, your list will prove invaluable."

Many writers just write the dialogue as best they can and then give it to someone from that character's walk of life to read. DiMaggio says she works on her dialogue, "until I'm not ashamed of it," and then turns it over to a person with whatever special knowledge her characters would have. In her TV movie script, *Belly Up*, for example, one of her characters was a man who gambled on the golf course.

"I wasn't about to take up golf to hear how guys gamble on the golf course," says DiMaggio. "So I sent the script to my brother. He gambles on the golf course all the time. In five minutes he told me things I couldn't possibly know unless I was out there. I made some changes in the script and all of a sudden it sounded absolutely real. One producer told me I wrote like a man, which, under the circumstances, I took as a compliment."

In one of my own scripts I created a character who was a professional crop duster, but I had never met a crop duster in my life. So I picked up the Yellow Pages, and a few lunches later I knew

Figure 17.2 *Read your script aloud to make sure it sounds natural.*

everything I needed to know to create a believable character. (And I now know to call them agricultural aviators.)

Go for Subtext

Syd Field, author of several well-known books on scriptwriting, including the now classic *Screenplay: The Foundations of Screenwriting*, calls dialogue one of the "tools of character." That's because what people say says a great deal about them. But what they don't say often says more. The best dialogue is not only about what your characters are saying, it's also about what they're not saying. This is subtext.

"Subtext is what's happening beneath the surface," says DiMaggio. "It's the key to truly great dialogue."

Examples of subtext abound in films like the 1944 film noir classic, *Double Indemnity*. One scene in particular comes to mind, in which insurance salesman Walter Neff (Fred MacMurray)

puts the make on a client's wife (Barbara Stanwyck). They fence back and forth in a conversation about cars and speeding, but driving is the last thing on their minds.

Subtext enlivens good writing everywhere—even commercials. We've watched that couple in the Taster's Choice commercials meet, woo and bed in Paris—all while talking about coffee.

Subtext probably isn't as important in most corporate video situations; still, you ignore it at your peril. Human beings talk around things. Dialogue that's too "on the nose" won't sound natural. Even the dialogue in infomercials has subtext.

Write the Right Voice Over

Voice-over narration isn't really dialogue, but many of the same principles still apply. This is especially true if the narrator is a particular character, as in the case of a host, or one of the actors, such as the Holly

Figure 17.3 *Congratulations! It's a script!*

Hunter character in *The Piano* or Walter Neff, who narrates *Double Indemnity*.

You write voice-over narration like you write dialogue—for the ear. It should sound conversational. Even if your narrator is an omniscient voice, that voice must conform to your audience's expectations of human communication.

"When I'm doing voice overs," says O'Connor Fraser, "I still get a character in mind and write for him or her. Of course this is really important if the narrator will ever appear on-camera, but I do it even if they won't. That way, the voice is consistent throughout."

Voice-over narration can be even harder to write than dialogue. "If you think dialogue has to be lean," DiMaggio says, "voice overs have to be the best of the best. It has to be very, very thrifty. The real gems. Otherwise it turns into an excuse for failing to write good exposition."

Practice, Practice, Practice

Writing authentic, believable dialogue is a special skill; it takes practice. But with some effort and more than a little patience, you'll get it. (See Figure 17.3.)

"You have to realize that the script you write today won't be as good as the script you write next year," says O'Connor Fraser. "And that's okay. I'm a much better writer now than I was a year ago. I learn something with every script I write."

"We're all students, really," says DiMaggio. "No matter how long you do this, there's always something to learn. I think that's the good news. It's one of the things that keeps this work interesting."

In the end, creating good dialogue is more about listening than it is about writing. Once you begin to hear the rhythms of human conversation, the dialogue you write for your videos will improve dramatically.

So keep your ears open.

18
The Storyboard:
Blueprint for Production

William Ronat

No doubt your mind swirls with little pictures, pictures that represent your ideas. Capture these "idea pictures" on videotape, and everybody can enjoy them.

But between your first idea—picture and your final cut there's a lot to do—and a lot of it involves people other than yourself. You must communicate your great ideas to these people, be they clients or camera operators, clearly enough so that they see the same pictures you do. How can you be so sure you're all on the same wavelength? By getting those idea pictures out of your head and down on paper where everyone can see them. And that is where the storyboard comes in.

Picture This Video

A storyboard is a graphic representation of your video. Like the comics you see in the newspaper, it consists of a series of pictures illustrating the scenes of your video. String these scenes all together, and these pictures tell your video's story.

The graphics need not be fancy— hand-drawn stick figures will do. Which is not to discount fancy graphics, from the most elaborately airbrushed fine art to computer-generated images.

Fancy or not, these graphics usually wind up pasted to a large board, the scenes pictured in little boxes corresponding to video's three by four format. Captions underneath each picture describe the audio story: what the characters say, what kind of music plays, what sorts of sound effects bang away in the background.

Obviously, detailed storyboards are not necessary on every video. Rough storyboards suffice for documentaries— there's simply no way to know exactly what scenes you'll capture beforehand. But on dramatic productions involving different actors and locations, a detailed storyboard can help you anticipate and avoid problems on the shoot. Using storyboards before producing commercials and computer animations can mean the difference between money in the bank and lots of costly reshooting.

Say you're shooting a short dramatic piece involving space aliens in the old West. You write the script, and then break the script into scenes, the scenes into shots. You prepare a production schedule. You think you're ready.

You're not. Go ahead and take that extra step. Prepare a storyboard. You don't have to be Michelangelo to draw these pictures. Stick figures and vague shapes will serve the purpose nicely (as long as you can tell what that blob on this side of the shot is, you're okay).

You draw the first shot. The aliens' space ship descends into Monument Valley. In the next shot, they get out of the ship. Cut to a stagecoach moving swiftly, clouds of dust trailing behind it. James, the driver, whoops wildly and cracks his long whip at the team of sweating horses.

The aliens watch from behind a butte. At the last moment, they jump from their hiding place and try to hold up the stage. Gabby, riding shotgun, lets them have it with both barrels. The aliens go down, splattering green slime. The doctor jumps from the stage and quickly examines one of the creatures. The driver watches expectantly. The doctor looks up at the driver and says hoarsely, "He's dead, Jim."

You look over your artwork. It's rough, but nonetheless it serves the purpose: you get your idea across.

But as you look closer, a nagging doubt creeps into your mind. Then it comes to you. You planned to use upside-down pie plates for the spaceship. This is fine for the long shot, but you also have a shot of the aliens getting out of the ship. The pie plate may not look so good in that one. You'll need to re-think it.

Also, using the pictures as a guide, you see you're still short a few props—Western outfits, a doctor's bag, alien make-up, some horses and, oh yeah, a stagecoach.

It's better to learn about these problems in the planning stages of your production, rather than when you're out in the field with your technicians and actors.

Okay, so you go out and get all these props, costumes and locations you know you need, thanks to your storyboard. You can now gather together cast and crew to discuss the details of the upcoming shoot.

Using the storyboard, you can show the camera operator how to get close to the ground for the shot where the stagecoach rolls over him. For another shot, he'll need to ride one of the stagecoach horses as he focuses back at the driver. And he'll also need some sort of harness for the overhead shot that simulates the space-ship crash landing. Of course, after reviewing this with him, you may also need a new camera operator.

You can show the actors how you'll cut the different scenes together, so they can get a better understanding of their characters and the motivations for their actions.

"So I shoot my gun at the aliens and not the doctor," the actor playing Gabby says after seeing the storyboard, a point that obviously needed clearing up. However, he still has a question about motivation.

From your perspective as the director of the shoot, the storyboard helps you make sure the shots you've planned will work well together. You have a shot of Jim followed by a similar shot of Gabby. From the storyboard you realize that this might look like a jump cut when you put the two together. You grab a pencil and quickly draw a point of view shot from Jim's perspective of the view from behind the horses.

Now when you cut to Gabby, you not only avoid the jump cut, you make a statement about the grumpy sidekick through the use of juxtaposition.

Tips from the Pros

The professionals have always used storyboards. Alfred Hitchcock planned out his films with storyboards to such degree that he found the actual shooting anticlimactic. He claimed he'd made all the artistic decisions long before the cameras rolled.

Steven Spielberg turns every single shot of his movies into graphics, which he then tacks to the walls of his office. He moves all of them around as he pleases to see how they will fit together. These directors don't leave much to chance, a good reason to use storyboards. But many professionals use storyboards for another good reason—money.

"Whenever we shoot live productions here," says Jane Carter, Studio Art Representative of Napoleon Videographics, a video and art studio in New York City, "we have our artist storyboard out certain scenes to show the client different ways to shoot the video—including a variety of shot angles and editing cuts. It's easier to make a shooting board to show a client what the director has in mind. It saves a lot of time and lets everyone know they are thinking the same way."

Napoleon Videographics specializes in producing highly polished work in very short turn-around times. The trick is to understand exactly what the customer wants up front.

"Most of the time we deal with advertising agencies," Carter says, "they're our biggest clients. They come in with their ideas and we go over those ideas with them. Once we know what they want, we transfer these instructions to the artist drawing the storyboard.

"Clients usually have an idea of how they want the board to look. They might want it drawn with a cartoon style or they might want it to be realistic. If the client wants a scene to look like it was shot with a special lens, such as wide angle, the artist can do that, too. And we can do quick pencil sketches to make sure that the camera angles are correct or that the people shown on the storyboard are the right ages.

"For these advertising agency clients, the storyboard is more than a production planner, it's an all-important selling tool.

"The agencies use the storyboards to sell their commercials," continues Carter, "to show how they would like to shoot it and where they would like to shoot it. They can have it set in Miami or in Norway. It's a visual tool to help their clients better understand what the finished product will look like."

Animated Storyboards

Storyboards also prove invaluable when planning computer animations. Anyone who's worked with an animation package knows what a time-consuming process it can be. If you're creating an animation for someone else, you want to make sure they understand what you will create for them; otherwise it could mean hours or even days of extra work.

Say you tell your client, Mr. Miller, that the logo you're making for him will be a huge three-dimensional steel structure and that you're drawing the dot on the "i" in "Miller" to burst into flames, he may just stare at you, glassy-eyed. But whip out a storyboard illustrating the process of the logo "growing" on the page while light plays off the metal of the letters, he'll follow your every step. If the dot on the "i" then burns like the sun, Mr. Miller will leave your session with stars in his eyes.

Storyboards for computer animation aren't necessarily drawn with classic artist tools such as pencils or brushes. Joseph Eagle, art director and animator at Traces, a computer animation firm in Bala Cynwyd, Pennsylvania uses a unique approach to creating storyboards. His clients include the Philadelphia Phillies and the Unisys Corporation.

"We sit down and talk about what they want," Eagle says. "We may start out by doing some real quick and dirty thumbnail sketches just to make sure we're speaking the same language. Now, say a client comes in and she wants a 3-D Wavefront animation. Instead of spending time on the Wavefront, which is an expensive machine, what I'll do is get on the digital effects system, and with the

2-D system, I can create pictures with a number of different tools.

"I get a product and then we set up the 35mm camera and just shoot straight off the monitor. Then we get the pictures developed, cut them out, paste them on a board and write little descriptions beneath them of the activity that's going on in the pictures. The images look very high quality and in some cases, they look almost exactly how they are going to look in the finished piece, so they get a real good idea of what it's going to look like. We started doing that because we found that storyboards drawn with markers and colored pencils don't convey as much as video does."

Make 'Em Move

You can make storyboards do lots of things, but if the storyboard consists of pictures pasted to a board, there's one thing it won't do... move. Using a caption, you can tell the person looking at the board that the camera is panning or dollying, but you are once again expecting the viewer to be able to "see" things the same way you do. There are other methods.

"A series of pictures will do that," says Eagle, "because it looks like some frames that have been pulled out of an animation. So there would be implied movement from picture to picture. If it's one picture and I want to say 'the logo flies down here' I would put a couple of trails to imply movement, even though they wouldn't show up in the actual piece."

To add movement to a storyboard, Napoleon Videographics takes the pictures and transfers them to videotape. Then they use a process called Animatics which uses "loose pieces" to convey movement. For example, say there was a cartoon man in one of the storyboard's scenes and this man needed to wave at the camera. The cartoon man's arm would be a loose piece—a graphic arm not attached to the picture of the man. Then when you shoot the man, you could move the loose arm back and forth as if he were waving. Two other examples:

A character's entire body, which you move toward the camera, and several different versions of a character's head, with each head wearing a different expression.

"Advertising agencies use Animatics a lot with testing," Carter says, referring to test marketing where groups view rough animated commercials to see how the audiences will react. "They use Animatics in these focus groups because it's a lot less expensive this way. If an agency has ten ideas, it's a lot more cost effective to do it in Animatic form and to test it in the focus groups around the United States then to make ten finished commercials. Then after testing they look at the results, and the storyboard with the highest scores becomes a real spot."

The Ever-Changing Storyboard

In the professional world, storyboards are just a part of the process—not an end in themselves. That's why artists aren't upset when clients ask for changes.

"They're always making changes," says Eagle. "Generally, it's understood that the storyboard is not a device for nailing down every last detail. Rather, it's a generalized look and we let them know that it's just an approximation of what they'll be getting on the 3-D system. If they come in and say, 'The storyboard is fine but the fifth or sixth picture isn't right' it's no problem, we'll do it over. Everything's still in the system. We can adjust them or even redo them from scratch.

"Storyboards are designed to be rough approximations of what will he produced. Some are real rough and some are pretty finished, but generally it's wise to look at them as a guide to what you're going to see.

"That's the point of the storyboard. Whether you're selling the job initially or you've got the job already, you're just trying to establish what exactly is going to happen. It's still only the initial rough step."

If you think all this preparation seems unnecessary, think again.

"There are a lot of people who don't think visually," says Eagle. "Even bet-ween artists, words don't really convey pictures. Two artists can sit there, and one artist can describe something and can write it down and get as verbal as you want with it, and the other will be listening and getting a visual image in his head, and it won't be the same image. It will be real different.

"The only way to find out is for one artist to make some kind of rough work, a quick sketch, and show the second artist. It's not always sufficient to describe something in words. You really need to look at a picture. Then you can say, 'this color needs to be that color or this needs to be bigger over here' Those are the kind of details no amount of words can convey."

It's a Wrap

Using a storyboard will give you communication powers you cannot acquire any other way. Your clients will understand your bizarre concepts immediately, showering you with money to carry them out. Your camera operator will pick up on your subtle imagery, helping you bring forth hidden meanings and subtexts in your compositions even you didn't know were there. And your actors will have a better than fifty-fifty chance of facing in the right direction during their scenes.

Obviously, storyboards can be very important to a video production. But what's more important is that I never used the phrase, "a storyboard is worth a thousand words" in this chapter. At least, not until now.

19
Budget Details:
Successful Video Projects Stick to Budgets

Mark Bosko

Creating and adhering to a realistic budget is important to the success of any videographer's project.

But just how do you compute that magical figure, arrive at an amount low enough to attract investors but large enough to get the job done? It's not easy. Thousands of people labor in Hollywood as budget wizards; not even they get it right always. So many variables and details can go wrong or astray; it's impossible to plan for every contingency. Most often, you just have to guess.

Still, in this chapter we'll offer a number of useful guidelines vital for budget preparation. Videographers who absorb these lessons will at least have a reasonable grasp of the basics of financial planning.

Reasons for Budgets

Video budgets both attract investors and allow you to exercise control over a production.

Since the budget is the foundation of any presentation to investors, it should be specific and accurate.

Realism is also a good idea. It's an admirable goal, applying LucasFilm-like effects to a dry-cleaning commercial, but hardly feasible when the video must come in at $499.

Most projects begin under-budgeted and under-scheduled. It's easy to understand why. A project will certainly seem more attractive to investors if you can convince them you'll finish the video for less money in less time than the competition.

But this shortsighted method of easy financing will eventually cause you suffering.

Projects under-scheduled and under-budgeted leave you with only two options once the show begins 1) the project goes over budget, or 2) the quality goes into the dumpster.

Say you tell a client $300 will do to create a training video. Then, during shooting, rain pours down; you must shut down the shoot and pay talent for a second day. You've now spent an extra $50, money intended for post-production. So will you skip the original scoring, budgeted at $50, choosing instead to give the client canned tunes? Or jettison the spiffy title effects for hand-lettered cards?

Sticking adamantly to an unrealistic budget forces you to continually compromise. This leads to a loss of quality.

There's a minimum budget for every project, a certain amount necessary to produce a video meeting reasonable standards of quality. Determine your video's destination, then calculate the smallest amount of money needed to reach it. If the available financing is less than this figure, change the project.

Keep budgeting until you have a video you can afford to make.

Step by Step

It's important to give equal emphasis to all stages of production, from writing and principal photography to music and editing.

It's easy to get excited about the shooting stage of a video project. Here is the place for lights, camera, action. Just don't make the mistake of creating an excruciatingly detailed budget for production only to carelessly slop but a few dollars to post production. You'll pay dearly.

Become familiar with the functions and costs associated with every step of the production process. Talk to the people responsible for the script, the shooting and the effects. Without such intense research, you may neglect such costs as B-roll tape, music copyright fees and catering charges.

Actual working budgets vary in size. Major Hollywood studio budgets may end up as two-inch-thick tomes, while an independent thirty-second cable spot can come in a tiny one-pager.

Regardless of size, most every budget consists of two sections: costs above-the-line and below-the-line. The former includes cash for producer, writer, director and talent. These costs are usually fixed, set amounts. Below-the-line costs include everything else associated with production. Each line item contains many separate details contributing to the total cost. I'll examine each in an attempt to explain what makes building a budget so tricky.

Over the Line

The first above-the-line item concerns screenplay and story rights (see Figure 19.1). If your production uses an adaptation of existing work, you'll have to purchase the rights. These can be costly for a known, popular author's work, or nonexistent if the story comes from a rookie simply seeking screen credit.

Unless you come up with it yourself, you will have to pay someone some amount for either an idea or an actual screenplay. Even for thirty-second commercials, people get paid to write scripts.

Hidden above-the-line costs can include photocopying, script breakdowns and rewrites, copyright registration and legal fees associated with purchasing work.

The producer is the one who generally runs the show; and, yep, often expects payment, too. A producer is responsible for finding story, actors, crew, equipment, locations, props, wardrobe and investors. This requires an enormous amount of time, even for a small, one-day shoot. A producer's talent lies in the ability to make and keep contacts; that's what they're paid for. Obvious expenses include phone charges, travel expenses, lunches, postage, contracts and legal fees.

In Hollywood, the director is paid for overall vision. On smaller productions, the director may be you, the camera operator or even the client. With very low budgets, you can skip this item; there isn't enough discretionary cash available to afford a director.

Talent includes lead actors, supporting cast, stunt people, voice-over artists and models. In budgeting talent, keep in mind daily or hourly rates. These usually vary according to how often the person works.

As an example, you might hire talent from the local actors' union for $200 a day, $150 for two days or more. Read the small print; often contracts demand that two or more days be worked consecutively. If you work your talent on Monday and Wednesday, you'll spend $400, not the planned two-day rate of $300.

For details on this intricate subject, consult Ralph Singleton's indispensable book, *Film Scheduling*.

Under the Rainbow

In an ultra-low budget affair you may spend little or no money above the line. But every production—large or small—incurs production expenses.

Depending on circumstances, production staff may or may not require significant expenditures. When the local car dealer hires you to shoot a thirty-second spot, you discover the script calls for a night shoot requiring a two-camera setup and live sound. What began simply now

Checklist Of Costs

above-the-line costs

screenplay and story rights
—purchase price of original work
destined for adaptation
—writing of original script
—script breakdowns
—script rewrites
—copyright registration
—photocopying
—legal fees
producer
—salary
—phone
—travel
—lunches
—postage
—contracts
—legal fees
director
talent
—lead actors
—supporting cast
—stunt people
—voice-over artists
—models

below-the-line costs

production staff
—camera operators
—sound recordist
—lighting technician
—makeup artist
—dialogue director
—script supervisor
—electricians
—dolly operators
—boom operators
—art director
—costume designer
—model builder
—prop maker
—set decorators
—hairdressers
—special effects technicians
—carpenters
—painters
—still photographer
—animal handlers
—security people
—first aid crew
—publicist
set operations
—camera
—lights
—tripods
—batteries
—tape
—nails, gasoline, books and more
—phone
—shipping
—catering

post-production costs

editing: equipment, for sale or rent
—music
—graphics
—titles
—special visual effects
—dubbing
—time-coding
—audio mixing
—looping
—sound effects
—tape
duplication expenses
—duplication costs
—packaging: artwork, layout, printing, photocopying, postage
advertising
—magazines
—radio
—television
—billboards
—classified
—flyers
—telemarketing
—direct sales
—conventions
—trade shows

And finally, the all-important fund for contingencies, a cool 10 percent of the total.

Figure 19.1 Checklist of Costs.

demands additional camera operators, lighting people, sound recorders and probably three or four grips to jockey cables and equipment.

On the other hand, your sixty-minute documentary on the mating habits of waterfowl requires only you and your camera.

Obviously, a project's length bears little relation to total costs. It's the script details that matter.

On most projects, regardless of size, the key staff members are the camera operator, sound recordist, lighting technician and makeup artist.

Support positions include dialogue director, script supervisor, electricians, dolly operators, boom operators, art director, costume designer, model builder, prop maker, set decorator, hair dresser, special effects technician, carpenter, painter, still photographer, animal handler, security, first-aid crew and publicist (see Figure 19.1).

Set operations, like staff, can demand either a large or small chunk of cash, again depending on the project. If you already own a camera, lights and sound equipment and all your locations currently exist; operation costs may be few. No need to buy or rent gear or build sets.

If your video seeks to portray a sci-fi world, get ready to dole out the dollars. The set builder needs wood, the wardrobe manager gold *lamé*, the makeup artist latex and the gaffer colored gels. The set operations segment of this budget requires great detail; include every imaginable associated cost.

It's not neurotic to include such costs as the tissues actors will need for their noses during chilly outdoor shoots. It's these little things that throw a project into disarray. Don't overlook such "obvious" items as nails, screws, bolts and glue; gasoline for automobiles and generators; rentals and permits; duplicate sets of clothing; bottles, cans, books and plants; and, of course, tape.

Again, talk to your people; learn what they require. Sometimes it's not a bad idea to ask your support staff to create their own budgets; these you can incorporate into your final estimate.

Postage and telephone fees can add up quickly. It's amazing how many long-distance calls people make in the middle of a production. And if you're on the road with a cellular, cash outflow can grow quite frightening. Same for postage and shipping expenses. If your client lives in another town and insists on dailies, you'll go broke if you haven't budgeted properly.

To cast and crew, the most important cost is catering. Believe me, you don't want to face down fifteen hungry people with nothing but a jar of peanut butter and a loaf of bread.

Post-production Funds

The last formal segment of the budget concerns post-production audio and video editing.

Those who own their own editing equipment will find costs fairly minimal. If forced to rent, the prices can get steep. Don't underestimate the amount of editing time. This will depend on such variables as length of script, timing, client approval and number of effects.

If you've shot footage for a thirty-second spot requiring only five or six edits, an hour may be enough. If that same spot requires a rapid assault of images and sounds, you may spend two or three days in the edit bay.

You usually reserve editing time by the hour, day or week, with price breaks for longer periods. Other post production costs include music, graphics and titles creation, special visual effects, dubbing, time coding, audio mixing, looping, sound effects and extra tape.

A final cost to consider is the promotional expense associated with selling and marketing. Final product should be presented in a professional form. All videos properly labeled and packaged. You don't want to hand a client a VHS copy in a cardboard sleeve with masking tape crookedly proclaiming the title on the spine. A nice hard-shell package with a printed label is the only way to go; include these costs in your budget.

If you want your production to reach the masses, think about full-color packaging costs like artwork, layout and printing. Any worthwhile marketing effort includes mailing promotional copies to potential distributors and buyers. Estimate photocopying, postage and duplication costs.

Don't forget advertising. Magazines, radio, TV, billboards, classifieds, flyers, telemarketing, direct sales, door-to-door, conventions and trade shows whatever

form you plan requires cash. Obtain quotes for ad rates during the time your ad will appear. Remember, your commercials will probably air more than a year after you've assembled the budget.

Because so much can go wrong, add a 10 percent contingency amount. This allows for unpredictable events, and lets investors know you're handling the project in a professional manner.

It's not easy to create budgets. It's harder yet to stick to them. But with a little preparation, forethought and diligence throughout the production, you could still find change in your pockets when the credits roll.

20
A Modest Proposal

William Ronat

Your phone rings. On the other end is a potential client. You like potential clients, as they represent potential profits. (Okay, so maybe you're not a professional videographer at this point. Stick with us anyway; as your skills and your reputation grow, you just might get such a call one day.)

The conversation is pleasant enough, with the potential client giving you a nebulous description of his potential video project. Then it's your turn. "What's it going to cost?" asks the potential client pleasantly.

That's the problem with some potential clients. They want to know exactly what you are going to do before you do it. And they want to know exactly what it will cost before they even know what they want to do.

Here's how I handle such a question. "Look," I say, "every video project is different. It's like asking how much a house is going to cost before you tell me what kind of a house you want to build. How

many rooms does the house have, does it have a water view, how many acres of land? Ceramic tile? A swimming pool? You see?"

"Ah, of course, I see perfectly," says the potential client pleasantly, "But *how much is it going to cost?*"

"A million dollars," I reply.

Learn to Earn

This is the time to get some details on paper, usually in the form of a proposal. A proposal is simply a document that outlines what the video is going to accomplish. How you plan to make it happen, and an estimate of what it's all going to cost. Ahhh—we're back to the cost issue.

How can you come up with an accurate estimate of a "potential" video? By learning everything you can about the project. Who will be watching the final video—CEOs of corporations or first graders at the local elementary school? How long is the video? Will you need to shoot on Digital Betacam or is S-VHS acceptable? Are you and your crew going to have to travel to Istanbul or will everything be shot locally?

Be sure the client understands what he or she is getting. If your price is for shooting and editing, then let the client know that scriptwriting will be an additional expense. Or if you do take the job from start to finish, then outline all the steps (selecting talent, scouting locations, production scheduling, shooting, editing, and dubs). Make sure the client understands that your price includes these items only.

Before you state a price for the project, see if you can find out what the client's budget is. It may be more than you thought, which gives you the freedom to add more elements to the production. On the other end of the spectrum, the budget may be so small that it's not even possible to accomplish what the client wants. It's better to learn this sad fact early, before you invest your valuable time.

Set the Parameters

Once you state a price, the client will try to hold you to it. Client-human nature is to lock you into a price and then add complexity that will cost you more money. "Did I forget to mention that you could shoot on the warehouse floor only between 3 a.m. and 5 a.m.? Must've slipped my mind."

This is why you want to be specific in the cost estimating process of your proposal. If you tell them exactly what they are getting for the price you are quoting, the client won't be able to add on more complexity without that price going up.

On simple jobs, I usually break my estimate down into two parts: 1) the Treatment and 2) the Estimate and Authorization.

Earlier, we looked at how you should ask questions to learn about the client's project. With the information you learned

through your questions, a natural method of creating the video will probably pop into your head.

For example, a client might be a builder of million-dollar homes. The client wants to show off the many features of the different models. Your treatment might look like this:

Classical music plays as the camera floats past the house with a breathtaking wide shot. The scene dissolves to a closer shot of the front of the house. As the camera floats forward, the front door opens and the camera (who is the viewer,) is greeted by a butler. This butler (a professional actor) proceeds to give the viewer a full tour of the house.

The treatment can be as simple or as complex as you like, as long as it serves the purpose of telling your client what the show is going to look like. If the client likes the concept and agrees on the price, you are ready to move into scripting. If there is something the client doesn't agree with, you know it before you discuss money.

Also, if the clients love your idea, they're more likely to go with you than one of your competitors. Of course, the client can always steal your idea and use another company, anyway. The unfortunate fact is you can't copyright an idea.

Author, Author

The second part of the proposal I send to clients is the *Estimate and Authorization*. On this sheet, I try to be as complete as possible, putting down my best guesses on what each part of the video will cost. There are two schools of thought on this. A buddy of mine, who also produces video, only tells his clients what the total cost of the production will be. He has found that some clients try to lower the price by eliminating parts of the video ("Look, we can save $200 if we re-use a stack of VHS tapes from home...").

Whatever method you choose, the important item is the last line on the page. This is where it says:

Authorization: _____ Date: _____

Have the client sign and date your document and you can get started.

Does this document protect you from a client who wants to rip you off? Nope. But neither do multi-page contracts. I know of a disreputable fellow who has run up thousands of dollars worth of video production bills (though not with me, thank goodness), refused to pay, been taken to court, ordered to pay by the judge, and still refused to pay. The last I heard, he had left town—without paying.

If you don't have a good feeling in your gut about potential clients and you think they might be in the sleazy category, back off. Talk to other people in the community who have worked with these people. Are your fears legitimate? A little homework can help you avoid major headaches.

So why get a signature? Honest people (the ones you *want* to work with) stand by their promises. But even these folks sometimes have short memories. ("I never agreed to that." "But you signed this document saying you did." "I did? I'll be darned.") Leaving a paper trail helps everyone remember these little details.

What Do You Propose?

Often, a proposal is much more than just a few descriptive paragraphs with a cost figure attached.

A proposal may become a long, involved chunk of paperwork, which explains in detail how you will create a specific video project. It is sometimes written in response to a *Request for Proposal (RFP)*. Government agencies and other large corporations often send out RFPs when they need specific services, be they video production or bomb shelter construction. What they get back from an

RFP is a mountain of proposals, each explaining why the proposer is the best choice to provide that service.

How can you get in on the fun of responding to an RFP? One way is to team up with a larger company which is responding to an RFP that calls for a video as a part of a larger contract.

For example, I once worked on a project where a company was creating a simulator for the Navy to train catapult officers on aircraft carriers. These officers stood on a simulated deck and looked at a large screen television showing F-14s, A-6s and other aircraft preparing to take off. The production company I worked for was the subcontractor responsible for capturing the aircraft on videotape.

As a subcontractor, the video company was only responsible for responding to a small part of the RFP. But it was important that the Navy was as comfortable with the information presented in the video portion of the proposal as the rest. If you can convince large companies in your area that you are the person to handle its video requirements, they might call you when they need a video subcontractor.

If the RFP is for the production of a video program, you could respond as the primary contractor on the job. But be aware that these RFPs go out to dozens of companies at the same time. If you don't feel that yours is the right company to do the work outlined in the RFP, you may want to save your energy for a project you *can* handle.

Why not respond to every RFP you can find? Because creating a proposal is a lot of work. You could conceivably spend all your time writing proposals and never win any of them.

For Example

What kind of information do most agencies expect to see in a proposal? The following is the actual wording of the proposal format from an RFP from the State of Florida.

1. Table of Contents

2. Tab 1. Executive Summary—Include a synopsis of the proposal prepared in a manner that is easily understood by non-technical personnel.

3. Tab 2. Certification and Reference—the proposer shall provide a list of not less than three (3) nor more than five (5) different previous clients during the past 3 years as references. This part shall include the dates of the previous projects and the name, title and telephone number of a responsible employee of the previous client who is familiar with the project. The proposer must include a certification that in the previous project it was the original provider of the services.

4. Tab 3. Resumes of Individuals Proposed to Work on this Contract—the proposer shall include resumes of the individual it proposes to assign to this project, specifying relevant educational and work experiences, and shall designate which individual will be the producer/director responsible for the coordination of work efforts of the other personnel assigned to the project. Availability of each individual shall be described, as well as the estimated number of workdays of commitment from each.

5. Tab 4. Description of Creative and Technical Approach—The proposer must provide a description of how it will produce the video programs. This description shall include the proposed production schedule of the estimated working days required to complete each part of each program, the degree of involvement by the Division, and the geographic location where the production will take place. It should also include general information about the talent (estimated number of professionals, semiprofessionals, and extras) and a general description of the proposed use of narrative, dramatics, animation and graphics.

6. Tab 5. Description of Video Equipment—The proposer must supply a list of production and post-production equipment intended for producing these programs.

7. Tab 6. Work Sample—The proposer must supply a sample in VHS format era previous instructional or training video program with production values similar to those offered in response to the RFP. The work sample will be evaluated for both production quality and creative treatment of the subject matter.

You Get the Idea

Also requested by the RFP were a Cost Proposal Form, a proposal Acknowledgment List and a Sworn Statement on Public Entity Crimes. If you think filling out one of these puppies sounds like more work than you are now putting into entire video projects—you may be right. This is why you should feel you have a pretty good shot at getting a contract before you go after it.

The sample above, from the State of Florida, was an extremely well written RFP. A video expert was called in to give the writer advice on how a video is put together. But sometimes an RFP is written requesting strange or unworkable video solutions. It doesn't matter. You must respond to these requests as they are, even if they are bizarre.

Responding to request for proposals is a skill. You have to answer every question, dot every i, cross every t. If you don't, your proposal can be thrown out for non-compliance. It's harsh, but true.

If you can find someone who has dealt with RFPs before, it might be worth it to "partner" with them. It doesn't really matter if this person knows anything about video; that's your job, as long as they understand the language of responding to proposals.

Check with local business groups to see if they know of any retirees who used to work for a corporation. These people might have been exposed to proposal writing and they might be willing to help you learn how. They might be happy to pass on their knowledge to a new generation. If you can't find a real human to give you advice, check your public library for books on proposal writing.

Is responding to an RFP worth the trouble? Winning a contract can be extremely lucrative. But it isn't easy. If you think you can fill the requirements, I propose you give it a try.

21
A Word From Your Sponsor

William Ronat

You have a great idea for a TV Show. Now all you have to do is produce it and wait for the accolades and money to roll in. Right? If only it worked that way.

Unfortunately, it takes money to put a show on the air. It takes money to shoot and edit a show and it takes money to buy the airtime so that the audience can receive it on their TV sets.

We can assume, then, that to go through this process you need... money. You can supply this money yourself if you have it, or you can find benefactors, patrons, or as we often call them, *sponsors.*

Sponsors might fund you for several reasons. They might take this step 1) because you look like a nice person who needs help, 2) to make money as a result of good publicity or the advertising value they receive from your show, or 3) because they're relatives and have more money than sense.

Addled relatives may be the best solution, but if you come from a family of poorly financed underachievers you will probably opt for an individual or business looking to make a buck. This is best. If a sponsor fails to prosper from your efforts, he will simply not fund your next project. A disgruntled family member may very well disinherit you.

On With the Show

You should pre-qualify potential sponsors before you contact them. What do I mean by pre-qualify? Find out the answers to questions such as these: will my potential sponsors be receptive to the idea of financing a TV show? Do they have the budget for such a project? Does the content of my show "fit" with the business of these sponsors?

The more you know about your sponsors, the better. If you approach a cigarette manufacturer to finance a show about lung cancer, you may not have much luck. Don't waste your time (or the time of the potential sponsor) by setting yourself up for failure. Rejections are mentally debilitating. Avoid them.

What kind of show might a sponsor pay to have produced? You can look to the television networks for the most prevalent type: one that attracts viewers who may buy the sponsor's product.

We call daytime dramas *soap operas* because companies like Proctor and Gamble—soap manufacturers—sponsor these shows. Traditionally, these shows were watched by the person in the household most likely to make the soap-buying decision.

Sporting events find sponsors in beer breweries, athletic shoe makers and car manufacturers, among others. Toy companies basically own Saturday morning children's programming. In fact, many of the shows have ties to specific toys and vice versa, an odd symbiosis. Think about who will want to watch your show and then match a business that has a product to sell to this audience.

Another type of show that a sponsor might pay for is the *infomercial*. Demonstrating a product can be an effective selling method. Door to door vacuum cleaner salesmen knew this technique. Throw some dirt on the floor and then use the product to clean it up while the customer watches.

Infomercials often go the same way. These shows are basically a long commercial for a product, but are disguised to look like a talk show or a news magazine format show. The channel-surfing viewer might get hooked before realizing he or she is being sold. If a sponsor can take direct orders (from an 800 number or through mail-ins) an infomercial might have appeal.

If one of your local stations is part of a home shopping club network you may be able to create a local version of their programming. Then, for X number of minutes (which you or your sponsor would buy) the station would be selling your sponsor's product instead of pots and pans or ceramic dolls.

If your goals include more intellectual pursuits, you may be able to get a sponsor for a drama or documentary if the quality of your show is superior. The leadership of a potential sponsor's company may understand that a fine drama adds to the quality of life of everyone. Or maybe they recognize that supporting a popular program could help change their "unpopular" status with a certain slice of the viewership.

Either way, the arts get funded. Without speculating on motive, here are two examples of corporations which sponsor artistic endeavors: the long-running series of dramas under the banner *The Hallmark Hall of Fame* and *Masterpiece Theater* sponsored in part by Mobil Corporation.

Psst, Buddy, Wanna Buy a TV Show?

How do you approach potential sponsors? First you have to find them. This means doing research to learn which companies in your area would even consider funding your show. But just because you have a great idea and your potential sponsors are the perfect candidates to benefit from it doesn't mean they will do so. You have to sell them on the idea.

Being in business, you have probably had to ask a bank for money. That's not an easy process, is it? You probably had to create a business plan and outline your life's history. Then you had to prove to the bank that you are a good risk and that the bank would get the money back that you wanted to borrow.

When you ask a sponsor for money, prepare your case as well as if you were asking a bank for a loan. You are asking a sponsor to risk capital without collateral or any guarantee that the sponsor will get the money back. How receptive will the potential sponsor be to such a plan? That depends on how good a plan it is.

It won't be easy to get an appointment to discuss a sponsorship. You will have to determine who the appropriate contact is. It may be a marketing person inside the company or the company may have an advertising agency, which would handle such a project. You may have to explain your way through several layers of people before finding the right contact. You may be ready to collapse by the time you reach the decision-maker, but this is the point where you must be your most convincing.

Be professional. You are conducting a business transaction. Dress and act appropriately. You may feel that you are an artist and should enjoy creative freedom—even in your choice of dress—but torn jeans and a ratty T-shirt may give a potential sponsor the wrong impression.

Prove the worth of the idea. You know your idea is a winner, but the decision-maker may not. Start at the beginning and take the potential sponsor through a step-by-step explanation of who the audience will be, what style you will use and why being associated with the show will be good for the sponsor. Don't try to do this off the top of your head. Make notes and *practice* in front of the mirror. If you lose the potential sponsor during the initial pitch they're probably gone for good.

Prove that the audience exists. If you plan to air the show on Channel X at 8 p.m. on Friday, get backup data from the television station that shows demographically who makes up the audience in this time period. Salespeople at the TV station need this information to sell commercials to their clients and will share it with you if you tell them your plan for purchasing a chunk of airtime.

Prove that the audience will watch. This is a tougher assignment. You can't really *prove* this point, but you can make a good argument if you do your homework. Show ratings for similar shows in similar time periods. Get as much hard data as you can. Facts often sell better than enthusiasm.

Prove to the sponsor that they will receive R.O.I. (Return On Investment). If 100,000 people watch your show and 1% of them buy the sponsor's product (a Rolls Royce Motorcar, for example) at an average price per car of $100,000, the sponsor will gross $10,000,000. (Actual numbers may vary.)

Prove that you can do the job. Have you ever done a TV show before? What was it? If not, have you ever done similar jobs before? Do you have a demo tape showing some samples? Can you give potential sponsors a *warm, fuzzy feeling*? You may have thought proving your competence in the video production field was the only step you needed in order to land a sponsor, but as you can see it is merely one of many important steps.

Because each show is different you will run into different obstacles for each of them, but if you can't go through the steps mentioned above with confidence, then you can go back to step 1. You remember step 1: look for a rich but not overly bright relative.

The acid test is this: is your idea good enough that you would use your own money to produce it and buy the air time if you had the money? If the answer is no, then why would you expect a sponsor to say yes?

Johnny on the Spot

Instead of finding one sponsor to fund your entire operation, you may instead want to sell pieces of time within your show. The sponsors can then use this time to run promotional spots about themselves. You've heard of these; we call them commercials.

You will still have to convince sponsors of the worthiness of your show, but because the amount they are spending is smaller, the decision may be easier for them to make. On the other hand, you will have to sell many different sponsors instead of one, which means your job will be more difficult.

There is also more risk for you because you will be buying the airtime and then recouping that money by selling commercials. Let's say you will run your show on cable.

The rate that a cable company can charge you for their airtime is based on a formula created by the FCC. The formula is tied to a cable company's markup and number of subscribers. For example, say your cable system buys HBO for $4.00 per month per subscriber and they turn around and sell it for $10.00 per month to their customers. But only 25% of the potential subscribers actually order HBO, which has an effect on the formula:

$10.00 subscriber fee

-$4.00 cost of programming

_____

$6.00 mark up

X.25 percent of subscribers purchasing a premium channel

_____

$1.50 implicit value of the premium channel, per subscriber, per month

These numbers are for a premium channel. For a channel included in the basic package, the cost would probably be closer to $0.50 per subscriber. But remember, that's *for the whole month*. If you only want an hour of time, you would divide that by 720, the number of hours in a month. The answer is $.00069 per subscriber per hour. If your cable system reaches 100,000 households you would pay $69.00 per hour. The cable system can also add fees for billing and collection, marketing and studio services, so ask for an estimate to be sure you are getting a good deal.

On top of that cost is the money you spend on production, promotion of your show, etc. If the total expenditure is $1,000 per week and you have room for 20 commercials, you would sell each commercial for $50 to break even. But you want to make a profit, and there will be times when you won't be able to sell all the commercials in a show. So you need to charge enough to make up for this shortfall. Supply and demand will have an effect on how much you can get for your spots. If the sponsor's target market loves your show and you can prove it, the sponsor will pay more. If nobody watches your show, you won't be able to give your commercial space away. But, hey, that's business.

Landing sponsors can give you the opportunity to create some great television. But before you start knocking on doors, be sure that your idea is as great as you think it is. Do your homework, back up your theories with facts and get your ducks in a row. Then, assume your best professional demeanor and go get 'em. It's not an easy gig, but you should be used to that by now. As you know, in video, nothing is ever easy. And that's the way we like it.

22
Recruiting a Crew

John Bishop

To the average camcorder owner, the notion of hiring a bunch of gaffers, grips and assorted techies may seem far-fetched. Probably because it is far-fetched. Most videographers go it alone, single-handedly performing every production role. Sound familiar?

To be honest, there are many times when you don't really need a crew. I recently set out to shoot a bunch of video in and around Los Angeles. A classic camcorder job; I grabbed two tapes and headed for the freeway.

The first thing I passed was a feature film crew setting up a shot. There were three trucks and dozens of people with walkie-talkies milling around. It wasn't on my shot list, so I kept going.

Six hours and 180 miles later, I'd filled my tapes and passed the film crew again. They were still working on the same shot. It felt good to be working light.

Keep in mind, however, that there's a lot of territory between my simple one camcorder job and a mega-budget feature film. Bringing a video project to the screen can easily involve juggling more details than a single director/camcorder jockey can handle (see Figure 22.1). Even the hobbyist who only tapes his family at holidays might need to recruit a bored relative to lend a helping hand.

Who knows? You may even find yourself in charge of a low-budget professional production someday. When that happens, if you pay an entire crew of pros what they're worth, your budget will rival the Pentagon's. But there are ways to get a crew to fit a small budget, if you know what you're doing.

High-Quality, Low-Budget Crews

While a crack crew may bring a lot of extra value to your work, most just-for-fun video productions don't really need a payroll full of professionals. If you have simple needs, you can often rely on friends and relatives to help out. This is a good way to work, especially when you

Figure 22.1 *Independent videographers are often both director and camera operator.*

need willing hands more than trained pros.

Any person willing to help and take direction is fair game to assist you on a video project. Once you learn that given individuals have good work habits, you'll start to rely on them more and more. Eventually, you could have people trained to handle most of the chores of video production. If they plan to go into a career in videography, or are just plain interested in video, all the better.

Schools are a good place to look for crew. Many cities have trade schools or colleges with plenty of film or video students eager for experience. Many cities also have media centers, which help to bring members of the creative community together. And most cable TV stations have access to a number of volunteers. All these places have bulletin boards. You can get names off the board, or post your own notice and wait for calls.

When putting together a volunteer crew, you must make it clear that it is a job. Make sure they know that they have to get there on time, stay with the job until it's done and bring a good attitude.

To get more experienced crew members, you have to offer them something. It's nice to pay a token fee, to show appreciation and defray the crews expenses. Sometimes I pay everyone a flat $100. But the real coin of the realm for low-budget indies is the credit. A screen credit on any finished work is a valuable commodity.

A crew person's career depends on credit for projects they've worked on. The promise of their name in the credits and a listing on the crew sheet holds great value, as does the chance to work with other crafts people who might provide good referrals for future work. (Crew sheets have the position, name and phone number of all members of a crew.) For crew jobs where the work shows on the screen, such as camera, sets, makeup or effects, the producer should make tape available for these crew members to put on their demo reel—a selling tool to help them get jobs in the future.

Some Typical Crew Positions

Now that we've discussed ways to find a crew, let's look at the "typical" production crew members themselves and what they do. Even if you're bound and determined to work solo for the rest of your life, you can learn a lot about the various aspects of making video by seeing how Hollywood divvies up the tasks.

A director relies on two essential helpers: the production manager and the script supervisor. These jobs relieve the director of the burden of organization. A videographer can benefit from the same kind of help.

The production manager breaks the script down and defines what each scene requires. He lists the actors, locations, props and special goodies that each scene needs. Using this data, he plans the most efficient way to shoot the program. For example, you should shoot all the beach scenes on the same day, even though they may be scattered throughout the script. If you need to borrow a friend's truck, rent a camera dolly, or commandeer a special prop, you should schedule all the scenes that require it at the same time. This is hard detail work, but essential for a smooth-running production.

It takes a person with a fine eye for detail and great organizational skills to be a good production manager. On a small production, any hard-working soul can perform this job just fine.

Scripts, Shots and Signals

While the production manager shepherds the work force, the script supervisor rides herd on the story. The script is a blueprint for the finished program. It is a list of scenes, each one representing a specific time and place, and comprising some actions and/or words. If you've got a script, you could probably use a script supervisor.

The script supervisor keeps track of the shooting of scenes. On dramatic productions, videographers often shoot each scene several times. Sometimes they do this simply to assure the quality of the scene. Other times, they do it to get a number of camera angles to have more choices when assembling the program. Sometimes, particularly when there's a special effect, you may want to shoot the scene with two or more cameras. And in the heat of production, lines of dialogue change, or even vanish.

As you can see, things can get pretty confusing for the director. Someone has to keep track of whether or not you've completely covered every scene. Likewise, the editor must deal with a bewildering melange of raw footage. The script supervisor maintains order and makes sure the footage shot jives with the script. He also makes sure there are good notes so it's easy to locate footage during the edit.

The script supervisor is also in charge of continuity. Since videographers shoot most fictional scripts out of order, some-

one must make sure that costumes and makeup match from scene to scene. Someone also has to keep track of the movement and placement of characters between scenes, so the audience won't get confused.

Script supervision is a hard job, but one that any dedicated person can pick up. Like production managers, script supervisors often start off as production assistants and move into the job after gaining some experience on shoots. If you've recruited someone to help you write or polish a script, why not make them your script supervisor?

Even if your video doesn't require a full-time production manager and script supervisor, it's often useful to have someone along to cover some of their functions and take pressure off the director.

Photo Opportunities

One crew position that for-profit videographers often neglect is still photographer. Photography is often an afterthought, if people think of it at all. But when the work is complete, you still need to promote it, and still photos of the scenes taken as you shoot them will help you do this.

On more complex productions, the still photographer often takes extra shots with a Polaroid™ instant camera to record setups. They shoot the placement of actors relative to each other at key points within a shot, details of costumes and makeup, placement of props and anything else that will help maintain continuity from shot to shot. For example, if you shoot an exterior of a person walking into a building through a door, and then on another day shoot an interior of the same person coming through the same door, you need some way to make sure their makeup and costume match up.

Another person who can prove useful on a small video shoot is a video techie who can keep track of cables, match mike impedances, tap into PA systems, read a waveform monitor or light meter to get the lights and exposures set up correctly, and otherwise check and troubleshoot the recording chain. For this job, I often hire people who work at college audio-visual departments or cable TV centers. These people know a lot about video and enjoy working out of the studio from time to time. If you've got a friend or relative with some technical knowledge, put them in charge of all the wires and signals.

DPs, Grips and Gaffers

On larger productions, the person in charge of the video's look is the director of photography or DP. The DP has the last word on the lighting, composition and movement of the image. Having a DP allows the director to focus on the acting and the story, while the DP worries about the technical side of creating the image. In practice, the DP and director work very closely together, because so much of the story unfolds through nuances in the image.

While most videographers do their own shooting and directing, Hollywood's use of a DP tells us something: it's hard to manage both your image and your talent. If you can find someone to run your camcorder, you can devote all your energies to getting good performances on tape.

Grips are the people who mount the camera. If you need a shot with the camera attached to a car, bicycle or model airplane; you need the tools and know-how of a good grip. It's the grip who sets up dollies and cranes and moves them during the shot. They're also in charge of some of the special effects devices, like wind, rain and fog machines. While many videographers perform this role themselves, it's not a bad idea to try and find someone gifted with wood and wrench to handle your camera mounting chores.

A gaffer or lighting technician's job is to help out with the lights (see Figure 22.2).

They balance the intensity ratio of different lights—for example, the key light, which sets the exposure and the fill lights which determine the contrast. They not only set up the lights in the proper places, but secure the stands with sand bags so they won't get knocked over, and tape the power cords down so people won't trip over them. This may sound like a simple role, but it's extremely important for safety concerns.

A great deal of lighting art involves working with the color and texture of the light, and keeping it from spilling in where you don't want it. The person in charge of lighting must know about the use of scrims to soften and reduce light output, barn doors and flags to control where the light falls, gels to adjust the color, a whole range of diffusion materials to change the quality of the light and reflectors to make one light do the work of several. Fledgling videographers working their first paying job would do well to get some help in the area of lighting.

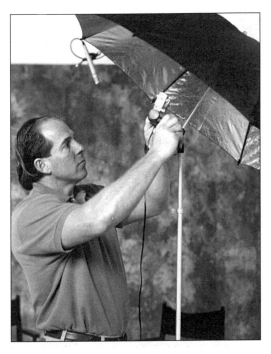

Figure 22.2 *The gaffer handles lighting elements.*

Sound, Props and Makeup

In documentary-style shooting, sound is the provenance of one person who both holds the mike and adjusts the levels (see Figure 22.3). In more structured shooting, a boom operator holds the mike and a mixer records the sound. These are important jobs, since the sound often carries more crucial data and keeps better continuity than the picture.

On a camcorder production, the sound crew has to get the mike in the best place without getting it into the picture. If you use wireless mikes, the sound crew sets those up and mixes their output into the one or two channels that will end up on the tape. When TV shows record on tape, they also make a backup of the sound.

A good sound crew can add a great deal of value to a program. They can record wild sounds from each site to use as sound effects in the editing. They will also record a minute of quiet ambience or room tone for each location that edi-

tors can use to cover audio glitches in the cutting. You probably understand the importance of audio, but have you ever considered getting some sound help when shooting? Having a person who worries about nothing but audio can really improve the aural impact of your productions.

On the set, a costumer will help the actors with their clothing and make sure the details are the same from scene to scene and shot to shot. A property manager is in charge of props, making sure they are ready at the right time and that there are plenty of those props that get used up during the shoot. Doing props and costumes on a low or no-budget video is a real challenge. It tends to send people on scavenger hunts through old clothing stores, junk shops and garage sales. The property manager may also spend lots of time in garage workshops coming up with unique devices. Often, having just one person in charge of props and costumes is a great help.

Makeup is a tricky business. It can range from very light makeup for realism to effects makeup for monsters and wounds. As with props and costumes, a good makeup artist can come up with fantastic solutions for very little money. Makeup for film or video is not the same as makeup for going out in the evening, nor is it like theatrical makeup that the audience will view from a distance. If your makeup person doesn't have experience, they should read up on video makeup and do some playing around with a model and a video camera before the production starts. If you have a friend who is gifted in applying makeup, have them visit the set at the beginning of a shoot. They will probably enjoy the challenge of learning about makeup for video.

How Much Help?

So, how much crew do you really need? You only need as much as it takes to get the job done. Smaller is better with free

Figure 22.3 *A sound technician will make sure that the sound track is crisp and clear.*

crews. It's better to keep a few people very busy than to have lots of people standing around and wondering why they've given up their time.

On the flip side, most videographers need to fight the urge to do everything themselves. After all, making video should be fun. Getting some help on your shoots will let other people in on the experience, and help you keep your sanity at the same time.

Crew up!

23

Location Scouting:
The Tricks of the Trade

Geoff Best

Whenever you're shooting outside the confined walls of a video studio, you're shooting on location. And whenever you're shooting on location, you must deal with site-specific circumstances—circumstances of lighting, power, noise, people, furniture, pets or anything else that might help or hinder your video efforts.

That's where location scouting comes in. Location scouting is the process of visiting a site before you shoot, and taking stock of all the conditions that might affect your shooting in order to prepare for potential problems or exploit the potential benefits of the site. It also involves the selection of locations when there's a choice to be made between several sites.

Location scouting doesn't have to be an elaborate process. It can be as simple as looking around, listening and jotting a few notes on paper. For both the weekend video hobbyist and the would-be professional videographer, this chapter will give hints and advice on the location scouting process. By the time you're done reading it, you should have a clear idea of how to improve your own videos with these time-honored techniques.

Two Kinds of Locations

For video producers, there are two distinct types of locations, and the way we investigate them is necessarily different. The first type, the one we encounter most often, is the predetermined or fixed location: weddings, graduations, musical events and sports, for example. These events are held at places that have been picked by other people. Our job is to figure the best way to deal with existing conditions. We may be hampered by poor lighting, or helped by a great sound engineer. We may find ourselves backed into a corner behind the potted palm, or discover someone has removed a whole section of chairs for our convenience. But knowing what to expect is where location planning pays off.

The second type of location is the one we choose ourselves to create a certain aesthetic look, or to meet some dramatic requirement. For this type of location shooting, there will often be a story to be told—possibly a drama or a documentary. Our job in this case will be to find places that will work to convey the mood and look we require.

Let's first consider the fixed location: assume we know a couple that would like to have their wedding videotaped. They ask their wedding coordinator, Elaine, to call Jim, a local freelance videographer, about doing the job. Although we'll be using a professional video producer for our example, the procedures outlined below are equally valid for an amateur—a friend who wants to videotape an event, for example.

Figure 23.1 A calendar of dates and times will allow you to scout locations under the exact conditions you'll be shooting.

Jim and Elaine talk and he assures her that he is capable of such an undertaking. Since Elaine is in charge of many of the festivities, she becomes his contact person. It is important to find a contact person right away. Your conversations will be more informative when you talk to the same person each time. Jim first asks for a schedule, along with the location, date and time of day.

While looking over the schedule, Jim notes that a rehearsal and fully programmed dinner party are planned. He realizes that being on hand for these events will add greatly to the finished tape. He marks these dates and times on his calendar also. The schedule tells Jim the wedding will occur on a Saturday at 4:30 pm and will be held outdoors at a country club. He wonders how the low afternoon sun will affect the camera, and knows that he must not be positioned facing the sun. He decides to go to the location on a Saturday at 4:30 pm. That way, he can see conditions similar to the ones at the actual event (see Figure 23.1).

This type of planning is the key to a relaxed and pleasant shooting experience. From this initial inspection, Jim will make two checklists. One will be a loca-

tion list and the other will cover the equipment. By now, Jim has a good idea of what to expect. He has found out where the bridal party will stand and where the processional will begin. Elaine tells him that in case of rain, the ceremony will be moved into the club's multi-purpose room, and Jim makes a mental note to give that location a thorough inspection.

The First Investigation

Jim packs his camcorder, a notebook, a compass, earphones, his wireless microphone and other items he feels he may need, then heads out on his investigation. His first priority is to locate electrical outlets, and determine whether they will be useful to him at his shooting location.

Jim finds that the nearest electrical outlet will be near the sound mixer—about 75 feet away. An extension cord will reach easily (see Figure 23.2). He decides that he will need to tape the cord to the outlet box to keep it from coming unplugged part way through the event. He jots down duct tape and an extension cord in his notebook.

Jim looks back across the lawn and realizes his car is a long way off. Bringing a wagon or handcart might be a good move.

Figure 23.2 *Scouting power sources will let you know what lengths of extension cord you'll need.*

He powers up his camcorder, plugs in his earphones (see Figure 23.3) and listens for noises that the ears normally block out. The camera's microphone picks up a constant rumble of traffic noise that Jim hadn't even noticed. He decides it's a good time to see how the wireless mike works. He sets the transmitter some distance away and listens through the camera as he walks around his shooting area. Wireless mikes usually work well outdoors but it's always smart to check for interference or dead spots.

Jim's investigation is almost complete. He stands near his shooting spot and takes out his notebook. He makes a sketch of the site noting the position of the sun, electrical access, location of the sound board and his own shooting spot. He then makes a note to get in touch with the sound man. If he can get an audio feed off the board, he might get better sound than he could with his own setup.

Indoor Challenges

Next, Jim heads over to the multi-purpose room. Shooting video indoors requires a similar careful approach. Inside a building, there is a stronger possibility that the wireless mike may not work properly, due to greater interference and dropouts caused by the radio transmission bouncing off walls and arriving at the receiver at slightly different times.

Jim rigs the mike transmitter at the podium and discovers a large dead spot near his preferred shooting area. But by moving several feet to one side, sound comes through loud and clear. He sticks a small piece of duct tape on the floor to mark his exact location.

The next item on Jim's agenda is power. He checks for the nearest power outlets, and notes that it only has two plugs. If he wants to set up a monitor other equipment, he may need to bring a power strip as well as an extension cord. Jim asks a maintenance person where the circuit breakers or fuse box can be found, to be prepared for potential power problems.

After telling him where the circuit box is located, the maintenance person tells Jim that the sound board will be very close to his camera, so Jim makes a note of the type of connector and cable he will need to pull sound right off the board. It always pays to check this out very closely.

Figure 23.3 *Use headphones to help hear sounds your ears normally block out.*

Accidentally pumping speaker voltage into the camera's mike-level input will be sure disaster.

Jim notes that the stage is bathed in a strange pink glow; the fluorescent lights illuminating the stage have a very strange tint to them. Our man Jim likes to come prepared: a trip to his car produces a small portable TV that he plans to use as a shooting monitor (see Figure 23.4). Running a cable from his camera to the TV confirms his suspicions. The white balance cannot cope with this weird light. He tries the outdoor and indoor presets and decides in this case he will have to manually white balance the camera. Zooming in on a piece of white paper will produce a true white and Jim is relieved to see natural colors reappear on his monitor.

Jim's location research is complete. He has created a checklist, investigated the sound and light characteristics of both the indoor and outside spaces and has talked with people in charge of the event. When he gets home, he will put it all down in order in his notebook and calendar. No surprises and a fun day are what a methodical investigation of location is all about.

A Location Of One's Own

The second type of location work involves finding a site for your own project. Here you will be on the lookout for dramatic backgrounds, colorful street scenes or mysterious old houses. When you first begin imagining what your video will look like, you can get an idea of what sort of mood or setting you want to create. This time, you get to pick your own location to match the look you're after.

Often, it's a good idea to have at least a written treatment of your script idea before you begin scouting locations. In a nutshell a treatment is a short piece that tells what the video is all about. So if the treatment says the camera follows three people through dry grass to an old barn, you know that scene will have to be found. Similarly, if the treatment tells you

Figure 23.4 *A color monitor will give you an accurate picture of what you'll be recording onto tape.*

that later on these people are involved in a heated discussion in a gloomy tavern, you will have to locate such a place and get the owner's permission to shoot your video.

So step one in location research is determining what actual places need to be found. But first—do you really need to go somewhere to shoot your video, or could you create a set in your own house or backyard? Maybe you already have footage from another time and place that you can cut into your production to suggest a location.

Out in the Field

When you have determined that you need to go out in the field, it's best to make a list of shots you need. In the previously mentioned story, suppose the three people walking through dry grass toward the old barn discover a perfectly preserved but dusty old car. The video centers on the discovery of the car, the subsequent disputes over its ownership and how the three agree to work together to restore it.

For this sequence, you need a barn, an old dusty car and a house with a garage. Let's fast forward to the barn, and concen-trate our research there. It goes without saying that you have first obtained permission to use this site. Since this barn has been deserted for many years, you'll need a generator to supply your own electrical power. You should also plan to bring plenty of fully charged batteries for the camera.

If your shooting script shows several scenes in the barn. Your checklist must then include changes of clothing that go with the different scenes. You want to get all the barn shots the same day. This means planning everything that will be needed at the location, including food and water. Since daylight will be streaming in the big doors, you will be faced with a mix of daylight and artificial light. Putting gel filters on the lights will equalize the two kinds of light. Or it could be possible to achieve an unusual effect by allowing the two kinds of light to interact. Whatever the circumstances, it's best to be prepared for any eventuality.

The same goes for all location shooting. Try to think of the things that will be needed on shooting day and keep your lists from one project to another. With these ideas in mind, your next location shoot will be a success.

24
Legal Videopoly: Playing the Game to Keep Your Videos on the Right Side of the Law

Stephen Jacobs

If you're making a tape that will never see the light of day outside your rec room, you don't have to worry much about the legal aspects of videography.

But when you start producing video for distribution and profit, you step onto a new game board. You must touch every space and pick up the right cards as you move along. To discover some of the legal pitfalls, problems and solutions, let's follow Stanley, a fictional videographer, as he tries to break into the big time.

Stan the Video Man

'Tis the dream of every videographer. Stanley's got a brand new SonySonic CCD V12345 camcorder in the trunk of his car. At home, a shiny desktop video post-production studio, complete with a NewTek Video Toaster 9000, is waiting to transform his raw footage into an edited masterpiece. Stanley has meticulously planned, storyboarded and scripted his parody-dramatic feature, *When Irving Met Soon Li*, down to the last detail.

Let's look in on Stan as he shows his script to the actor who will play Irving, his writer friend Sam.

"Hey Stan," says Sam, "some of this dialogue sounds pretty familiar. Are you sure you're OK on copyright here?"

Copyright Friend

Copyright law stands by you, the producer of original material. It protects your exclusive right to profit from distribution and/or performance of your original work, whether written, videotaped, audiotaped, painted, photographed or chiseled. It also protects you from the creation of derivative works based on yours.

The United States entered into the Berne Convention for the Protection of Literary and Artistic Work in 1989. This strong international copyright agreement implies that you need not place a copyright notice on your work. This treaty protects your copyrights to a piece if you can prove that it's your work.

To make sure, however, you may wish to follow the provisions of the U.S. Copyright Act of 1976. Under this act, you secure your rights to your work for your lifetime plus 50 years.

If you want to display a copyright notice on your work, it should follow the format described below:

The word "Copyright," its abbreviation "Copr." or the copyright symbol "©" must appear first.

The year of the work's publication/production should appear next.

Last, the name of the company or individual that owns the work should end the statement.

Putting your personal equivalent of "©1994 Stan the Video Man, Inc." at the end of your tape and on your tape labels will give you common law copyright protections.

Still, to get the full protection of the law, take one more step. Register your piece with the copyright office in Washington. Fill out a form, pay a small fee and send two copies of the work to D.C. It's that easy. And worth its weight in gold if you ever need to use it.

Copyright Foe

Copyright stands against you when you want to use someone else's original material in your own.

If you want to include footage, audio, text or lines from another copyrighted work in your own piece, you've got to get the rights to do so. If you don't get the rights, their holder can bring you to court, and sue for copyright infringement. You could fork over between $250 and $10,000 just for unknowing or unwilling infringement. You could lose up to $50,000 for deliberate infringement.

Those are just the fines. The judge could also force you to pay all your profits and/or damages of lost income to the plaintiff. The court can also order all

copies of the piece in question destroyed. It gets worse; you could land in jail. (See Figure 24.1.)

If you want to use the work of others, you've got to track them down and get permission to use it. The owner will want to know what, specifically, you intend to use, including the run time if it's a recorded piece; what you intend to do with it; what specific rights you want; how you will modify or edit the piece if it's your intention to do so; and finally, what compensation you plan to give him. In other words, he'll ask you all the questions you would ask someone wanting to use *your* work.

Fair Use

In some rare instances, you can use copyrighted material without obtaining the rights to use it. Your usage must fall within some rather ambiguous guidelines also defined by the Copyright Act of 1976.

If your production is educational or informational and uses a relatively small section of the work that will not impact on the market for the copyrighted piece, you may be able to invoke the fair use act.

A good example of fair use in action is The Empowerment Project's Academy

Figure 24.1

Award winning documentary *The Panama Deception*. The documentary examined the military action taken against General Noriega and his forces in Panama. It focused, in part, on the media's coverage, or lack of coverage, of the ensuing events. The big three networks could charge hundreds of dollars a second for news footage in which their news anchors appear. The Empowerment Project, a nonprofit organization, couldn't afford to buy the footage from the networks. Instead, they were able to obtain copies of the news programs recorded off-the-air. Because the producers used the news footage in critical and educational comment on the medium itself, fair use protected them.

Like anything else regarding copyright issues, fair use can cut two ways. George Holliday could have lost his suit against the networks and CNN due to fair use. Holliday sold his footage of the beating of Rodney King to KTLA, CNN and NBC. Soon after, the footage graced airwaves around the world. Holliday maintained that he sold the footage exclusively to those three entities, without giving them the right to distribute it. The judge observed that theoretically fair use entitled the news stations to broadcast it whether they paid Holliday or not.

In the end, however, the judge did not make a fair use decision. He ruled instead that Holliday sold his rights to the buyers along with the actual footage. He thereby applied the rules for professional news photographers to Holliday, an amateur.

Public Domain

If the copyright lapses on a work, you can use it freely. Such work falls under public domain. Most, but not all, of this work is very old. Since Frank Capra copyrighted his *It's a Wonderful Life* under the pre-1976 law, he had to renew his copyright every 28 years. He forgot to renew and the great hit film fell into public domain.

Never assume others have made Capra's mistake or that a work lies in public domain for any other reason. Research it until you confirm it.

Derived Works

As with fair use, there are no fixed guidelines for derived works. If a court decides that your story bears "substantial similarity" to those in another work, you could be guilty of copyright infringement. Stan's lucky that *When Irving Met Soon Li* is a parody; courts typically hold that parodies do not infringe on marketability or copyright, since their markets are so different from those of the originals. Still, Stan should play it safe and mimic only the storyline—not the dialogue—from *When Harry Met Sally*.

The Shooting Permit

We rejoin Stan several days later, after he's completed his script rewrite. Now Stan's in the middle of Wall Street, about to tape the first meeting of his soon-to-be-lovebirds. As he raises his camcorder to his eye he feels somebody tap him on the shoulder. (See Figure 24.2.)

"Good afternoon sir," says one of New York's finest, "may I see your shooting permit please?"

Shooting permit? Yes, this may be the land of the free, but it's also the land of the bureaucracy. Not only do you have to obtain a shooting permit, you probably have to obtain several. Depending on where you shoot, you may need state, city and county permits. Will you shoot in a park? You may need additional permits from the state or federal park authority to shoot there. How about the subway, or on a bus? You may need a permit from the transit authority. How do you know what you need? You make a phone call, and then pay a visit, to the state (U.S.) or

Figure 24.2

provincial (Canada) film office, bureau or commission.

Almost every state and province has such an office, usually located in the capital city. If it's a large state, it may have one or more regional bureaus as well. Contact the statewide office first and they may refer you to the regional office afterwards. If there is no state or regional office, contact the mayor of the city you plan to shoot in before you begin production.

These offices have a dual responsibility. First, they're responsible for promoting their state, or region, as a "great place to make film and video." (They want those big bucks generated by film and television production.) To that end, they provide a variety of services. In addition to lining up all your paperwork, they'll help you pinpoint shooting locations, secure production insurance and even provide police for traffic control or street closings while you shoot.

The state film bureau is also responsible for making sure that you conduct your production safely. That's one reason for all the permits. Permits typically request the following information:

- The names of the principals.

- The dates, times and locations of the shoot.

- A list of vehicles used by the production.

This list pays off, as you will often receive special parking permits for your vehicles. In a city like New York, these special parking permits save hours of hassle.

A summary of the scenes to be shot and a list of traffic control requirements. Again, the support you receive in crowd control can be life saving.

Evidence of liability insurance. The state doesn't want to pay for injuries that happen during your shoot.

This may sound like quite an ordeal for a guy with a camcorder who just wants to shoot a few scenes. Most offices will deal with you at your level of production, and may not require the same kind of paperwork from a student or small, independent production that they would from a big Hollywood production. However, you should go ahead and work with the local bureau or government offices no matter what your level.

If caught without a permit, at the very least you'll have to stop shooting for the day and reschedule. You may find yourself liable for substantial fines as well. If you're actually endangering passers-by, you may even face criminal charges.

Save yourself the money and the hassle. Go get your permits. It may cost you a few hours time and some processing fees, but it will be time and money well spent.

Written Releases

Back to Stan the Video Man. Stan sends his actors home for the day and wanders over to the Governor's Office for Motion Picture and TV Development on Broadway. The folks over there get him all he needs without taking too big a bite out of his budget. He asks a lot of questions; in return, he gets his answers and those handy parking permits. However, the folks over there have pegged Stan as a first- timer. As he gets up to leave, the staffer calls him back.

"Hey Stan," she says, "you've got releases for your locations and your actors, right?"

"Releases?" asks Stan.

"Americans have a right to privacy. This includes not being videotaped if they don't want to be. It even includes not having their property videotaped. People who find themselves on videotape they don't want to be on can sue the producer of said videotape."

"But they wanted to be in my movie!" says Stan.

"But how can you prove it?"

Learn from our friend Stan. When you contract with professional actors, the contract typically covers releases. Some states allow for videotaped, oral releases. Most of the time, however, it's best to have a written, signed release (see Figure 24.3) from everyone who appears in your production.

A release defines your rights regarding the use of the subject's image. It usually specifies the purpose of the footage, the length of time the producer can use the footage and so on. Make sure it also stipulates that performers are free to appear in your production without violating any other contracts they may have. Finally, it should state performers' compensation— even if it's only one dollar. If the performer does not receive some form of compensation, the release is not a binding contract.

A location release defines the dates and times you will be using a property for a shooting location. It protects the property owner by: 1) making the producer liable for any damage to property or personnel during the shoot and 2) spelling out the producer's right to the footage and the necessary fees. It protects the producer against charges of invasion of privacy and illegal intrusion.

If you're a news producer, releases can prove a problem. Your subjects may not want to sign releases, or you may not want to approach them about releases before you shoot them. You can bet that the reporters doing hidden camera reports for *60 Minutes* and *20/20* aren't asking those folks committing crimes for permission to take their pictures. The trick is to make sure that your footage and the "facts" you present are correct. Otherwise you may find yourself embroiled in slander or libel suits.

Music Rights

Stanley's finished the rough cut of *When Irving Met Soon Li*. As this is a romantic film, Stan has gathered some CDs of his favorite mushy songs to lay down onto the soundtrack. He's brought in his pal Jim, the Rock and Roll Animal, to help him mix down the soundtrack.

"Whoa, dude," says Jim, "where'd you get the bucks to get the rights to these tunes? You win the Lotto?"

Stan must check out the two organizations that track music copyrights for songwriters and performers: the American Society for Composers, Artists and

Figure 24.3

Publishers (ASCAP) and Broadcast Music, Inc. (BMI). They track music use for royalty purposes and help arrange for the purchase of rights to music.

Copying a track from a CD onto the soundtrack of your video is the most expensive way to get music onto your tape. This method could cost you thousands upon thousands of dollars.

If you must have a particular song on your tape, buy the rights to record someone else performing the song. This is much cheaper, a question of hundreds, rather than thousands of dollars, plus the cost of the performer you record.

Better yet, buy a licensed music collection. There are hundreds of companies out there that sell music for video production. You pay a couple hundred bucks for a collection of CDs, and you own the rights to use the music in your productions. This is a much simpler answer to the problem of music rights.

The simplest answer: do it yourself, or find someone to do it for you. There are hundreds of struggling songwriters and bands in every major city in America. If you don't know anybody, ask around. Get the word out; post announcements at your local college, radio stations and clubs. If it works out, you'll have inexpensive, original, exclusive music for your video. What's more, you give the performer a big break—and you know that "what goes around comes around."

It worked for Stan. A quick call to ASCAP convinced him that his dream soundtrack would require a third mortgage on his house. He cut a quick deal with Jim for a soundtrack and completed his dream feature.

It's a Wrap

The law is a tricky business; this article hardly represents the last word on the subject.

The author is not a lawyer, but has contacted film commissions, ASCAP and the like (see contact information at the end of this chapter). You should do the same.

Call your state's film commission and get acquainted. Often they will provide special producer's kits designed to educate you in the ways of permits and requirements. You'll find the time and money you invest in getting "Media Law Literate" well worth the investment.

Where to Go for Help

United States Copyright Office
(www.loc.gov/copyright)
Register of Copyrights
Library of Congress
Washington, D.C. 20559
(202) 707-3000

American Society of Composers, Authors and Publishers (ASCAP)
(www.ascap.com)
In New York:
1 Lincoln Plaza
New York, NY 10023
(212) 621-6000

In California:
7920 Sunset Boulevard
Los Angeles, CA 90046
(323) 883-1000

Broadcast Music, Inc. (BMI)
(www.bmi.com)
In New York:
320 W. 57th Street
New York, NY 10019
(212) 586-2000

In California:
8730 Sunset Boulevard 3rd Floor West
Hollywood, CA 90069
(310) 659-9109

Legal Advice From the Pros

Media Law for Producers, 2nd Edition by Philip Miller. Knowledge Industry

Publications, Inc. c/o Focal Press
(www.focalpress.com)
225 Wildwood Avenue
Woburn, MA 01801-2041
1-800-366-2665
ISBN 0-240-80303-5

Volunteer Lawyers for the Arts (VLA)

If you're making a non-profit piece, or are a non-profit company, VLA may offer you specific legal advice. Contact your state bar association to find the office closest to you.

PART III
On the Road

Production Techniques

You've got the machines you need and they're running well. A copy of your production plan is in your camera bag. Your talent stands poised before your lens. You are ready to roll tape. This section of the *Handbook* will help you acquire the best possible pictures and sound with the equipment and people you have.

If there is one gem of wisdom underlying all the various techniques described in this section and the next, it is this: making a video is not a passive activity. It is not a matter of simply recording what's "out there;" it is a constructive act of creation. You take control of elements in the world and features in your camcorder to create images and sound that are informative, compelling, entertaining or amusing. Even if your goal is to simply document a real event as it takes place, you nevertheless are engaged in controlling the light and sound entering your camcorder. Though your subject is the real world,

you are painting an image of that world no less than a painter would. You are composing a sound track no less actively than a composer creates music. You become engaged in artifice simply by pointing the camera at this rather than that, by setting the exposure to f/22 rather than f/8, by placing the mike a foot from the talent rather than five feet away. The truthfulness of your video comes, not from trying to escape this artifice, but by using it to accurately convey the realities of the situation or the truths it reveals. You might as well embrace it. Approach the making of every video as a creative act.

To a great degree, this entails making choices consciously. The first place you set up your camera may not deliver your best shot. Try different angles, tripod heights, camera moves. Then *choose* the shot you will take according to the purpose, intent and style of your video.

Recording whatever sound hits the camcorder's built-in mike may not capture the sound you really need. If you are trying to capture the words of a speaking subject, you may try handing her a stick mike and then try a lavaliere to *choose* which sound and "look" best suits your video. Letting your subject stand where you happened to meet may not give you your best shot. Try moving her out of direct sunlight for a better picture and to get her to stop squinting. Place her in front of the building she's discussing so your viewers can see it. Then *choose* where she will stand in relation to the camera. Good videos are the results of hundreds of these choices made consciously and well.

The chapters of this section will introduce numerous techniques for capturing quality moving pictures and sound. Learn them by practicing them. In the end, it will be the way you use these techniques and the choices you make that reveal your artistry and yield high quality footage.

25
Reflecting on Reflectors

Michael Loehr

As good videographers, we're always searching for new ways to improve our production techniques. Too often we believe improvement means adding new gadgets to our gear. Lighting is no exception.

Lights, gels, scrims, stands, flags, reflectors, clamps, clips, gobos, sandbags, diffusers . . . the list keeps growing. No one can deny these tools are excellent additions to any light kit. But simply owning such lighting luxuries won't necessarily better your technique. Tools like these are in fact more than most videographers will ever need. Some cost more than most videographers are willing to spend.

One lighting accessory any videographer can afford and easily master is the reflector, the most versatile lighting tool available. Whether you're using one light, two lights or ten lights, a set of good reflectors is always an excellent addition to a lighting kit.

Reflectors serve many purposes. They can raise or lower the light level. They can make a light source appear larger or smaller. They cast natural, pleasant light. And they do it without extension cords, batteries or costly bulbs. They're also lightweight and easy to carry.

Best of all, reflectors are cheap. If you can afford it, consider investing in professional reflectors. For the budget-conscious, a few bucks and a trip to the art supply store will yield good custom reflectors.

Light Theory

Understanding the principle of reflection is essential in learning how to supplement light with reflectors.

Light is a complex physical property lacking one concrete definition. From the videographer's perspective, light is a form of energy known as electromagnetic radiation. Light moves through space as an electromagnetic field. When light collides with an object, the electromagnetic field changes and energy is reflected away or absorbed.

The eye translates this reflected energy into mental images. Your camcorder translates it into video images.

Just how the field changes and where the energy reflects depends on the object's color, size, shape and surface texture. While specific reflective properties vary, there's a general law of physics that describes how light behaves: the law of reflection.

The law states that light striking a surface at angle A will reflect away at angle B. If a surface is smooth, like glass or polished chrome, angles A and B will be the same. This is called *direct reflection.*

The most common direct reflector is the mirror. A mirror maintains the pattern and intensity of incoming light; it simply changes the direction. Other common direct reflectors are water, glossy paper, plastic, glass and polished metal.

When incoming light strikes an uneven surface, angles A and B vary. The variation disrupts the light pattern, causing the light beam to scatter in *diffuse reflection.*

In diffuse reflection, beams of light scatter in all directions away from the surface, casting a soft, even light over a large area. The light leaving the reflector is less intense than when it left the source, though it spreads more evenly over a much larger area.

Diffuse reflectors for video applications come in paper, cloth or wood. Most are white, silver or slightly grey to promote the greatest reflection of incoming light.

Most videographers won't need more than one or two reflectors in a lighting kit.

Manufactured items dwell in almost any professional video retail shop. They're collapsible, very compact and easy to pack into a small bag or light case.

Light Umbrella

The most common commercially available diffuse reflector is the umbrella. Light umbrellas come in sizes ranging from 12 inches to 72 inches in diameter. The most common shapes are round and octagonal.

Essentially coarse concave mirrors, light reflectors resemble a typical rain umbrella. They're made of either white or foil-coated flame-retardant fabric.

Umbrellas are an effective way to control diffuse reflections. Since an umbrella is a concave reflector, to obtain the greatest reflection you should position the light source somewhere along the shaft. The light source points toward the underside of the umbrella. Soft light reflects away from the umbrella and toward the subject in a large, even circle.

A light source positioned closer to the umbrella's reflective surface will project a larger, less consistent blanket of light. Moving the light away from the reflective surface results in a dimmer, more controlled reflection.

With a light positioned somewhere other than along the umbrella axis, the resulting reflection is uneven and unpredictable.

Mounting an umbrella to a light is fairly easy, since most manufacturers include a small bracket on their lights for this exact purpose. However, it's important to verify an umbrella will fit your particular light before purchase.

If your light doesn't offer an umbrella mount, consider buying or building a different type of diffuse reflector. Or invest in another brand of lights.

When using an umbrella, watch for umbrella reflections in eyewear, windows or other smooth glossy surfaces. The small white octagon reflected in a window or someone's glasses is something to avoid.

Flexible Rings

Another popular commercial diffuse reflector is a round piece of reflective fabric surrounded by a thin plastic ring to provide support and retain shape. These rings also come in a range of sizes, from 24 inches to 72 inches in diameter.

The great thing about flexible reflectors is they shrink to less than half their full size with a turn of the wrist. This makes them as compact and lightweight as umbrellas.

Flexible reflectors also feature two separate reflective surfaces. They come with standard white on one side and either gold or silver foil coating on the other.

One thing these reflectors lack is rigidity. Anchoring them down on windy shoots can be tricky. They're also a sort of flat-plane reflector, compared to the concave umbrella. This means reflections are less controlled. Light can strike a flat-plane reflector from any angle and still produce an adequate reflection, however.

Flexible reflectors are usually cheaper than comparably sized umbrellas.

If you're a budget-conscious videographer, or if you simply can't justify spending the money on professional reflectors, consider building your own. You'll save anywhere from 50 to 80 percent of the cost of commercial reflectors. And the end results should be just as good.

Custom reflectors are a bit more cumbersome to work with, simply because they don't collapse like commercial units. Still, they weigh almost nothing and fit easily into any trunk, back seat or hatchback.

Card Tricks

The easiest reflector to make is the white card. The white card, a diffuse reflector, is exactly what it sounds like: a stiff card with a surface of pure white. The white card is also a flat-plane reflector. The best material for making white cards is foamboard.

Foamboard is a stiff, narrow piece of foam laminated between two heavy sheets of white paper. It's extremely rigid and weighs almost nothing. Art supply stores stock foamboard—also known as foamcore—in a wide range of sizes.

You can cut foamboard into any size or shape with a utility knife. Most white cards are rectangular; common sizes are two feet by four feet and four feet by six feet.

As a diffuse reflector, foamboard is ideal. Its bright, glossy finish provides excellent brightness when reflecting even small light sources. Small cards are excellent for illuminating small areas with soft, bright light. The large cards effectively spread even a small light source across a large area.

When building a reflector using foamboard, remember to use both sides. Consider putting foil on one side, leaving the other pure white.

A more diffuse white card consists of foamboard or stiff cardboard and plain white bedsheets. Wrap two layers of the sheet around the cardboard and attach it with staples at the edges and corners. This card doesn't reflect as brightly as the bare foamboard, and the light is much softer.

Fun with Foil

Another inexpensive reflector consists of cardboard or foamboard and aluminum foil. More direct reflectors, foils are perfect for adding bright highlights to a subject.

Standard grocery-store aluminum foil works well. Examine both sides and you'll notice one side is bright and shiny, the other slightly dull. The shiny side is an adequate, lightweight direct reflector, useful for reflecting a large, slightly dimmed spotlight from the source towards the subject.

The other side is more diffuse, but still acts like a direct reflector. The dull side reflects a larger, less intense spotlight toward the subject.

Affix foil to foamboard or cardboard with duct tape or strong masking tape. Cut the foil to the same size as the card. Apply strips of tape on all four sides of the foil. Leave about half an inch of tape extending from each edge. Then lay the foil across the board, gently fold the tape down and stick it to the backside of the board. This yields a taut surface that won't wrinkle. You can fold the card inward without damaging the foil.

Some art supply shops stock colored foils. These offer creative alternatives to standard silver foil, and are usually more expensive. They don't reflect as brightly as silver foil, but can change the apparent color of a subject. For example, an orange or golden foil reflector will add a warm, pleasant color to fair-skinned faces.

Foil will reflect more diffuse light if carefully crumpled and restretched before it's taped to the card. You can also splatter foil lightly with white or grey paint to further diffuse the reflector.

With such lightweight and relatively small reflectors, another crew member can hold them while you are shooting. If you're shooting solo, prop the reflector against a nearby heavy or stationary object. The ideal situation is a separate light stand. Use a large pony clip, available at any hardware store. If the stand feels flimsy, anchor it down with a small sandbag or other weight.

Let Freedom Ring

With lights and reflectors in hand, it's time to put them to work for you.

Using reflectors is more an art than a science; the techniques are based on personal preference, not hard rules.

A particular lighting scheme might seem innovative and stunning to one videographer, tacky and overstated to another. Reflectors are a tool that can help you express your own visual ideas more effectively.

For reflectors to improve your lighting techniques, you need to experiment with them. Even before your next project, spend some time getting comfortable with your new toys. Try staging simple scenes that allow for lots of variation. A friend reading while relaxed in a chair, writing a letter while seated at a table or fixing a bicycle on the porch.

Good practice before the next real shoot will help you understand how each type of reflector can affect scenes in future projects.

Although tremendous freedom exists for experimenting with reflectors, there are a few general ideas that apply to nearly any lighting scenario.

Reflectors light, or fill in, dark, shadowed areas, a technique commonly called "creating a fill."

The most common fill reflectors are the white card and white flexible reflector. These offer the even, balanced reflection necessary to created an adequate fill. The aesthetic purpose of a fill light is to enhance the perception of contour and shape by eliminating visual competition between areas of light and dark.

Casting Shadows

Fills are usually positioned at a 90 degree angle to the main light source. A properly positioned fill doesn't eliminate shadows, it simply makes them less distracting.

Shadows are an important part of any image, because our brain uses shadows to interpret shape and dimension. Our cameras and our minds perceive a subject that casts no shadows as flat and two-dimensional.

If a subject casts sharp, dark shadows, it will appear three-dimensional. But the contrast between bright and dark areas might distract the viewer's attention away from the subject and toward the shadow itself. A proper fill light should brighten these dark areas while still allowing the subject to appear shadowed. This effect can be difficult to achieve if you're using an additional powered light source as your fill.

As a rule, fill lights should never be as bright as the main light source. Shadows will disappear; the result will be a flat, two-dimensional image. Using a powered light source as a fill is difficult because the fill's intensity often matches the intensity of the main light.

Reflectors, however, are ideal fills. Their intensity never approaches that of the main light source. Consequently, reflector fills allow greater control of shadows and contour.

Start by casting shadows on a subject. Shadows are created by positioning the main light at an angle across the subject. Main lights are commonly positioned on one side of the camera at a 45 degree angle. The larger the angle away from the camera, the more prominent the shadows.

Position the reflector near the subject on the side opposite the main light as in Figure 25.1. Aim the reflective surface halfway between the main light and the subject. Stray light beams from the main source and from the subject itself will bounce from the reflector back toward the dark side of the subject.

Rotate the reflector around the subject and observe the effect. As the reflector moves, different amounts of light will bounce back toward the subject. Position the reflector closer to the subject, and the dark areas will get even brighter. Move away and the shadows grow darker again.

Illustrated Fill

To illustrate the use of reflectors as fills, imagine shooting an exterior scene at noon on a sunny day.

At noon, the sun is at its highest point in the sky. It casts long, dark shadows that fall straight down. This effect is very distracting on faces, because the eyebrow ridge frequently casts a shadow hiding the eyes. Noses also cast unappealing shadows on mouths and chins.

Position a white card below the subject's face and out of camera range. Angle it slightly upward. This will lighten the shadows and reveal the subject's face and eyes clearly, still retaining the contour and features of the face.

Creativity and personal taste are important factors in using reflectors as fills. Only you can decide how much shadow is appropriate.

Another common use for reflectors is backlighting. This procedure is similar to fill lighting, except that reflected beams move from the main light toward the back side of the subject.

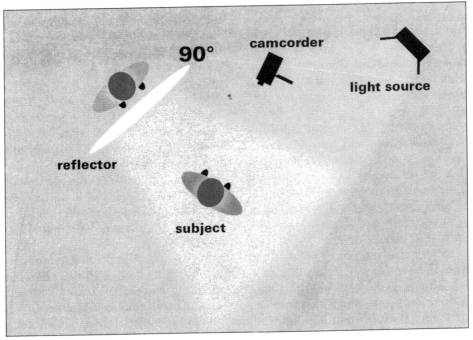

Figure 25.1 *Positioning a reflector opposite your main light helps modify shadows.*

Aesthetically, backlight creates a subtle edge highlight, enhancing depth by separating the subject from the background. This edge also enhances large contours.

Backlight reflectors are usually either white or foil cards. White cards yield a softer haze; foil cards provide more intense, focused coverage.

To create a backlight, position the reflector out of camera range behind the subject. Aim the reflective surface toward the main light until the light bounces onto the back side of the subject.

Move around the subject with the reflector pointed toward the main light. Notice the variation in pattern and intensity as you move the reflector above and below the subject. The backlight is stronger when the reflector is slightly above and behind the subject. The effect is more subtle as you bring the reflector below the subject's back side.

Be careful to avoid bouncing light back into the camera lens. This light will generate an annoying flare or bright spot in the image.

Light in Back

The backlight technique is also useful in other situations. If the subject you're shooting is silhouetted by a very bright backlight, reverse the technique to lessen the silhouette effect and brighten the subject.

For example, if you're shooting a cyclist poised against a bright orange sunset, use the reflector to bounce sunlight back toward the cyclist. The best reflector would be a golden flexible reflector or foil card. You'll reveal more color and detail, while still retaining the sharp outline produced by the sunset.

By locating a foil card somewhere between the typical fill light and backlight positions, you cast sharp, bright highlights along one edge of a subject. This is ideal for adding hair highlights, or accenting the contour of a face.

Finally, reflectors can enlarge and soften main light sources. This technique is excellent for videographers with one powered light source.

Reflecting a primary light enlarges its apparent size by scattering the light beams across a larger area. This can make an otherwise meager light source seem substantial. The scattered beams also cast soft, even light on subjects, making an otherwise harsh, focused source more pleasant.

Primary Reflections

The most common tools for reflecting a primary light source are the white card and the umbrella.

To reflect a main light using a white card, position the card where you normally position a main light. Aim the reflective side toward the subject. Position the light source between the subject and the reflector. Point the light away from the subject and toward the card surface.

By adjusting how light strikes the surface, you can alter the direction and intensity of the light reflecting toward the subject. Even though a white card is a diffuse reflector, the law of reflection still determines how the reflected beams will behave.

When light strikes the card from a low angle, the law of reflection states it will reflect away at a high angle. The reflected beams in this example will illuminate a higher area of the subject.

When light strikes the card from the right, the law states it will bounce away to the left. This can be quite useful if you want to dominantly light one side of a subject.

Using an umbrella is much easier than using white cards, simply because the umbrella is attached to the source itself. The umbrella and source as one unit eliminates the need for additional stands or props.

Position the main light normally, then attach the umbrella. Spin the light around so the underside of the umbrella faces the subject.

An umbrella reflects a smaller, more controlled light than a white card. The light is still very soft; it just covers a smaller area. Since the umbrella is attached to the light source, it's much easier to control.

Light from the umbrella is softer than light from a bare light source, so you have greater freedom in positioning the umbrella around the subject. The softer source will cast softer shadows on the subject. This allows you to position the umbrella wide to one side without casting dark, distracting shadows across the subject.

26
The Power of Three Point Lighting

Robert Nulph

See video clips at www.videomaker.com/handbook.

Everybody has to start somewhere. Camera operators learn the basic shots and movements and graduate to more difficult and exciting work. Audio operators begin with a basic knowledge of microphones and recording devices and move up into the creation of soundscapes. Lighting designers, called "directors of photography" or DPs in the film world, learn the art of lighting by perfecting the most basic yet important tenet in light design: three-point lighting.

In this chapter, we will look at the various components of three-point lighting. We will discuss different lighting instrument options and describe how to use them. Finally we will explain how you put it all together to create lighting that will give your videos a professional look.

To begin, you need to ask the director (who might very well be yourself), some very important questions. First, where will the camera be in relation to the subject? Is the shot going to be outdoors or indoors? If outdoors, how will the sun affect the shot? If indoors, is there a win-

dow or door that has sunlight filtering through it? Will the subject be moving? Is it day or night? What mood are you trying to achieve? While this may seem to be a fairly extensive list of questions, it is only the beginning. Once you have the answers to all of these questions, you are ready to begin designing the lighting scheme. You begin by deciding how to set up the components of three-point lighting: the key light, fill light and back light.

The Key Light

The key light is the main light. (See Figure 26.1.) It is usually the most intense and direct light shining on your subject. For basic lighting such as that used for interviews, imagine your subject in the middle of a clock face on the floor with 12:00 behind the subject and the camcorder at 6:00. You would place the key light beside the camera at about the 4:30 position. Elevate to an angle of 35 to 45 degrees above the subject. This placement creates some shadows on the face, giving it a three-dimensional shape. The height is enough to keep it out of the subject's eyes, yet not so high as to create dark shadows under the eyes. For those subjects with glasses, lower the key light slightly to eliminate the shadow across the eyes caused by the frame of the glasses. Be careful of glare reflected off the glasses. Typically, you would use a softlight to soften the lines and texture of the subject's face. If you require a more dramatic look, you would select a harder, smaller light.

When designing for a dramatic story, look at the surrounding set and try to duplicate the lighting that would usually exist. If the scene takes place in a living room at night, set the key light to approximate the look of a floor lamp, table lamp or ceiling light. If the scene is a daylight scene in an office, the key light might be the sunlight coming through a window or a desk lamp or top light.

Whatever the situation, the key light is the starting point from which the rest of the light design evolves.

Once you figure what kind of key light you need and where you want to put it, the next step is to determine the style and placement of the fill light.

The Fill Light

The primary purpose of the fill light is to fill in the shadows, thus its name. The fill light is typically larger, softer and more diffused than the key light. (See Figure 26.2.) While the key light gets all of the notice, the fill light is more critical to establish the mood, time of day, and quality of light in a scene. Because of this, the

Figure 26.1 *The key light is the primary light source in a scene.*

fill light is perhaps the most important of the three lights used in basic lighting.

In a basic setup such as an interview, you would place the fill light on the side of the camera opposite the key light at about the 8:00 position. It is elevated at an angle of 20 to 30 degrees above the subject. This placement fills in the side of the face away from the key and reduces or fills in the shadows created by the key. The brighter the fill light, the fewer shadows and the less dramatic the scene will be.

This difference in brightness between the key light and fill light is called high-key and low-key lighting. In high-key lighting, the fill is very strong and the contrast between the fill side and the key side is reduced. This is typically how you would light news, most interviews, talk shows and game shows. In low-key lighting, the fill light is very dim, which cre-

ates a high contrast between the key and the fill lights. This lighting is very dramatic. You would use this type of light for movies, dramatic interviews and other dramatic scenes. It is the fill light that determines how dramatic a scene will be.

To distinguish between an indoor daylight scene and night scene, you would again turn to the fill light. For daylight scenes, you would use a strong fill light and reduce the contrast between the fill light and the key light. For night scenes, you may decide to use no fill at all.

The lighting instruments used to create fill light are also the most fun and creative. Besides diffused lamp light, you can use a host of other lighting accessories. A white piece of foamcore or white posterboard makes a wonderful bounce card for fill light. Place the white bounce card so the light from the key light

Figure 26.2 *The fill light softens shadows caused by the key light.*

bounces from the card to the subject's face. You will be surprised at the amount of light you can bounce off the white surface. This is very useful when doing interviews in the field and you need to carry a minimum amount of equipment or have very little power available. For a more intense reflected fill light, you can use a variety of silver reflectors. If you want to add a little color to the scene, use a gold reflector or a colored bounce card.

After you have established the fill needed to create the appropriate look for the subject, you have one more light to deal with—the backlight.

The Back Light

The back light's primary function is to separate the subject from the background.

(See Figure 26.3.) By lighting the back of the subject's head and shoulders, the resulting rim of light helps establish that the subject is located some distance from the background.

For the basic three-point lighting setup, you would place the back light at a higher angle than the key light, behind the subject. Make sure that the back light does not shine on the lens of your camera. As with the key light, the back light can be hard or soft, depending on the look you want to achieve. Soft light is more difficult to control and will tend to cause problems with lens flare.

As with the fill light, the amount of back light will determine the dramatics or glamour of a scene. The more intense the back light, the more glamorous and dramatic the subject. However, you must determine the amount of back light

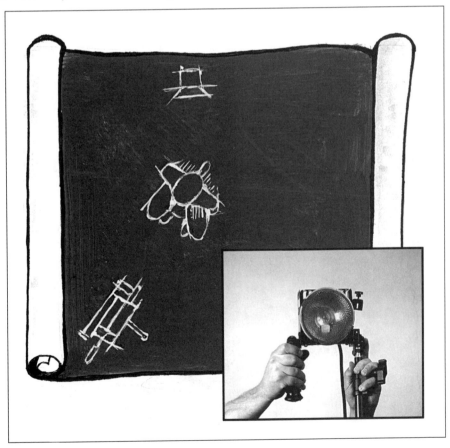

Figure 26.3 *A back light adds depth, and separates a subject from the background.*

needed by noting the hair color of the subject. Blonde-haired people require less back light than brown-haired people do. A good rule of thumb is the lighter the hair (or head in the case of a bald man) the less back light that is needed.

Fine Tuning

Once you determine the lighting needs of your scene and set up the key, back and fill lights, you are almost ready to shoot. Always look at your scene through a viewfinder and on a color monitor if possible. You can fix weird shadows with fill light or changes in the key light placement. You can also fix reflections off glasses or bald heads by slightly moving the key or fill lights. Remember that light moves in a straight line and bounces at opposite angles. If you see a strange reflection, determine which light source it is coming from by placing your hand in front of the reflection and looking for the shadow, or turning off one light at a time until the glare is gone. The shadow will be exactly in line with the culprit light source.

The last thing to keep in mind while setting up your three-point lighting is that this is just the beginning (see Figure 26.4). Use your imagination, and adapt your lighting setup to your surroundings. Recreate the light that you would see in real situations. By controlling the light, you are controlling the mood, time of day and dramatic intensity of a scene. This is an awesome power in the hands of a creative videographer. Use this power to your advantage to create professional quality productions.

Figure 26.4 Basic three-point lighting set up.

27
A Dose of Reality:
Lighting Effects

Robert Nulph

See video clips at www.videomaker.com/handbook.

The firelight flickered against the cabin wall, warming the cool blue light of the full moon filtering through the tattered curtains. Suddenly the ominous blue then red flash of police lights filled the small room and Carson knew his game was up.

Suddenly the director yells "Cut!" and the camera pulls back to reveal two Hollywood flats painted to look like cabin walls and a squadron of techies moving a myriad of lights and other equipment to new locations. Nowhere in sight in the cavernous sound stage is there a squad car, a full moon or a flickering fire.

For years Hollywood and independent filmmakers as well as corporate video producers have used lighting techniques to make us believe things exist that aren't really there. You can too! It is all a matter of collecting the right lighting instruments and accessories and adding a large dose of imagination. Mix them all together to give your scene a large dose of

reality. Throughout this chapter, we'll look at a variety of ways to bring reality to your scenes. It is all in the power of lighting.

Mr. Sun and Mr. Moon

It's a good idea to always plan the outdoor and daylight shots first for your productions, because you have more control of indoor lighting than you do over the weather. All you need to make sunshine or moonbeams is a small, powerful light source and some colored gel. You can create sunshine, even at night, by placing a powerful light (1000 watts or so) outside your window. (It is not advisable to do this if it is raining.) Make sure you place it at an angle similar to that of the sun at the time your scene takes place and is out of the camera shot. It works best if you use a small, intense light to create the light of the sun or moon because you want to imitate their qualities. If you think about it, the sun and moon are very small intense lights that throw very hard shadows. A big soft light will not do the trick.

To recreate the sun, you have to determine what time of day your scene is taking place. If your scene is in the early morning, you may want to place a single blue gel in front of the light. For midday, use no gel and for evening, use a yellow/gold, orange or red/orange gel, going towards the red as the day progresses. A light shining into a hard gold reflector and reflected through the window makes a fabulous evening light.

To recreate the moon, place two Color Temperature Blue (CTB) gels together in front of your light. Dim the lighting in the room to pick up the color of the moonlight and create the feeling of nighttime.

Figure 27.1 *Shine a light through a set of mini blinds to imply the existence of a window.*

If you are creating the sun or the moon on a sound stage or other big room, you can also create windows through which they can shine. Place a window frame just out of camera shot so that its shadow falls across the floor and the background wall. Set up window blinds (see Figure 27.1) and let the light filter through the slats. You instantly have a wall with a window.

Cars and Cops

With a little mechanical skill and a good sense of pacing, you can easily imitate car headlights, city streetlights, the flashing lights of a squad car or a searchlight being used to find the bad guy. You'll also need a couple of small, focusable lights that you can gel.

One of the easiest, yet most effective lighting effects you can use is the imitation of a car's headlights. Using a four-foot long 2X4, mount two narrow beam lights about two feet apart. Slowly sweep the beams of light at an angle across the darkened back wall of your set. Instant car lights. If you are shooting a scene in a car at night, you can use the same technique both for cars passing you from the other direction as well as those coming up from behind.

In the same driving scene, you can imitate the passing of city streetlights, by rhythmically passing the beam of a powerful flashlight over the hood of the car, avoiding the camera lens. A flashlight works well because its lamp has a yellow color temperature and should look different from the lights you are using for headlights.

If your characters get in trouble with the law, you can fill the car or house with flashing blue and red lights by rhythmically passing a double or triple blue gelled light then heavily gelled red light past the background or interior of the car. The Lowel Omni light has a comfortable soft rubber grip that allows you to move it around without being burnt. Focus your light's beam to the tightest setting possible and pass first the red then the blue past the set. You can flash the set, tilt the beam to the floor and pass it again. With two people, it is a bit easier, but one person can handle it. Take the gels off one of the lights, put on a yellow gel, widen the focus on the beam and you have just created a searchlight.

If your scene occurs on a city street or in a seedy motel room, you can add the pulse of a red neon light. Reflect a diffused red-gelled light onto the background or into the interior of your car. By turning the light off and on or moving a flag to cover the light occasionally, you can imitate the stuttering of an old neon sign. Add a few sound effects and your characters are infor a long and dramatic night.

Living Rooms

Fireplaces, televisions and lamps that you see used in video and movie scenes more often than not, don't really work the way we think they do. You can create it all through the magic of lighting.

If your character is supposed to be watching television yet you don't see the front of the set, you can create a very believable TV light. Get an old TV set, remove the picture tube and tack a double CTB gel to the front. Inside, place a lighting instrument that has a good quality switch on its cord. Quickly turn the light off and on; pausing at times for longer lengths of both light and dark. A television is never always bright so the flickering makes it look more realistic. Of course, you could always plug in an actual TV set, but hey, that would be too easy.

If your character is sitting before a warm fire, you can create the effect by setting up a small, diffused light, angled up from floor level. In front of the light, hang inch wide strips of red, yellow and orange gels on a broomstick. Gently shake the gels in front of the light to create the feeling of firelight movement, as in Figure

27.2. Another method uses a round wheel (like an old bicycle wheel) covered with various orange, red and yellow gels cut with holes and layered to provide a variety of combinations and the occasional flash of real light. Turn the wheel slowly in front of the light to create the movement of the flame. Again, add sound effects and bake to perfection.

For lamps that you will see on the screen, the first thing you need to do is remove the regular bulb. A sixty-watt bulb will cause the lamp to glow on camera and look much brighter than it should. Place a 15-watt bulb in the lamp to provide a soft internal glow and supplement the light with a diffused 600-watt or more lighting instrument. Be sure to flag the light so that its beam does not fall on the lampshade of the light you are trying to use. If you place the lighting instrument just off-line from the real light, you can light your character in a warm glow that will look like it is coming from the lamp beside them.

Water Water Everywhere

Sometimes, the script calls for water ripples reflecting in your characters eyes or on her face. Often, it just isn't very convenient to set up lighting to get this effect using a real water source like a creek or

Figure 27.2 *Red and yellow gel strips waving in front of an orange-gelled light create the illusion of firelight.*

Figure 27.3 *Light reflected off of water and broken pieces of mirror create a shimmering pool side effect.*

lake. Don't worry, it is really a quite simple effect to recreate. All you need is a deep pan like a roaster or a painters roller pan. Carefully break up a mirror into two to three inch pieces and place them in the bottom of the pan, face up. Cover the mirrors with about three inches of water. Shine a small, intense light into the water so that the light reflected from it falls onto the face of your character. (See Figure 27.3.) Gently lift one end of the pan up and down to create a soft ripple effect. You should see water ripples in your characters eyes. If your scene occurs at night, add a CTB gel to your light. Add a few seagulls, some water sounds and your ready for a day or night in paradise.

Reality

Always be aware of the world around you. Look at the light that makes up our world, its reflections its colors and the shadows it casts. If it occurs in the real world, you should be able to re-create it for the camera. A bit of knowledge, a dose of imagination, and a touch of lighting magic can create any reality you wish.

28
Outdoor Lighting: What you Need to Know to Shoot Great Footage Outdoors

Michael Loehr

How do you light the outdoor scenes in your videos? Do you plan and stage each shot carefully to make the most of the sun's glow? Or do you just switch to outdoor white balance, call out "Action!" and roll tape?

Even if you choose the latter, chances are your videos still look pretty good. Today's camcorders work well enough in daylight to make very acceptable pictures, even with no attention to lighting.

Maybe that's why videographers don't worry too much about outdoor lighting. Perhaps they think making the best use of sunlight requires expensive instruments and tools they can't afford. Perhaps they just never learned the tricks of managing sunlight in a video project.

That's where this guide can help. In this chapter are some of the popular outdoor lighting techniques. They can help subjects look more natural on video, and improve the overall look of your projects.

You'll learn what tools and gadgets you need to make the most of sunlight. You

can build many of them with inexpensive stuff from art supply and hardware stores. We'll even teach you how to create the illusion of a dark night in the middle of the afternoon.

So start taking advantage of what may be your greatest asset as a videographer: the sun.

Principles of Light

The fundamental principles of good lighting apply whether you shoot video indoors or out. However, sunlight presents unique challenges to videographers. On almost any given day, there is more than enough light outside to shoot a scene. At first, an abundance of light seems like an asset. However the hundreds and thousands of lumens cast by the sun can actually cause problems for your camcorder. Not technical problems, but aesthetic ones.

At its brightest, the sun can shed more than 10 times the light of one typical indoor instrument. When it shines brightly, it also casts very dark shadows.

In video lingo, the difference between these light and dark areas is commonly called the contrast ratio, or *contrast range.*

Our eyes can compensate for the high contrast range of a bright day. Our camcorders, however, don't react as well. They require a much lower contrast range, especially to capture detail accurately. (Of course, our eyes also see better when we lower the contrast range, which is why we often wear sunglasses on sunny days.)

On bright days, the contrast range is usually too high for your camcorder to make good pictures. If you shoot without any lighting equipment or assistance, the sunlight won't flatter your subjects. Dark shadows may leave unpleasant or unnatural accents on facial features. Your images may also look washed out.

A high contrast ratio also affects your camera's automatic iris feature. You may have noticed that when the auto iris is on, its position changes constantly while you shoot.

As you move into a shadowed area, the iris opens to allow more light into the lens. As you move back to the bright areas, it closes again to avoid overexposure. That means you might get even, natural lighting from one angle, and harsh, overexposed lighting from another. The constant movement of the iris makes maintaining continuity between different camera angles difficult. It's also very distracting mid-shot.

The goal of outdoor lighting design is to lower the contrast range without damaging the natural look of the subjects and the outdoor setting. You want a lighting setup that looks the same to your camcorder, no matter where you put it. To do this, you need to brighten the dark, shadowed areas, and perhaps even lower the overall light level, depending on how brightly the sun shines.

Tools and Tricks

If you're shooting indoors and need more light, the standard practice is to plug in a light and point toward the dark areas. Outside, you do practically the same thing, only with different tools.

You only have one light source—the sun. It doesn't need extension cords or power outlets. Even better, it will usually give you more than enough light to work with.

All you must do is redirect some of that excess light toward the shadowed areas of your set and your subjects. The best, most affordable tools for redirecting light are reflectors and diffusers; they will point light in different directions, and alter the way it falls on a subject. Light will bounce off a reflector, and pass through a diffuser.

Learning to use reflectors is easy. Their behavior is somewhat constant, given the fact that light bounces in predictable angles.

Reflectors vary, however, in three ways:

1. how much light they reflect,

2. how large an area their reflection covers and

3. the color of light they reflect.

Foil or mirrored surfaces reflect the most light over a small area. Pure white surfaces usually cover larger areas, but with less light. Some reflectors have a gold foil surface; these bounce light with a warm, rich quality that really flatters skin tones.

Diffusers filter direct beams of sunlight, spreading them evenly over a large area. Like reflectors, they're easy to use and fairly predictable.

A material's porosity and transparency determine its diffusion characteristics. Dense or very cloudy materials allow less light to fall onto the subject. Highly porous materials allow more.

Diffusing sunlight is probably the most effective technique for taming unpleasant shadows and reducing contrast. It does an excellent job of brightening dark areas, while retaining much of the outline and contour.

To use a diffuser, simply suspend or position the material between your subject and the sun. Where you place the material and how you angle it depends on the look you want. To create shadows on the face, place the diffuser close to the subject and off to one side. To spread light evenly and minimize shadows, place the diffuser above and away from the subject, angled down slightly as in Figure 28.1.

Experiment with the diffuser to determine the most effective position for your particular scene. No matter where you put

it, your camcorder will make better pictures with diffused light.

There are, however, a few drawbacks to using diffusion. It simulates the light you might see on a slightly overcast day, especially when you suspend the diffuser overhead. This lighting tends to be flat—some subjects may look bland under diffused light.

One solution: bounce more sunlight toward the subject. This highlights the subject and slightly increases the contrast range.

Or you can abandon diffusion altogether and bounce light around the scene with reflectors instead. By using reflectors, you can maintain the look of a summer day and still reduce contrast.

Position the reflector to bounce excess sunlight toward shadowed areas, as in Figure 28.2. This lets your camcorder use more incoming sunlight without washing out your subjects.

If you're shooting at midday, unpleasant shadows may appear on your subject's face. The simplest solution is to move your subject out of the direct sun, if possible. (See Figure 28.3.)

Another solution is to use a reflector. Try putting the reflector below the subject's face; this should help eliminate the shadow. Be careful to avoid the "monster look," however. Strong light from below the face is a classic horror film technique, hence the name. Unless you want your subject to look frightening, make sure the reflected light flatters the face. Reposition the reflector as necessary to eliminate the monster look.

One last tip: videographers on the go may prefer reflectors to diffusers. Diffusers can sometimes be cumbersome to set up. Reflectors offer better portability, and still solve many outdoor lighting problems.

Simple Solutions

Other very effective and inexpensive outdoor lighting techniques involve simply

Figure 28.1 *An overhead canopy of diffusion material (a) can help to control excessive contrast from the sun (b).*

Figure 28.2 *When shooting in direct sunlight (a) a simple reflector can bounce light back into dark areas to improve contrast (b).*

staging a scene in the proper place with respect to the sun.

You've heard the saying that the sun should always be behind the camera when you shoot. True enough, but it doesn't tell you whether the sun should be to the left, the right or directly behind the camera. Many videographers default to the center position, where the sun sits directly behind the camera. This is a bad idea for two reasons:

It puts the sun in the subject's face, which almost guarantees squinting eyes in

the shot (see Figure 28.4) and the shadow that you and your camcorder cast is likely to wind up in the shot.

You can avoid these rookie moves by adopting an outdoor version of the classic three-point lighting setup, which is used to add more light to indoor shooting situations.

In the three-point setup, one light serves as the main or "key" light. It provides most of the light for a scene. It's positioned to light one side of the subject, angled approximately 45 degrees horizontally from the subject.

A second, less intense light shines on the opposite side of the subject. Called a "fill" light, it balances the shadows that define contour and shape. It's often somewhere between one half and two thirds as bright as the key light.

Sometimes a third light adds backlight. A *backlight* separates the subject from what's behind it, and provides shoulder and hair highlights on people.

Here's how you can adapt this three-point setup to outdoor lighting situations. Instead of standing with the sun directly behind you, change your position so that the sun shines from behind you over either your left or right shoulder. In this position the sun becomes a key light, shining light on one side of the subject's face.

This takes the light out of your subject's eyes and lowers the chance of your shadow appearing in the shot. When the sun shines at an angle similar to a key light, the shadows will fall away from you and the subject, and hopefully out of the shot.

You can also add a very inexpensive fill light by just using another reflector. Once you've established the sun as the key light, either clip a reflector to a spare light stand, or have an assistant hold a reflector near the subject on the side opposite the sun. Rotate the reflector back and forth to bounce light onto the "dark" side of the subject. Move away to lessen the intensity, closer to raise it.

If you have another spare reflector, particularly one with a foil surface, you can

Figure 28.3 If you must shoot outdoors, avoid backlighting with direct sunlight (a). Instead, place him in the shade of a tree (b) or building (c).

simulate a backlight. Stand just off camera behind your subject, on the side with the key light. Point the foil side toward the sun and rotate it until the reflection lights up the back of your subject. Presto! Instant backlight.

Occasionally you will encounter outdoor settings where the background is as brightly lit as the subject. This is another aesthetically unpleasant situation.

When the subject and background are both very bright, they conflict with each other, creating an image viewers will find difficult to watch for very long. To solve this problem, you must highlight the foreground subject. Instead of trying to reflect more light onto the foreground, try shadowing all or part of the background.

The technique is subtractive lighting, or "flagging." It involves using a card called a "flag" to block sunlight from hitting certain areas. You can buy ready-made flags from video stores, or build your own from black foam boards. In a pinch, a reflector will work as a flag, but black foam board is better. The reflector's white surface sometimes bounces light where you don't want it.

To shadow the background, position the flag behind the subject, just off camera on the key light side. Angle it so that it casts a shadow on the background. You

Figure 28.4 Direct sunlight hitting your subject from the front causes squinting eyes.

may need to move the subjects away from the background to avoid casting a shadow on them as well.

Position Problems

Because the earth rotates in space, the sun's position, intensity and color balance change through the course of a day. This can create problems for uninitiated videographers. Understand these changes, however, and they can become tremendous assets.

If you shoot a series of scenes during an entire day, you'll notice the lighting changes from scene to scene. Shadows gradually change position, density and direction, and the contrast range changes. Color temperature also changes throughout the day.

For example: a scene shot very early in the morning will have long horizontal shadows, a slightly orange glow and a lower contrast range. A scene shot in the same location at midday will have dark vertical shadows and a much higher contrast range.

You will experience difficulty when you try to edit these scenes together. Differences in shadow placement and color balance will reveal that you shot the scenes at different times. (See Figure 28.5.)

A diffuser is an excellent way to prevent such problems. Diffusing sunlight hides the movement of the sun across the sky, and disguises the time of day. Sometimes the earth's atmosphere provides its own diffusion in the form of cloud cover. If the forecast says the clouds will hang around all day, you may not need to set up a diffuser at all.

Many projects call for dramatic use of light and shadow to convey specific moods or emotions. If yours is such a project, avoid using diffusion; it lessens the impact of shadows. Also avoid shooting in the middle of the day, when shadows make subjects look less than their best. Instead, shoot your footage either late in the day or early in the morning, when the shadows are most flattering.

Figure 28.5 *Shooting in the evening or early morning results in a soft light (a). Mid-day sun is brighter and harsher (b).*

When the sun is near the horizon, its color temperature is different from when it's high in the sky. At noon it casts a white light high in color temperature, usually around 5600K (Kelvin). Your camcorder's outdoor filter works best with this type of sunlight.

At dawn and dusk, however, the sun is lower in the sky, and its glow is a warm, golden-orange color. Videographers often call this period the "golden hour," since it usually lasts right around an hour. Its color registers much lower than the 5600K light of mid-day—usually around 3100K.

Consequently, your camera may react differently when switched to the out-door setting. If you white balanced early in the day under regular 5,600K light, the video will turn more and more orange as the evening progresses. If you don't want this look, simply white bal-ance your camera at the beginning of every shot.

While this change in color tempera-ture may prove inappropriate, it can also be perfect for certain types of shots. The golden hour's long shadows and warm lighting make it an ideal time to shoot dramatic or romantic scenes.

Be aware that the moment only lasts a short time. You can extend the golden hour a little by reflecting sunlight off a gold-surfaced reflector. However, once the sun either disappears in the evening, or reaches a 45 degree angle above the horizon in the morning, the golden look will be difficult to maintain. If you know exactly when the golden hour will hap-pen, you can plan to take advantage of it on your next project.

On very rare occasions you may need to add artificial light to make an outdoor scene suitable for shooting. This is most common when shooting under either very dark clouds or heavy shadows. In these cases it may be appropriate to use your indoor lighting instruments instead of reflectors to light a scene.

Remember, the color temperature of sunlight is much higher than that of indoor studio lights. To use indoor lights outdoors, you must put a blue gel in front of them. Also be aware that indoor lights shine a very small amount of light when compared to the sun. You may need two or even three instruments to light a subject adequately outside.

Night Lights

Shooting outdoors at night can be trouble for professional and amateur videogra-phers alike. Even with low-light cam-corders, it's still very difficult to get good pictures without adding artificial light.

To solve this problem, use a technique called day-for-night shooting. It involves shooting a carefully staged and controlled scene during the day, making it look as if it were shot at night.

Day-for-night shooting isn't easy, and it isn't always effective. To make it work, you must create an illusion of nighttime that will fool your audience. To do this, pay close attention to how your eyes see at night.

Pay close attention to colors. At night our eyes don't see colors as well because of the lower light level. The same is true for our camcorders. Dress your subjects in muted colors to keep the color intensity down.

If your camcorder has a monochrome mode, consider using it instead of the color mode; this will help reduce the amount of color in the scene. Some edit-ing VCRs have chroma controls or mono-chrome switches, which can also mute color intensity.

Consider buying a blue filter for your camcorder. This helps create the illusion of moonlight by turning sunlight blue. Most video stores carry a selection of fil-ters to fit your camcorder. Be sure to get one that fits your model's lens.

When combined with the other tech-niques, the blue filter greatly enhances the nighttime look.

If there are any ordinary lights in your scene—car headlights, porch lights, win-dow lights—switch them on. Indeed, before you shoot you should turn on any and all lights normally on at night.

You also must know how to disable your camcorder's auto iris and auto white balance circuits—if it has them. When activated, an auto iris circuit lets the opti-mal amount of light into the lens to make pictures.

With day-for-night shots, you want to limit the light entering the lens. You can only do this when you turn off the auto iris.

The same applies to auto white balance. If active, the feature will try to get an accurate white balance, even with the blue filter on the lens. The goal is to fool the camera and ultimately the audience, so switch off the auto white balance.

With the circuits off, white balance the camera without the blue filter. Place the filter on the lens, and manually close the iris until a small amount of light enters the lens. Let enough light through to distinguish your subjects, but not any more than that.

The result is the nighttime look: a grainy bluish image with muted colors and contrast. If your editing VCRs allow it, try lowering the black level and raising the luminance during post production.

This increases the contrast enough to match what our eyes typically see at nighttime.

Wrap It Up

Enhancing your outdoor shoots with reflectors and diffusers is more art than science. The techniques reflect personal preference as much as rigid rules.

So use reflectors and diffusers to express your own visual ideas more effectively. The best way to learn is to experiment with them.

Stage a simple scene outside, and then create four or five different moods by just changing the lighting design. This'll teach you how sunlight works, how to make the most of your tools and how your camcorder reacts to sunlight.

Experiment, too, with different materials and techniques. You may discover a style that becomes the signature element in your videos.

29
Miking Basics

Loren Alldrin

When it comes to recording sound, the microphone has a greater effect on sound quality than any electronic gizmo in the signal chain. If you've got a crisp, clear recording from your mike, it takes a pretty good effort to mess it up later. Likewise, if the mike signal reaching your camcorder is poor, there's not much you can do about it after the fact.

Image Isn't Everything—Distance is

Even if you forget everything else, there's one rule of miking you should permanently burn into your mind: with microphones, distance is everything. To elaborate, the distance between the mike and the sound source is the number-one factor affecting that mike's performance. In most cases, a $50 mike 12 inches from the source will sound much better than a $500 mike at 20 feet.

This is due to the nature of sound. Unlike light rays, which we can filter very selectively through the lens, there's no tidy way to "zoom in" on specific sound waves. There's no microphone equivalent to the center-field camera at a baseball game, the camera that captures the subtlest movement of the catcher's hand from 300 feet away. Even high-tech surveillance microphones can't block out unwanted sounds to that degree.

So what can you do to get some aural distinction between your subject and the competing noise? The answer is simple: get the mike as close as practical to your sound source. If you're deciding between a lavalier attached right to your speaking talent and a directional mike placed several feet away, go with the lavalier. If you're trying to get clean, full sound from your voice-over talent, do what the pros do: place a high-quality directional mike 6 to 12 inches from his or her mouth.

The "distance is everything" rule explains why the camcorder's built-in mike rarely delivers good sound—it's just too far from the subject. If you have to use the on-camera mike, try to do your shooting close to the subject (by using wider zoom settings). Have your subject(s) speak loudly, and move to the quietest location available. And because the on-camera mike is closer to you (the operator) than the subject, stay as quiet as possible.

Picking a Mike

Getting good sound on tape requires selecting the right mike for the job, and placing it correctly. With all the different physical styles, directional patterns, and element types out there, choosing the best mike can be tricky.

As mentioned earlier, directional mikes aren't as directional as you might think. The most common directional pattern is called "cardioid," due to its heart-like shape. The mike's sensitivity is almost as great at its sides as directly in front. This means that a cardioid mike will do a good job picking up sounds above, below and to the sides—it's only directly to the rear that the cardioid pattern has a high degree of sound rejection.

If you're trying to record a large sound source (a group of several people, for example), a cardioid mike's wide pickup pattern will be a plus. If you're trying to "focus in" on a relatively small or distant sound source, the cardioid mike may not be directional enough. Mikes with tighter patterns are available (such as hypercardioid, supercardioid, shotgun, etc.), but they have some drawbacks as well. First, tighter pickup patterns often result in a less "natural" sound. Second, the tighter patterns actually have areas of increased sensitivity radiating around the back of the mike. If the sound is coming in at a certain angle, a highly directional mike might pick up more sound from behind than a standard cardioid.

And while a directional mike will have the aural effect of "closing the gap" somewhat between mike and source, it's not the ultimate solution; closer is still better. An omnidirectional mike (which picks up sound evenly from all directions) placed close to the source will often sound better than a more distant directional mike. Some of the most natural-sounding mikes are omnidirectional lavalieres, because they capitalize on the mike's close proximity to the sound source.

Don't be taken in by so-called "zoom" or variable-pattern mikes. These mikes go from wide stereo pickup to a tighter (mono) pattern, and some of them actually track the motion of the zoom lens. But remember that no mike can match the "pickup pattern" of your camcorder's zoom lens, and these mikes often create strange-sounding artifacts if you zoom while taping. If you can control the zoom mike's pattern manually, your best bet is to find the pattern setting that gives the most natural and uncolored sounds, and leave it there. If you can't disable your zoom mike, you've got further incentive to not zoom mid-shot (which is a good habit, anyway).

Something else to watch for when selecting mikes is their impedance rating. High-impedance mikes are usually quite inexpensive, but have a serious drawback—the longer the cable run, the more dull they sound. Low-impedance mikes usually cost more, but can drive a good-sounding signal through hundreds of feet of cable. Though not directly related, the type of connector is often a giveaway to the mike's impedance. A three-prong (balanced) connector on the mike itself usually means the mike is low impedance. A permanently attached 1/4-inch cable often indicates that the mike is high impedance. This type of mike is generally suitable only for very short cable runs (20 feet or less).

Every type and model of mike has its own particular sound. The science of converting sound waves into electricity is not cut-and-dried—the design of the mike has a huge bearing on the resulting sound. Even in the same price range and class, you may notice that one lavalier mike sounds "boxy" and "dark," while another sounds "open" and "clear." Lay some sounds on tape from each mike before you shoot, and listen to the results. Switching one mike for another can mean the difference between ho-hum sound and a track that practically jumps out of the speakers.

The mike's intended application also has a major bearing on the resulting sound. Some directional mikes are meant to be placed quite close to their sound source. This position creates a buildup of bass frequencies due to what's called the "proximity effect." Most handheld vocal mikes compensate for this by generating a similar boost in the high-frequency "presence" range.

If you use a vocal mike to record a sound source a few feet or more away, you won't have any proximity effect or bass buildup. You will, however, still have boosted high frequencies. The result? Thin, tinny sound.

Instrument mikes, on the other hand, don't usually sit so close to their sound source. Hence these mikes don't have a high-frequency boost to compensate for the proximity effect. Place an instrument mike a few inches from your talent's mouth, and you'll get dark, bassy sound. Proximity effect is working against you, with no compensating high-frequency boost from the mike.

Placing your Mikes

The second half of the miking art involves where you place your mikes, and how you apply the "closer is better" rule. As we mentioned at the top of the column, mike placement is the biggest factor influencing a mike's sound quality. Mike placement controls the balance between ambient sound (or noise) and your subject.

Sometimes, if you're trying to accentuate the surrounding clutter of sound, you'll actually want to place your mike a good distance from your sound source. Let's say two actors are struggling to hear each other over the roar of a stadium crowd. Putting mikes on or near the talent would block out a great deal of the ambient sound, making their voices undesirably clear. Placing a mike five or 10 feet away would allow the surrounding noise to compete nicely with their voices. Trying to lift the voice of a training video host over the nearby machinery demands the opposite approach: get the mike as close as possible.

The majority of on-screen dialogue heard in television and movies is picked up with a directional boom microphone suspended just out of the frame. This offers two advantages to the director: freedom of movement for the actors, and a relatively short mike-to-subject distance. You can set up your own simple boom with any number of different items: a broomstick, a mike stand without its base attached, a sturdy fishing pole, whatever. Simply find a way to attach a mike to the end, run a cable down the length of the boom, and train someone to keep it close to the action and out of the frame. You may want to mount the mike with foam or elastic bands, to isolate it from mechanical vibrations.

Try to avoid picking up the same sound with multiple mikes, especially when they're located almost, but not exactly the same distance from the source. This can cause cancellation. You're better off moving one mike closer to the source, or eliminating the second altogether.

For recording numerous different sound sources in mono, you have two options: place one mike an equal distance from all sources or use one close mike for each individual source.

When setting up your mike (or mikes), don't be afraid to experiment with "unusual" placement. A lavaliere mike, for example, is not just for lavalieres. Why not conceal a lav mike in a plant between two actors, or hang it overhead just out of the frame? A handheld vocal mike, because of its presence boost, may sound great picking up the muffled giggles of two kids playing in makeshift sleeping-bag fort.

When you're struggling with difficult mike placement, understand that it's better to be slightly off-axis from a directional mike than too far away. In other words, if you can get a directional mike close to the sound source but not pointing directly at it, this is usually preferable to placing the mike a great distance away. Directional mikes aren't all that selective, and many deliver very good off-axis sound.

Other Considerations

Microphones turn vibrations in the air into an electrical signal. Unfortunately, the element inside a microphone converts physical movement into sound as well. This is why some camcorder's built-in mikes record as much transport, zoom, and finger noise as they do desired sound.

Good handheld mikes suspend the element in foam to minimize handling noise. Cheaper handheld models (or those instrument mikes not designed for hand-held use) may have little or no suspension, a design shortcut that can turn even the tiniest hand movement into the aural equivalent of an elephant stampede. Again, it pays to consider the original application of the mike before plugging it in. Most lavaliere mikes, because of their small size, have no suspension system. This makes the lav a poor choice for any application where someone will be touching the mike directly.

It's important to note that you can (and should) break all these "rules" under certain circumstances. The final authority for any miking decision should always be your own two ears. Record some test footage, and listen back with headphones. Experiment with different mike placement. Try different mikes. If you're having no luck with a stand-mounted directional mike, try substituting a boundary (pressure-zone) mike. Is your handheld mike not cutting it? Try a lavalier instead.

Remember—when it comes to microphones, if it sounds right, it is right.

30
Audio on Location

Michael Loehr

Camcorder audio signal quality has come a long way. Today's PCM and hi-fi audio tracks rival those found on the professional gear of a decade ago. Now videographers can easily pair sophisticated images with stunning soundtracks. With basic location audio techniques, superior sound is as easy as "Check, 1, 2, 3."

In the beginning, camcorders offered only one linear audio track. Because videographers didn't have much to work with, they often assembled soundtracks with little or no consideration for quality or fidelity.

These days, camcorders record in stereo, boasting audio specs approaching the quality of CDs.

Unfortunately, the built-in camcorder mike still has its limitations. Though often fairly sensitive, the mike's location and pickup pattern are appropriate only for events like birthdays and reunions, where setups are simple and audio integrity less than critical. When your video productions get serious, you'll want to invest in location audio accessories.

Sound Tools

As inherently creative people, videographers are always searching for new ways to ignite the imagination. Too bad most new tricks and techniques center exclusively on video.

While good images are essential to any video project, audio is just as important. Dynamic range, ambience and tonal balance are as important as zoom, pan and focus. In fact, poor location recording can easily drain the life from an excellent visual sequence.

It's important to approach audio with the same care as video. When storyboarding a scene, think a bit about the audio. Imagine the sound of the scene inside your head.

Then, when out on the shoot, strive to match the sound in your headphones with the sound in your imagination.

The tools of audio field production link real world sound to the magnetic tape in your camcorder. Each tool serves a different purpose.

Microphones react to raw sound waves moving through the air, converting changes in air pressure to voltage changes. These are sent to the camcorder through cables.

Cables allow you to position the mike away from the camera, a great asset for audio control.

Audio mixers combine the sound from two or more sound sources into one unified audio signal. Equalizers and filters alter the signal by shaping it to meet certain technical standards or personal tastes.

You should wear headphones on every shoot (see Figure 30.1). Listening through headphones is really the only way to verify the quality of the signal you're recording on tape. Although a signal meter might indicate proper audio level, it won't reveal echo, buzz, hum or overload distortion.

The best headphones are the over-the-ear type. These insulate the ear from sounds in the real world, making it easier to determine what you are actually recording.

When on location, always tape down audio cables with duct tape or gaffer's tape. Or bring a couple of throw rugs to place over the exposed cables.

You want to protect yourself, your crew and any onlookers from tripping over

Figure 30.1 *Using headphones is the only way to ensure you're getting good sound.*

cables, as well as keep the place looking tidy and organized.

Leveling

It's important to know how to set proper camcorder audio levels. Manual audio input level controls are most often found on high-priced camcorders, especially Super VHS and Hi8 models. Input levels are set based on the input level meter, which uses either a standard moving needle or LCD bar graph display.

An acceptable audio level means the meters only occasionally hit the -2dB or 0 dB mark. Here, "occasionally" means once or twice every few minutes. The average level should run around -5dB. This allows enough room for signal peaks without overloading the circuit.

Some high-end and professional 8mm camcorders use Pulse Code Modulation (PCM) circuits for audio reproduction. These sound great provided the signal level doesn't exceed a certain limit. Otherwise, the digital circuits overload instantly and heavily distort the audio.

To prevent distortion, many camcorders include automatic gain control (AGC) to constantly adjust the input signal level. Although these limiters are theoretically a good idea, they rob the audio signal of dynamics and presence, adversely affecting realism. Your audio will almost always sound better with the limiter switched off and levels adjusted manually.

On VHS camcorders, the linear audio track record level is set via AGC, even if the unit boasts hi-fi input level controls.

If you fear overload, run the levels slightly below normal.

This shouldn't present any problems, especially with PCM and VHS hi-fi circuits, as both schemes feature very low noise floors and almost no tape hiss.

Another excellent aid in better location audio is a separate crewmember dedicated to running the audio gear. With someone else devoted exclusively to the audio, you can focus full attention on the images.

Outdoor Ambience

When shooting outdoors, pay special attention to your location's ambient sounds. In everyday life, our minds filter out much of the noise around us. A quiet city park might actually buzz with sounds we don't notice. Voices, aircraft, automobile traffic and wind often sneak right past untrained ears.

Before shooting outside, stop and close your eyes. Let your ears determine what's happening. Decide if the sounds you hear are appropriate for your project. Plug a mike into the camera and listen through headphones for an even better indication of surrounding ambient sounds.

If you're recording a political rally in a civic park, you can live with urban undercurrents on the soundtrack.

But if you're trying to stage a romantic walk in a quiet scenic park, best find another locale.

Wind often creates serious audio problems. Wind modulates a mike's output signal in a way that makes it difficult to set proper recording levels. Even worse, wind can affect a mike's entire frequency range. The result is a basically useless soundtrack.

Wind noise most often occurs as pulsing low-frequency rumble. Gales blowing directly into a microphone can break up the entire audio signal.

The videographers's only defense is a windscreen, a thin foam jacket or sleeve that fits around the mike pickup. A windscreen provides acoustic resistance against the low-frequency waves generated by wind. Yet its light weight and high porosity pass high frequencies through to the pickup unchanged.

The result is near elimination of annoying rumble and a dramatic decrease in the loud pops and booms associated with high wind.

While windscreens don't completely eliminate the effects of wind on audio, they do make wind noise manageable.

Sounds Inside

When shooting indoors, ambient noise is still a concern. There you'll contend with a different batch of unwanted sounds.

Fluorescent desk-mounted and overhead light fixtures are common in many office buildings. These lights actually switch on and off thousands of times every minute. This constant switching generates a faint buzz or hum easily detected by sensitive mikes. The best solution: simply shut the things off and light with supplemental sources.

Or you can install a mixer with a low-cut filter on your mike cable between the microphone and the camcorder. Fluorescent lights turn on and off between 30 and 60 times a second, or 30Hz to 60 Hz. The low-cut filter on a mixer corresponds to this range.

The air conditioner is another common indoor audio enemy, particularly frustrating because its sound covers such a wide frequency range. This makes it difficult to remove the noise without removing important audio as well. The best solution, obviously, is to shut the air conditioner down for the duration of the shoot.

Sounds from adjacent rooms or floors might go past your ears unnoticed, but your microphone will capture them faithfully.

Even conversations in the next room can travel through walls and disturb your shoot.

Speak Into the Mike

When recording an interview, the video is constant, focused primarily on the person speaking. Good audio quality will add authority and credibility.

If you shoot the interview one-on-one indoors, use lavaliere mikes. These are tiny microphones designed to clip to a tie, lapel or collar. The cable is concealed inside a shirt or jacket, running to a small housing in a pocket or attached to a belt. In many models, the housing holds the batteries and a connector for cable or camcorders.

Proper positioning is important. The mike should lie in the middle of the chest. Point the pickup toward the person's mouth. Make sure the mike doesn't sag or dip. Narrow or flimsy ties may not support even tiny mikes.

You can also clip the mike to the upper part of a shirt or blouse, either between buttons or folds in the fabric. Hide the cable by feeding it underneath the shirt or blouse. If the subject is clad in a suit coat or sport jacket, try clipping the mike to a lapel.

A lavaliere mike will also work well outdoors. Remember to use a windscreen to eliminate any wind problems.

For an audible interview in a crowded location, use a hypercardioid or shotgun mike, so named because it focuses only on sound directly in front. Sound from the side and rear is rejected. These mikes are ideal for isolating a specific person or activity.

Hold the shotgun mike in your hand and point it toward your subject. If you notice a popping sound on harsh consonants, move the mike back a bit.

Shotguns on Fishpoles

Scenes in dramatic videos are rarely shot in consecutive order. Days or weeks might elapse between shoots, and tens or hundreds of miles might separate locations. Yet maintaining audio consistency is vital.

An excellent way to ensure a smooth soundtrack is to record each scene with a shotgun mike mounted on a fishpole. And what is a fishpole? A long telescoping tube designed specifically to support a shotgun mike. The length of the pole varies with adjustment of the separate sections.

This arrangement requires somebody to act as boom operator. The operator must hold the mike high in the air, pointed down toward the subjects to reject any noise coming from above. Sound waves from the actors will float up toward the pickup. Floors and walls can also help reflect sound up toward the mike.

When actors move within a scene, the boom operator must move with them. The operator should wear headphones to hear and react to changes.

You can enhance consistency by placing a stationary omnidirectional mike somewhere on the set. Feed it to the second channel on your camcorder. This will supply a stable sound source in case something goes wrong with the shotgun mike.

Feeds

Stage and theater works present formidable videography challenges.

Whether a local production of *My Fair Lady* or a company CEO addressing conventioneers, these situations require extra attention.

Visit the site well before the performance. Meet someone affiliated with the production and ask about the setup.

Inquire about camera positions, setup time, cable runs, mixer feeds and AC power connections.

The first question to ask yourself is, "Do I mike it myself, or do I trust the house feed?" Try the house feed first, the signal fed from the auditorium's master mixer directly into the camcorder. In many professional and semi-pro presentations, the audio will feature a public address system with a mixer to balance and adjust sound levels. With a little detective work, you can determine exactly how to join the house mixer to your equipment.

Know what audio input connectors and signals your camcorder accepts. Most will take either a line-level or microphone input; their connectors are almost always located separately on the unit. If the connector reads Mic/Line, you can adjust its sensitivity with a small switch located near the connector.

Determine how the house mixer distributes its output signals. Most mixers offer a range of output levels. For videographers, the ideal output is either Line Out or Program Out. Be sure the feeds you select include all signals passing through the mixer.

Don't use outputs marked Speaker or Monitors. The high-level outputs designed to drive loudspeakers can damage camcorder audio circuits.

Common connectors on all camcorders are RCA phone jacks for line-level inputs, and 1/8-inch phone plugs for microphone inputs. Most mixers use professional, balanced XLR plugs or RCA phone jacks, or occasionally a 1/4-inch phone plug.

If you use XLR outputs, you'll need a transformer to convert the balanced XLR to the unbalanced camcorder input connector. Adapters can convert the XLR signal to mike or line level. Use balanced XLR cables to preserve audio quality when taking feeds from a house mixer. Install the XLR transformer at the camcorder input.

Once you have the proper adapters, match the output and input levels as closely as possible. Feeding the line-level input on your camcorder with the line-level output from the mixer is the ideal combination.

Connect the cables and mikes and adjust the recording level. If distorted audio fills your headphones, the house mixer output signal level is too high for the input circuit. If this happens, you need to install a limiter or attenuator between the mixer output and the camcorder input to limit the input level.

If the sound in your headphones is faint or non-existent, connect the feed to your camera's mike input. If the result is distorted audio, you're overloading the mike input. Use an attenuator to conform the house mixer output to the input circuit.

If a house mix feed isn't available, you'll have to run the cables and the mikes yourself. You'll need to set aside more time to fine tune the setup. Remember, preparation is the key.

PZM Pickup

Theaters, auditoriums and large conference rooms possess unique acoustic properties.

A good microphone for recording in areas like these is a Pressure Zone Mike (PZM). These sensitive, accurate mikes capture natural balanced sound free of echoes or room reflections. In a PZM, the actual pickup element sits suspended above a metal or plastic plate. The distance between the pickup and plate is very small. This arrangement cancels the annoying reflections commonly associated with regular omni mikes.

PZMs work well as long as you locate them somewhere near the center of the action. If the event is a discussion with people gathered around a table, simply place the PZM in the middle of the table. PZMs are also excellent for recording plays or stage performances. Place the mikes anywhere along the edge of the stage for an even, balanced recording.

An alternative to PZMs is the standard close-mike technique, which means placing a mike as close to each source as possible. To close-mike a show, you need a mixer with enough channels to handle each source. Close miking results in a more dynamic soundtrack.

In a situation where a PA system is used, but a feed isn't available, simply put a good omni mike in front of one of the PA speakers. If you're recording a presentation by various speakers to an audience, a good omni mike attached to the podium will yield excellent results. If members of the audience will ask questions, either point a shotgun toward the audience or use a wireless handheld mike.

Going Wireless

For extra flexibility, a wireless mike is an excellent solution—perfect for miking people who move around a lot.

When working properly, wireless technology is the ideal way to connect mikes and camcorders. Unfortunately, wireless systems frequently fall victim to radio interference. The best technique to avoid interference is to maintain a line of sight between the receiver and transmitter. Walls, doors, windows and light fixtures all reduce signal strength and increase interference. Light dimmers, two-way radios, cellular phones and power transformers also impede performance.

Interference usually creates a buzzing or popping sound. Signal drift is the result of poor reception, and comes as more of a whispering sound, or delicate hiss. Like interference, it's also affected by movement.

If you use more than one wireless mike of the same brand, tune them to different frequencies. If the two mikes occupy the same band, neither will transmit or receive properly.

Some videographers record projects with stereo mikes, which are actually two microphones housed side-by-side in one unit. A stereo mike adds the illusion of depth and space. Be aware, though, that inconsistencies in stereo soundtracks are more noticeable than in standard, single-channel soundtracks.

If you elect to record a particular project in stereo, leave the mike in one place to ensure a consistent soundtrack.

If you're recording in unusually loud circumstances, condenser or battery-powered mikes will present more problems, simply because they're more sensitive than the passive dynamic types.

Finally, always record two or three minutes of each location's natural, ambient sound (see Figure 30.2). Use this sound to cover the silent audio gaps between edits that occur when you assemble the program. The result will be smoother audio tracks within each scene.

Figure 30.2 *Record a few minutes of ambient sound for smoothing your soundtrack during editing.*

31
Great Shots:
The Art of Composition

David Welton

See video clips at www.videomaker.com/handbook.

Composition is the art of framing your video world. The visual frame is the artificial border that limits and defines the image you shoot—a border you set, for better or worse. Learn to set that border in a manner that pleases the human eye and you make that all-important leap from good video to great video.

In this chapter, we'll discuss this most artistic aspect of videography—from the basic rules of composition to tips from the pros.

Anatomy of an Image

Before we look at the mechanics of composing an image, let's focus on the receiving side of the visual communication process: the human eye.

The human eye scans an image from many directions at incredibly fast speeds. Certain patterns, colors and shapes attract the eye, but other elements confuse it.

It requires considerable brainpower to interpret a visual message. The more complicated and unfamiliar the image, the harder the brain must work.

A printed page is a good example of a familiar image; the brain has learned from experience to direct the eye to the upper left-hand corner, then proceed across and down the page.

Most video images present a greater challenge to the brain. Typically, it must use a different strategy to digest the visual information of each individual image. Poorly composed images make the brain work harder, directing the eye to scan and re-scan images in search of meaning. Whatever message the image is meant to communicate is lost in the confusion; the brain tires and simply ignores the message. A properly composed image, on the other hand, is easy for the brain to digest.

Rules to Shoot By

Everyone makes aesthetic choices every day; we know what styles and colors of clothes we like to wear. Likewise, we can usually muster a yea or nay opinion on the artistic merits of any given video image, though we might not be able to explain why.

A closer look at these subjective opinions reveals some simple rules of composition, rules we operate by—whether we realize it or not.

Play by the rules. Artists have been aware for centuries that placing objects at certain predictable points results in greater harmony and a more pleasing appearance. As a result, a mathematical formula evolved that filmmakers and videographers use today. It is called "the rule of thirds" and states that if you mentally divide your frame into thirds, both horizontally and vertically—like a tic-tac-toe board—and place important elements on the lines, you create a more visually pleasing image. (See Figure 31.1.)

Give me some room. The rule of thirds can be applied to any object, but given that we generally spend a lot of time shooting people, the *talking head*—a closeup shot of a person's head and shoulders—deserves a

mention. You may have no choice than to center your subject in the frame, but you can use the rule of thirds to help conquer another people shooting problem—that of headroom.

The space between the top of the frame and the subject is called *headroom*. Too little and your subject appears partially decapitated. Too much and the subject appears to be dropping out of you shot. Using the rule of thirds to place your subject's eyes—on the top horizontal line—will ensure proper headroom. (See Figure 31.2.)

The nose knows. Another visual imbalance occurs when someone looks or points in a direction other than straight into the camera. A full profile of a person is one example of this situation.

Well-composed shots leave space in the direction of the looking or pointing, as in Figure 31.3. This extra space, called *noseroom* or *leadroom*, is particularly important when a person or object is in motion and the camera is panning to maintain framing. (See Figure 31.4.) Whether you're

Figure 31.1 *The rule of thirds is the key to good composition.*

Figure 31.2 *Place your subject's eyes on the top third line—no matter how wide your shot—to ensure proper head room.*

shooting a profile of a marathon runner or a passing Indy racing car, the rule applies equally—leave space in front of the subject.

The human form. In face-to-face interactions we can usually see the other person's entire body. In television, often it's necessary to show only a portion of the human anatomy. There are guidelines to help in the composition of these types of shots.

Don't compose video frames that cut people off at the eyes, nose, mouth, chin, bust, waist, knees or ankles. Instead, frame your shots just above or below these features. This still leaves you plenty of options; standard shots range from extreme long shot to extreme close-up.

For instance, the medium shot will frame the body above the knee while leaving ample headroom on top. The extreme close-up frames just below the chin and slightly above the eyes; in this example, no headroom is necessary.

After you've properly framed your subject, check the background for objects that may interfere with your image. You don't want a telephone pole sticking out of a person's head, or a branch of a potted plant looking like the third arm of your interviewer.

Whatever your message, knowing the rules will help you communicate it more effectively.

Camera Moves

Whenever you're shooting something in motion you must consider the directional axis of action, that is, which way your subject is moving.

Say you're shooting a football game. In your quest for the best camera angle, you shoot from both sides of the field, during the same quarter. This proves kind of confusing when you play back the tape—without warning your team rushes towards the wrong goal line. You've forgotten the directional axis of motion, which changed a full 180 degrees as the camera switched across the axis, from sideline to sideline. You can break this rule, as long as you inform the audience.

When you move the camera to follow action, mimic the movements of the human head and eyes. This way, such moves as panning and tilting a camera

Figure 31.3 *Placing a subject on the left third line (A) makes the shot feel crowded. Placing her in the center (B) is better, but too uniform. For proper noseroom, place her on the right third line (C).*

seem natural to the viewer—like typical movements of the head.

Make sure to follow the rules of good composition throughout the entire pan or tilt. Rehearse the camera movement first. Determine whether you'll need to alter focus or make other midstream adjustments. Make sure your tripod is level; a high quality fluid head will give you the best results.

Establish where the camera movement will begin and end, and hold the camera still for a few seconds in both places. Experiment with different pan speeds and camera positions; you can decide which one is right later in the solitude of the edit suite.

Be sure there's a reason to move the camera; it's best if an object or event is the motivation for a pan or tilt. A passing train, descending glass elevator or off-screen scream justifies movement, but camera movement without apparent reason appears contrived.

Let's examine the most obviously contrived camera move of all: the zoom. I know, it's hard to resist using a variable-speed 15:1 power zoom lens when you've got one. The zoom lens is a great tool for setting up framing before you roll tape, but remember that you don't need to use it during every shot.

Unless you're the Six Million Dollar Man, your eyes don't have a zoom feature. Since zooming is not a human phenomenon, it often appears unnatural to the viewer. So use zooms sparingly. Watch a professional production carefully; you'll discover that on-air zooms are quite rare.

In fact, the handheld look dominating music videos and television commercials now illustrates the point well: the human point of view is the most natural.

Instead of zooming to follow a subject retreating from the camera, try physically moving the camera closer to the action—following the movement step for step. Unlike zooming, this technique produces an image and feel that is similar to the human experience, and therefore more real to the viewer.

A dolly or boom works best for smooth results. Many videographers don't have access to these devices. If you don't, make do with a simple yet effective alternative: your feet. Try to keep your knees bent to absorb shock, and slide your feet instead of picking them up. A wide-angle lens setting will minimize picture jiggles. Technological advances in mechanical, electronic and optical image stabilization systems help videographers get remarkably stable shots while in motion.

If an on-air zoom is necessary, be sure to maintain proper composition throughout the entire zoom range. It's often necessary to tilt up or down as you zoom in or out. If you zoom out from an extreme close-up of a person to a medium shot, you must tilt down to maintain the proper amount of headroom. As we have seen, too much or too little headroom is very disconcerting to the viewer.

As with pans and tilts, rehearse the zoom before you roll tape. Whenever possible, use a tripod for steady shots during the zoom.

Artistic Considerations

It's always a challenge to recreate a 3-D world with the two dimensional reality of a television screen. There are a few tricks that enhance the perception of depth: using shadows, placing objects in the foreground, manipulating the depth of field and experimenting with camera angles, to name a few.

Shadows are good. This may come as a surprise to those of you who learned to design lighting that eliminates shadows. Don't be afraid of them; they can help mold objects with light and dark, subtly adding dimension.

Placing objects up in the foreground also gives the viewer perspective on the depth of the scene.

Videographers can also alter the depth of field—the portions of a scene that are in sharp focus. A large depth of field keeps everything—from foreground to infinity—in sharp focus, thus enhancing the perception of depth. Manipulating the shutter speed, aperture and lens focal length all affect an image's depth of field.

Don't be afraid to keep the camera still, allowing the action to move towards and away from the camera. By doing this, we let the moving objects get bigger and smaller without using the zoom lens,

Figure 31.4 *When panning, allow room for subjects to move into. In shot (A) the subject bumps into the edge of the frame. In shot (B) she has room to move forward.*

recreating the way the human eye naturally tracks objects.

Give yourself lots of time to experiment with camera angles. Even a minute change in camera position may radically change the resulting image. Whatever you do, don't just let the camera rest on your shoulder. Remember, too, that for the best impact and clarity on the small 3:4 ratio television screen, you must get as close as you can to the action.

Color composition is also an important artistic consideration. Much of the beauty and excitement of television comes from color; unfortunately for the videographer, many camcorder viewfinders display in black-and-white.

If possible, connect a small color TV to your camcorder while shooting. The new

LCD varieties are quite good for this. Even better, most camcorders today are equipped with a color LCD viewfinder. These cost a bit more, but they allow you to compose your shots in color without hooking up to a monitor.

Two final viewfinder tips:

Check your viewfinder for framing accuracy. Viewfinders are notorious for overscanning, or cutting off the edges of the frame. It's very frustrating to find out that your viewfinder lied and your composition paid the price.

Newer camcorders often clutter viewfinders with on-screen displays and prompts. Turn off all but the essential ones—don't let these annoyances make composing a shot more challenging.

When to Break the Rules

A certain amount of aesthetic tension is good. It makes images more interesting to look at, less boring. Don't feel constrained by the rules we've outlined here; once you've learned them, you can break them with good purpose. In fact, there are times when breaking the rules is the most effective way to make your visual point.

Composition is more art than science. You determine which visual elements to include in your shots, and which to exclude. The choice is yours.

32
It's Your Move

Robert Borgatti

See video clips at www.videomaker.com/handbook.

Remember that famous shot in *Gone With the Wind*, the one where Scarlet O'Hara wanders into the rail yard looking for a doctor? The camera begins on a standard wide-shot and then slowly lofts skyward, revealing row upon row of wounded and dying soldiers, ultimately coming to a halt on a tattered Confederate flag flapping in the wind.

Not all camera moves are as grand and eloquent as that one. But whether you're shooting a cinematic epic or a home video, add a little movement to your shots and you'll bring a whole new dimension of creativity and visual energy to your efforts.

In this chapter, we'll look at some equipment you can use to move your camcorder. These will influence the mood, look and emotional intensity of your productions. Such movement often means the difference between dull and dazzling video.

Moving vs. Zooming

Today's small, lightweight camcorders offer virtually unlimited mobility. Even so,

videographers sometimes let the zoom lens do all their moving for them.

They rely so heavily on the zoom because it offers a quick and easy alternative to the typical static shot. Novices, in particular, go zoom crazy, charmed by the notion of doing a smooth, professional-style camera move.

The zoom is a great feature and has many useful applications, but it's no substitute for the real camera moves. In fact, the pros rarely use zoom shots. Zooms are conspicuously absent from most movie and TV productions, with the exception of electronic news gathering, documentaries and other formats where all shots are not pre-planned.

Most big-name filmmakers use fixed focal length lenses, making zooming impossible. The professionals mainly use the zoom to change framing between shots or to maintain composition while they perform some other type of camera move.

Why this reluctance on the part of professionals to use the zoom? Because zooming alters the focal length of the lens. As you zoom in, you gradually reduce the angle of view from wide to telephoto, thereby magnifying a selected portion of the original shot. This effect is quite different from the effect you achieve by actually moving the camera itself toward the subject.

For example, using the zoom-in to simulate the view of a person walking down a hallway does not convey the impression of real movement. Since the camera does not change position during a zoom, the perspective of the shot—that is, the spatial relationship of the objects within it—remains virtually unchanged.

To our eyes—which do not normally zoom—this sort of movement appears unnatural. To obtain a convincing shot, you need to move the camera down the hall.

The artificial quality of the zoom is further emphasized by the fact that going from wide-angle to telephoto compresses the apparent distance between foreground and background objects.

Also, as you zoom in, you gradually reduce depth of field, thus throwing more of the foreground and background out of focus.

The effect you get by physically moving a camera through space more closely resembles our own human way of viewing reality. Unlike a zoom, a shot obtained with a moving camera continuously reveals new perspectives and emphasizes space and depth. The movement itself is more interesting.

Pans and Tilts

First let's consider the most basic of camera moves—the *pan* and the *tilt*. Like the zoom, pans and tilts are camera moves that involve shooting from a fixed position. A pan is a left-right or horizontal rotation of the camera; a tilt, an up-down or vertical rotation.

Shooting in just about any situation requires a fair amount of panning and tilting; you do much of this panning and tilting intuitively as you follow the action.

However, you can plan those moves and spice up your shots dramatically.

Say you're shooting Grandpa's birthday party. Why not begin with a close-up of the cake and then slowly tilt up to capture Grandpa's face reeling from the heat of all those candles? Pan across members of the family as they sing "Happy Birthday." Time the shot so you're back to Grandpa gladly blowing out the candles by song's end.

Wedding ceremonies present many lulls in the activity where you can perform interesting camera moves. Try panning across the bridesmaids and ushers, or from the minister giving his sermon over to the bride and groom. For a shot of the congregation, you might begin by tilting down from a stained glass window.

Often you'll find that good composition requires a simultaneous combination of panning, tilting and zooming. That's why it's always a good idea to rehearse the move before committing it to tape.

You can also use pans and tilts as transition devices to introduce or conclude a scene. Let's say you're shooting an on-camera narrator discussing the dangers of over-exposure to the sun's rays. You start your shot with a pan following a swim-suited couple as they stroll along the beach. You stay with them until they cross in front of your narrator; at this point you stop panning and your talent begins the monologue.

Use the pans and tilts at the end of a shot to reveal some poignant or dramatic visual element. For example, in a scene where a husband and wife discuss divorce, you could pan over to a photograph of the couple in happier times—after they leave the room.

Pedestal Shots

Another type of move you can perform with a tripod-mounted camera is a *pedestal*. This movement—often called *pedding*—involves the raising and lowering of the entire camera on its pedestal or elevator.

Professional tripods, however, feature pneumatic, hydraulic or counterbalanced pedestals, which allow extremely smooth raising and lowering of the camera.

You might use a pedestal move to jazz up a shot of someone writing a letter. You start at table level with a close-up of the pen moving across the paper. Slowly, you pedestal up. As you do, you also widen out your zoom and tilt down to maintain good composition.

Dollies, Trucks and Arcs

Equip a tripod with dolly wheels, and you can roll it in any direction you want. Unless you are on a smooth, seamless surface or employ some form of image stabilization, however, your moving shots will probably suffer from excessive shakes and jolts.

You find dolly wheels primarily in TV studios, where the cameras need to move around freely and easily on floors as smooth as glass.

In TV production jargon, there are three distinct rolling moves. A *dolly* is a move toward or away from the principal object in a scene, making it appear either larger or smaller in the shot. Since the video camera itself moves through physical space during a dolly shot, there's a more dramatic change in perspective than you would achieve by merely zooming in or out.

Roll the camera laterally to the left or right, and you've performed a *truck*. This move is commonly used to follow talent across the set.

An *arc* is where you roll the camera on a curving path around an object while maintaining a consistent distance from it.

In dramatic productions, an arc move provides an interesting way of changing your perspective during lengthy dialogue shots. It can also work well for showcasing products in promotional videos.

Getting a Grip

Outside the studio, a tripod with dolly wheels is not very effective for performing moving shots. So you use special camera mount devices instead.

In film and video production, these devices—known as grip equipment—often require the use of an extra person to assist the camera operator with the moves. You call this person—what else-a grip.

The most common piece of grip equipment is the dolly. Not to be confused with the tripod wheels we already discussed, the dolly is a low-riding cart with chubby pneumatic tires; a pull handle steers the device in much the same way you steer a child's wagon.

Dollies often feature a fully functioning tripod head mounted on a central column. Dollies without this central column require you to mount an entire tripod on the cart. The camera operator rides on the dolly with the camera, while the grip pulls, pushes and steers.

When ground conditions are too poor for tires, use a tracking dolly. Tracking dollies ride on rails laid out in advance along the desired shot path, providing a perfectly smooth ride regardless of terrain.

Dolly moves or tracking shots are used in any number of situations, but most often to follow talent as they move across a set or through a location. The idea is to keep the framing of the shot consistent during the move by moving the dolly at the same speed as the talent.

In big-budget movie productions, heavy-duty dollies boast cranes, permitting elaborate and complex camera moves. For example, you can get a shot that starts at ground level and, as it pulls back, arcs around the talent while rising to a point high above the set. The camera operator goes for a ride with the camera on the crane, while a grip handles the move.

There is a smaller version of the camera crane, called a jib arm. Mounted on a dolly, the jib arm is controlled from ground level by the camera operator, who views the shot on an external monitor.

Affordable Alternatives

The equipment outlined above is big and heavy and costs hundreds of dollars a day just to rent. But videographers eager to perform these exciting film-style moves can find some relatively low-cost alternatives.

You can get great tracking shots using PVC pipe and your standard tripod with dolly wheels. Bolt two lengths of pipe together to form each track. Then position your tripod so that two wheels ride on one track and one wheel on the other.

If you can't afford PVC pipe, use wheelchairs, wagons or shopping carts as a dollies. The camera operator rides on board while a grip navigates. This is effective on surfaces that are reasonably smooth. Even a small pebble can cause a big bump in your shot. To minimize shaking and jolts, hold the camera freely in the air.

Do not rest your elbows on your knees or any part of the dolly. The same applies when shooting from a moving car. Hold your camcorder like you would a cup of coffee to avoid spilling it while you drive.

The Right Moves

Like any special effect, use camera moves only when there's a logical reason for them. This reason may be as simple as keeping up with your actors as they walk across a set or following an elaborately choreographed sequence of actions.

The best camera moves are like a movie in miniature—they have a beginning, a middle and an end.

The beginning should attract the viewers' attention, the middle should hold their interest and the end should prove revealing and satisfying to them.

Now go ahead and make your move.

33
It's All About Direction

William Ronat

See video clips at www.videomaker.com/handbook.

When a video project fails, there's no one to blame but the director. That's the bad part.

The good part is that when it succeeds, the glory is the director's alone.

We've all seen poorly directed films. They wander aimlessly, camera pointed at nothing in particular, actors sleepwalking through their roles. You find yourself wondering, "Where in heck were they trying to go?"

It's the director's job to know the answer to that question. If you are unclear about a project's direction, the crew, the actors and the audience will be, too. You'll end up with a sloppy mess no one can sit through.

In the paragraphs that follow we'll explore some basic directing mechanics. The more you know about these, the less time you'll spend worrying about them.

The People

The word talent refers to your actors and spokespersons. It's an amusing word, since these people sometimes don't have much

in common with it. Still, they're your responsibility as director.

Sometimes you get to audition and pick your show's talent. Sometimes circumstances force them on you (a company president wants to be the talent in her marketing video, for example). However they get there, it's your job to make them look and sound as good as possible. We'll delve into this farther when we're on the set.

The archetypal director is a guy in jodhpurs yelling "action" and "cut" through a megaphone. The costume has changed, but the director still uses these phrases. The first couple of times you yell "action" on a set you'll feel goofy, but it soon becomes second nature.

But the director doesn't just walk onto a set and start barking orders. The step before the shoot is the most important aspect of directing. Skip it at your peril.

Plan, Plan, Plan

The planning stage is where you draw the blueprint, create the battle plan, decide the direction the project will take. Without this step you may as well forget it; your ship already has sunk (mixed metaphors, anyone?).

Directors of major films often begin planning a year or more in advance of a project. You may have neither the luxury of nor the patience for this much planning time.

Start by visualizing your script—anything from a bare-bones outline to a full-blown document complete with dialogue and stage directions. Break your script into shots. Visualize them. The vision will change when it bumps up against the reality of your budget, but for now, let your imagination go.

List every shot you think you'll need. You may want to underline and number the sentence in the script corresponding to a shot. For example, the first half of sentence one is *Shot 1*, the second half of sentence one is *Shot 2*, and so on. On a separate sheet you'll write 1.) Wide shot of Kasbah, booms down to reveal hero. 2.) Close up of hero, and so on.

Create a storyboard. Each shot becomes a picture—it can be fine illustration or stick figures, as long as you can tell what the pictures represent. Under each picture write the dialogue or voice over. These pictures show how scenes cut together. If someone else is paying for the production, the storyboard is valuable for showing a client what the video will look like. You can also use it to show crew and actors what you're trying to achieve.

The Production Schedule

When you get to Hollywood you won't have to take the next step on your own. An assistant director/production manager/unit manager will handle the details of production schedules. But it's a good idea to understand how they work.

Shooting each shot in sequence, as called for in the script, is inefficient.

Say you start out with an establishing shot in the busy Kasbah, using hundreds of extras. Next, your hero enters a build-

ing for a one-on-one conversation. Then he comes out—again, into the Kasbah. Would you leave all those extras waiting around while you shoot the interior scene? Of course not. You'd shoot out of sequence, picking up both Kasbah scenes at the same time.

A production schedule groups shots with common elements. If your script calls for a kitchen scene near the beginning and another at the end, shoot them at the same time. If a particular actor appears in only two scenes, schedule them on the same day (especially if you pay union scale). If you need a beauty shot of a building, schedule it for the time of day when the sun hits perfectly.

Your production schedule could include a second shot list grouping similar shots: Location: Kasbah, *Shot 6*.) Wide shot. *Shot 12*.) Hero runs out of shop. *Shot 13*.) Wide shot, car forces its way through the crowd to pick him up. Later, when shooting in this location, you can check off shots as you get them. This way you're less likely to forget a shot.

Nothing is worse for the crew's morale (or your mental state) than hearing, "Uh, I forgot we needed another shot of the hero back in the Kasbah," especially if the Kasbah set has been struck and the hero's gone home.

Whenever possible, scout the location. Visit the place where you plan to shoot to anticipate problems. For an outdoor shoot where talent speaks on camera, you'll need to know if your location is near a highway, construction site or other noisy, disruptive area.

Check interior locations for power requirements. Are there outlets for plugging in your lights? If you plug them in, will they trip a breaker, crashing all the computers in the building?

A location may sound perfect on the phone, but a scouting trip may reveal it just won't work. It's better to find this out before you show up with equipment, talent and crew.

The planning stage is not everyone's favorite part of directing. It's meticulous, challenging work. There's no flash, no glamour. But if you cover every detail before going into the field, you'll have more energy for making your show its best.

Plan to the *nth* degree. Then plan some more.

Who Does What?

If your production involves several crewmembers, meet with them before starting the shoot. Go over the script, or provide a synopsis. Discuss technical concerns and work out solutions. (Are you shooting a night scene? Your lighting director will know that blue gels on the lights suggest moonlight.)

This does two things. It involves the crew in the decision-making process, which makes them feel good. It also tells the lighting director to include blue gels in his light kit, which he might not otherwise do.

Once you're on the set, encourage your crew to participate with suggestions, but be sure they understand that you're the boss. As director, you are responsible for the final product. You have the final say on all content.

If you're working with talent and a crewmember has a suggestion; he should suggest it to you, out of talent's hearing. If you feel the suggestion has merit, you can suggest it to the talent. This keeps the talent focused on one source of direction. When suggestions fly from every crewmember everyone gets confused.

Pick your crew for technical ability and for personal compatibility. They need to be good at their jobs and easy to get along with. Be aware that the first shot of the day takes the longest to set up—the equipment must be prepared, power located and sets dressed.

As a general rule, the crew's energy (and yours) will be low in the early morning, peak at mid-morning, lag after lunch, peak again mid-afternoon and then drop off until the end of the day. Try to schedule your shots so the most difficult ones hit at peak times.

Once the set looks right, the lighting's in place and all the gear appears to function, it's time to bring in the talent. How you interact with these people determines how they appear on camera, so take care what you say.

Be Nice, Be Positive

Be gentle with your talent. Nonprofessional talent may have no video experience. This can be a nerve-racking experience for them (and possibly for you).

Say Ms. Brooks barely makes it through her first rehearsal. You yell, "Cut! Ms. Brooks, you're blinking too much. Your enunciation is sloppy. You're slouching. And try to smile more. Ready and *action!*" What is Ms. Brooks likely to do?

This is her first video experience. What you're saying sounds a lot like criticism. She may become more nervous, or become wooden as she tries to remember all the faults she needs to correct. Or she just might get angry and walk off the set. You don't want any of these reactions.

Instead, look for positive comments to make at the end of a take. "Ms. Brooks, that was an excellent smile. Now on the next one, please hit your words a little crisper. I'm a little hard of hearing and it will help me. Thanks." After the next take: "Very nice enunciation Ms. Brooks. Now try one with your shoulders flat against the back of your chair. Yes, perfect."

Tell your talent what *to do,* not what *not to do.* It's easier to remember a positive suggestion than a negative one. Every time you offer positive feedback to the talent, you increase their confidence. By the time you correct the problems, your talent should be ready to give their best performance.

How Many Crewmembers Do You Need?

The size of your crew will affect your directing style. If you work alone, you'll have to wear more hats than you would if you had a 10-person crew. A large crew presents further challenges.

Even if you're the only person on the crew, you're still the director. The problem is how to keep an eye on lighting, properly set up the camcorder, monitor audio, and consider countless other technicalities. At the same time, you must coax a good performance from your talent and watch for aesthetic problems.

If the show is a documentary or a very simple shoot, you may be able to handle all the chores and stay in control. But all the distractions increase the chances of making a mistake.

Be careful. At times like these you'll record over and lose forever perfectly good takes when the operator (you) forgets to re-cue the tape.

With two people on the crew, you have a fail-safe method. While your head swims with the details of directing, your crewmember can keep an eye on levels and lighting, and watch the tape and take notes.

This person can also point out poor audio quality or picture problems you might have missed. Because only the here and now concerns this extra member (as opposed to the show's concept, or how shots will go together later), he can concentrate on little things that might escape your notice.

A three-person crew boasts extra help, perhaps even a lighting expert. This person adjusts lighting instruments and wields reflectors for a constant level of lighting quality from shot to shot.

The director can concentrate on what's *in* the shot without being concerned about unwanted shadows or reflections.

The crew can grow to any number. You can add assistant director/production manager, makeup person, assistant gaffer, focus puller, grips, etc.

The more bodies running around on the set, the more important good communication becomes. Your best bet is to fill key positions with the most competent people and meet often before the shoot.

Be sure the crew understands what you're trying to do. Most important, make sure every crewmember has a specific job. If 20 of your friends show up to "help," you may be in big trouble. Each may spend the day asking, "What can I do, now? "—almost immediately annoying.

If they want to help, be sure they're serious. Video production is more work than most people realize. You don't want key crewmembers disappearing when the going gets tough.

Choose crew members according to how well you work with them. On a long and difficult shoot even the best of friends can end up at each other's throats. And only the best of friends will forgive each other the next day.

Keep It All Covered

During the shoot, it's a good idea to grow eyes in the back of your head and develop the ability to be in two (and possibly three) places at once.

As director, it's your responsibility to ensure that shots will cut together; that all shots called for in the script actually are shot; that talent not only has good line delivery, but that they don't change lines and alter the meaning. You have to keep an eye on background action for lingering curiosity seekers that can ruin your shot. And if a plate glass window reflects camera and crew, who's to blame? The director, of course.

Give yourself some options as you go along. You must plan out every shot and know exactly where it will go in the final program, but you should also get some extra shots during the shoot itself. Why? Because things slip through the cracks.

Say your sequence begins with the hero getting in a car, lighting a cigarette, driving a short distance and then exiting the car. During editing you notice the hero has the cigarette in his mouth only in the first shot. Not a huge continuity error, but the sequence is awkward.

If you've shot some cutaways of the street during production you can put a couple in. Then when you come back to the hero, no one will notice the cigarette has gone up in smoke.

If you're shooting a documentary and can't know in advance what shots you'll include in the final show, you'll need a lot of cutaways.

Say the mayor holds up a sheet of paper with a proclamation on it. He talks. You keep the camcorder on him. When he's finished, you have an assistant hold up the document for a close-up. In editing, you put the close-up in as if it happened in real time. Watch for opportunities like these.

Planning: The Sequel

After the shoot and before you cut the show together there's yet another planning stage.

Look through all your footage and make notes. Know where to find all the shots you need. This means jotting down what reel the shot is on and where it is on that reel.

Consider transitions (dissolves, fades), music, sound effects, titles—every aspect of your show. Create an edit decision list, made up of all the shots in order with their start (and sometimes stop) times, for the person doing the edits. You don't want to search for shots in an edit session.

You might even want to do an off-line edit. This is where you cut your footage together to see how it's going to look, but without any special effects, titles or music. Also called a rough cut, this step is especially helpful in spotting potential problems and eliminating them before taking your show into the on-line suite.

The Cutting-Room Floor

The actual edit can be the most enjoyable part of a production. As you put the shots down on the master, you watch your vision come together and form a cohesive whole. If you're pushing the buttons, the process should be smooth. If someone else does your edits, you'll need to communicate.

Try to meet with the editor before the session. Tell him what you want to achieve. If he's a professional, he'll have suggestions for improving your show. He may encourage you to avoid effects or sequences he knows from experience have little chance of working, or that will take too much time to achieve. Listen to the advice, use your own judgment and your edit should be enjoyable.

We've Only Scratched the Surface

Of course, there is a lot more to directing than all this.

Background and special areas of interest also shape a director's style. If the director is a former actor he may focus on performances, entrusting technical areas to his crew. If the director came up the ranks on the production end, he may be active in the show's every aspect.

As long as the director has a clear idea of where he's going and can convey this vision to cast and crew, chances of success are good. Whatever the outcome, good or bad, it reflects the director's effort.

Directing a complex video may be the biggest challenge you ever face. It may also be the most satisfying creative project you'll ever take part in. You won't know until you yell, "Action!"

34
Sets, Lies and Videotape

Jim Stinson

When you think of movie sets you probably imagine structures built from scratch on Hollywood sound stages—structures that need too much big-time money, skill and space for most video graphers.

True, full-scale sound stage sets can cost thousands or even millions of dollars, but professional videographers routinely "build" effective video sets that cost little or nothing in under an hour or even a very few minutes.

But wait, you say. I shoot weddings (or family events or vacations or training guides or business conferences), so I don't use sets.

Oh, yes you do. If you move a distracting picture, fill a dull corner with a ficus plant, pull a desk away from a wall to remove a shadow—in short, if you change the shooting environment to improve the look of your video, you're creating a set, however modest in scope. And if you know the tricks of the trade, you can make more elaborate sets—whole new environments, in fact—with a minimum of time, effort and

expense. This chapter explores those tricks of the trade.

To show you what we mean by a "set," let's start with a quick review of the various types.

Types of Set

We begin, of course, with the classic built-from-scratch sets you find on a Hollywood sound stage. I don't recommend that you construct one of these, even if you do have some place to put it. Real sets are heavy, expensive and time-consuming to build. Interiors are generally finished in actual drywall (though it's much thinner than normal), complete with taped and "mudded" seams and full paint jobs. Windows, doors and moldings are real.

In short, a set requires all the building and finishing skills you need for a real room. A more practical alternative: a sort of location set—a part of an actual environment that you can customize for your video.

For instance, I once needed four offices for an industrial video I was directing. We couldn't work in the maze of cubicles where the client's actual personnel toiled without disrupting operations, so we commandeered a large conference room. Once emptied of furniture, what we had was a 25-foot square room with wood paneling on one wall, draped windows opposite it and neutral-toned painted walls at either end.

Perfect! By arranging furniture in one corner, I made an office with windows to the left and a painted wall at the rear. The opposite corner yielded the painted wall on the left (with different artwork) and the fancy paneling to the right. Upscale furnishings turned this corner into an executive office. The third office set used only the full expanse of the painted wall, again dressed with different artwork and a potted plant.

And the fourth? We obtained two panels of the modular office dividers used in the client's actual offices and set them up at right angles inside the conference room. By propping them up on wooden boxes (called "apple boxes"), we raised them high enough so that the camera could not see over their tops and reveal where they actually were. And since we videotaped the actors seated at their desks, the camera did not show that the divider tops were abnormally high.

Improving Reality

More often, you don't have to create a set from bare walls; you just want to customize an actual location to fit your needs.

A desk is a desk, after all. Put a "GO, PANTHERS!" poster on the wall behind it and you have a principal's office. A couple of medical school diplomas and it's a doctor's office. A performance chart for Acme Industries and it's an executive office.

The relatively low resolution of video makes improving reality all the easier. Countless sets have been constructed or enhanced with nothing but pieces of foam, glue and spray paint. A cheap poster hung behind a window in the set wall becomes a beautiful view. Some TV news studios dot their sets with photocopies of equipment to look more impressive. Our saving grace is that the video camera is not that discerning, particularly in the consumer realm, and especially from a distance.

Often you create a set by subtracting items as well as adding them; that is, you frame off unwanted elements so that they never appear on screen. Once, for example, I needed to shoot into the corner of a room with a door in one wall. The trouble was, the adjacent wall was actually an archway into a much larger room. But I had three feet of solid wall before the arch began, so I made sure that my camera never panned quite far enough left to reveal it. As a result, the background "became" two full walls of a small room.

Major Principles

This example demonstrates an important principle of set making: *if it's not in the frame, it doesn't exist.* Remember that the sets you build or the locations you use need only be finished as far as the camera will reveal.

The flip side of this principle is: *if you can't see it, you can't tell that it's not really there.* I once shot a scene in what seemed like a large and busy restaurant. It was actually a wall with a door in it—plus just three tables, three actors and video crewmembers as extras.

Imagine the door screen on the right and the three tables to the left of it, along the wall. We put upscale paintings on the wall, added the tables with snowy white linen and bentwood chairs, lit the area to simulate intimate down-lights and installed the crewmembers as patrons.

Action: over-the-shoulder of the maitre d' as two diners come in the door. He greets them, picks up two menus (from a tacky orange crate beside him but out of sight of the camera), gestures left and escorts them to the far left table as we dolly past the other "diners" to hold them. Add ambient restaurant sound in the editing process, and voila! a blank wall becomes a complete restaurant populated by scores of staff and customers.

As long as we're talking about principles, here's another biggie: *the world on the screen has no fixed geography.* For example, the two sides of a door are not always the same door. I taped my diners as they walked down an actual street, turned and entered an actual restaurant. In editing, I then cut from that exterior to the shot of the diners entering the restaurant set. By matching the action of the door opening—actually two different doors—I "attached" the set to the real exterior, which, in fact, was 10 miles away. In addition, the actual restaurant exterior helped sell the reality of the set.

For another video I zoomed in on a window high up on a great glass skyscraper. Then, in my location office set, I placed a pane of glass in front of the camera, deliberately lit to show light reflections. A piece of fabric hanging behind the glass on one side simulated office drapes.

Then my actor stood behind the glass, staring pensively down at the "city" below, then turned and walked away, revealing the office behind her.

When making a cut from the exterior zoom shot to this shot, I effectively placed the office inside the skyscraper.

Making Do With Less

Sometimes you can get away with a very small shooting area. In the video with the four offices, an actor in one of them makes a phone call. I placed the actor who answered that call in a swivel chair in front of a painted wall on which I'd hung a framed print. This wall was in a little-used office corridor.

I then framed loose head-and-shoulders profile shot of the actor as he answered the phone. There were no desks, no cabinets, no other decorations; the actor had to hold the telephone base on his unseen knees. But by rocking and swiveling gently as he talked, he sold the idea that he was sitting in a complete office.

You tend to think of a video set as falling behind the actors, but it can be very useful to place it in front of them instead. I once needed a shot in the stacks of a library, but it was impractical to create long rows of six-foot-high shelves full of books. Instead, I placed a three-by-four-foot open-backed bookcase on a table five feet out from a blank wall and filled most of it with books. The spines of the books faced the wall.

Next I positioned the camera on the far side of the bookshelves so that it faced the books, which filled the entire frame. Strategic gaps in the ranks of books let the audience see parts of the wall beyond.

The actor walked on-screen between the wall and the book case, paused with her face framed in one of the gaps between books, selected a book and walked out of the "library stacks." For that scene, three-by-four feet of set was all I built.

If you can, though, choose a room that gives your set the most "throw." Throw is the distance from your camera to your subject, and it's crucial for achieving certain visual effects. If you want the compressed, flat look of a long zoom setting, you need to get your camcorder back from your subject. A 10-by-10 room won't allow you this option.

Remember that you'll never be able to make a wide shot in a small room look like a wide shot in a large room, but it's easy to make a large room look small.

Moving Outdoors

We tend to think of sets as indoor critters, but you can make them outdoors as well.

Just remember the two key principles: if it's not in the frame it doesn't exist, and in the world on the screen there is no fixed geography.

To illustrate the first principle, I once had to show a vigorous senior citizen playing golf. The trouble was, we had to videotape her at her home, far from any golf course. Here's the sequence of shots I made on her front lawn, to solve the problem:

1. HIGH-ANGLE CLOSE-UP: a patch of cropped grass. A woman's hand enters the shot, deposits a golf ball on a tee and exits. CUT TO:

2. WORM'S EYE ANGLE (sky and tree branches fill the screen): the woman wearing a golf outfit, eye shades and golfing gloves, enters the shot and addresses the camera with a club, as if it were the ball on its tee. CUT TO:

3. CLOSE-UP (an out-of-focus hedge fills the background): The woman raises her look from the ball to the distant, off-camera green. She swings. CUT TO:

4. HIGH-ANGLE CLOSE-UP: the ball being whacked solidly out of the shot. CUT TO:

5. CLOSE-UP: The woman follows the ball (off-camera) as it sails (presumably) toward the green. She smiles at the result and walks out of frame.

See? Because the telltale details of her residential neighborhood were never in the frame, they didn't exist.

I warped screen geography to solve a problem in another program, in which a passing citizen comes upon a bank robbery.

As the bad guys rush out of the bank, the passerby looks around, spots a phone booth, runs to it and calls the police.

Just one hitch: there was no phone booth anywhere near the bank exterior. To solve the problem, I put the citizen in the foreground, watching the perps in the background.

He looks off-screen right, sees something and rushes out, screen right.

Pause, while the felons clear out of the shot; then the citizen runs back in from screen right. An hour later and three miles away, I centered a phone booth in full shot. The citizen rushes in from screen left, enters the booth and makes his call. Then he runs back out the way he entered, screen left.

In editing I inserted the phone booth sequence into the bank shot, between the point at which the man leaves and the later point at which he returns. The effect: to move the phone booth next door to the bank.

The Frankly Artificial Set

So far, we've talked about sets that work hard to convince you that they're actually offices or restaurants or golf courses. But other sets are just, well, sets. News, game show and talk show sets are examples of video environments that make no attempt to disguise themselves as something else.

Like other studio sets, these are probably too elaborate for modest productions. (Imagine trying to build the set for *Jeopardy!*) But you can use two simple alternatives to create "frankly studio" sets.

Your first option is a seamless paper backing. Available on rolls from photo supply houses, seamless paper comes in a wide range of sizes, colors and patterns.

Simply hang the roll high in the air and pull down as much as you need to create a "wall" behind your performers. If you select a chromakey blue color seamless, you can use properly equipped special effects generators or computer editing systems to replace the blue color with matted-in images (like the TV weather reporter's maps).

The simplest set of all is no set whatever. Instead, you create your set by lighting the actors and furniture in the foreground and leaving the background dark.

This works best when 1) you can exclude ambient light from the shooting area and 2) your camera permits manual exposure control.

For simplicity, we've discussed ideas for sets one at a time, but you can combine several of them.

For instance, one of my video students wanted to insert an actor into an actual event. Happening on a serious auto accident on a busy boulevard at night, he'd taped it with the Hi8 camcorder he carries everywhere.

It was perfect for the public service video he was making on safe driving. Now the trick was to place his on-camera narrator at the accident scene.

Here's how he did it. He took the actor to another busy boulevard, where the ambient background sound was similar, but shooting was safer because of wide sidewalks. He placed the actor under a streetlight so that the camcorder's auto exposure system would reduce the background to darkness.

Then he positioned an assistant just out of frame, holding a battery-powered revolving light. As the actor read his lines in medium shot (down to the waist), the revolving light splashed his face with an intermittent red glare. Intercut with shots of the real accident, the narrator seemed to be standing beside a police car at the scene.

In faking his narrator into the scene, the director used the following principles:

1) If it's not in the frame it doesn't exist (the technician holding the light),

2) if the videographer says it exists outside the frame, then it does exist (the police car) and

3) screen geography is not fixed (combining the accident and narrator locations).

If you keep these fundamental principles in mind as you plan and shoot your programs, you can put together almost any set you like.

35
Looking Good: Makeup and Clothing Tips for Video

David Welton

See video clips at www.videomaker.com/handbook.

Bright lights glaring, camera in tight, tape rolling. . . cue talent.

Oops... what's that glare where your subject should be? This is no time to realize the talent's forehead reflects light like a beacon, or the color of the talent's clothing clashes with the upholstery.

Makeup and wardrobe are often overlooked in pre-production planning. You ignore them at your peril. Even the simplest production suffers when makeup and wardrobe concerns are not properly addressed. The CEO loses credibility when he perspires under the studio lights; so does the diesel mechanic dressed in a pale blue cardigan sweater and white slacks.

Master the basics of wardrobe and makeup, and your videos will improve dramatically.

Facing Up to Makeup

When you think about someone, what do you remember most? Most people remember the face. Like it or not, it's the face that serves as the basis of first impressions and lasting memories.

When you listen to someone, either on TV or in person, you naturally direct your attention to the speaker's face. That's because the face plays a significant role in non-verbal communication. It stands to reason that the face deserves careful aesthetic consideration, especially for videographers. As the center of attention, it's the face that requires careful makeup.

There are three basic ways to use facial makeup: to improve appearance, to correct appearance and to change appearance.

The everyday makeup applied by many women and an increasing number of men improves the appearance of the face. With the help of makeup, you can cover up minor skin blemishes, accentuate the eyes and even emphasize preferred features.

Makeup can also perform minor miracles, visually correcting small facial imperfections such as circles under the eyes, sagging chins and misshapen noses.

Sometimes you will need to completely change the appearance of talent—say, an actor portraying a character. These transformations may include changes in age, race, gender or even species.

With makeup, you'll find it fairly simple to improve appearance, a bit harder to correct appearance and quite complicated and time-consuming to completely change appearance. Fortunately, most videographers don't often need to transform a 20-year-old actress into an 80-year-old grandmother—a job best left to the professional cosmetologists.

But all videographers need to know the basics.

You start with skin tone. If you have any doubt how important skin tone can be, just think of a poorly adjusted color television set where tint, color saturation, brightness and contrast levels appear as if set at random, and skin tones turn green or red or purple.

Since most viewers don't have access to a color bar generator to calibrate their TV, they resort to skin tones as the reference point for color and tint adjustments. Makeup allows you to ensure that your talent's skin tones appear natural on your video.

Begin with foundation, which evens out skin tone and serves as the base for makeup to follow. Choose the shade of the foundation carefully, matching the natural skin tone of the talent as closely as possible.

A general rule of thumb: warm colors are best for television makeup. Avoid cooler colors; since television lighting already accentuates these colors they often appear exaggerated on camera.

Consider the effects of the lighting on the set where you shoot. If you apply makeup to talent off the set, you should do so under lighting similar in color temperature. Why? Because makeup applied under fluorescent lighting looks much different under studio lighting; the change in the color temperature of the light source alters the appearance of skin tones. Always use a color monitor to check your talent; what looks good to the eye might prove unacceptable on video.

Fair warning: it's possible to go too far with makeup. The painted face of the clown is appropriate for the circus, not for the up-close world of television, unless of course you make circus videos.

Remember the difference between makeup techniques for the stage and those for video. Stage makeup dramatically highlights facial features, so you can see them from the last row in the theater.

For video, you want everything to look good in close-up. The lights tend to wash people out making them corpse-like. Properly applied makeup brings them back to life.

These makeup differences present a dilemma for videographers taping theatrical performances. Should you design the makeup for the theatergoers or the TV viewers?

The best answer: compromise.

For most video, however, the right makeup is transparent, undetectable to the camera. The idea is to make people look natural. As a viewer you shouldn't even know it's there, but rest assured, it's there. The only time a viewer should notice makeup is when it's purposely applied poorly.

Figure 35.1 *The daily makeup used by most women is generally fine for video.*

Making Up the Talent

Women are usually adept at applying their own makeup (see Figure 35.1), but not all of your subjects will turn their faces over to you for makeup without a fight. Men may be particularly difficult to convince, especially if they're not professional talent.

If, after looking at your male talent on a monitor, you decide makeup is necessary (see Figure 35.2), be tactful. Suggest the need for makeup but don't appeal to his vanity. Rather, appeal to his desire to perform well on camera. Explain the need for makeup in technical terms; say that cosmetics will overcome some of the shortcomings of the video medium, thereby making him appear more natural.

Most large drug stores can supply you with the materials needed for basic makeup techniques. Theatrical supply retailers offer a more complete line of professional products.

With a little practice, your talent can apply their own simple makeup. Don't forget to direct them to apply makeup on hands, arms and other areas exposed to the camera's scrutiny.

Hair also plays a significant role in the on-camera appearance of talent. Both the style of the hair and overall grooming must complement the performer's "look."

Figure 35.2 *Men will usually need help with makeup, at least at first.*

Usually the best strategy is to choose one hairstyle and stick with it. Make sure your talent maintains this same hairstyle throughout the shoot: if your talent's hair was impeccable in the morning, but tousled after a windy break outside, fix it. You don't want such an unexplainable change to befuddle your viewers.

It's true that not every video situation requires makeup. Still, a little attention to makeup can make the difference between amateurish and top-notch professional work, especially in close-ups.

Clothes Make the Video

The clothing worn by talent is as important as their makeup. In big-budget productions, a fully staffed wardrobe department provides the clothing, or costumes. A wardrobe person remains on the set during the shoot attending to every clothing detail.

Corporate or industrial productions can seldom afford such luxuries. For these productions, the talent's clothing usually comes from the talent's personal clothes closet. Here are a few tips to help your talent make better clothing choices.

Remember, this is a two-dimensional medium. Often this means the television camera doesn't flatter the human form; TV has a tendency to put a few extra pounds on a performer. Avoid needlessly baggy clothing since this only compounds the problem. Horizontal striped clothing makes people look broader and shorter; vertical stripes make people look thinner and taller. Be sure to use this phenomenon to your benefit, not detriment.

Know your colors. TV is a color medium, but a very biased one. Certain colors don't fare well when processed through television; you must be careful to avoid overly bright, highly saturated colors like reds, oranges and yellows. You'll achieve better results with earth tones, especially browns, greens and blues. Try using clothing with pastel colors.

Be careful of contrast. The television signal has a difficult time accurately reproducing large differences in contrast. Avoid clothing that combines both very bright and very dark colors. It's possible to wear relatively light or dark clothing if the material isn't reflective, as long as you don't wear both extremes at the same time. A good rule of thumb: stick to colors near the middle of the brightness spectrum. These colors also have less effect on skin tones.

No tiny patterns, please. Avoid tight or close patterns like small checks, herringbones and narrow stripes. These patterns strain video's ability to reproduce them accurately. The result: an unintended *moiré,* effect, a vibrating rainbow of colors. This effect distracts viewers. Wide stripes and other large patterns work well on camera, as long as they're not overpowering or too high in contrast.

Consider the set. Choose clothes that work well with the setting; they shouldn't blend into, or clash with, the background or the furnishings.

Keep it simple. For most non-dramatic productions, the talent's clothing should appear attractive and stylish, but not too conspicuous or flashy. Remember, the goal is to focus the viewer on the production's message, not the talent's wardrobe. Accessories such as shiny, dangling earrings that dance with light under studio illumination are an example of something to avoid. Remember, too, that today's fashion is tomorrow's *faux pas.* The trendier you dress your talent, the more dated your show will appear in a few years.

Dramatic productions may demand period costumes—from turn-of-the-century pioneer wear to space-suited aliens. For these special needs, try theatrical costume shops. Don't worry if there's not one nearby; many will ship apparel right to your door.

If budget constraints preclude professional costumes, don't despair; other options do exist. Check your local fabric store for clothing patterns; many pattern books include designs for costumes. With a little help from someone handy with a sewing machine, you can obtain first rate costumes on a budget.

Don't know a seamstress? Ask the sales clerks at the fabric shop; chances are they'll direct you to someone who can help. Don't forget to try thrift shops and stores specializing in Halloween garb or other sources of costume help, like high school and college drama departments, or local theater groups. But remember this when choosing theater wardrobe items for television work: the construction of outfits—especially the detail—designed for stage often proves too coarse for the close-up scrutiny of video.

Clothing Hazard

Certain video situations present their own wardrobe challenges to the videographer. Using a chromakey background complicates wardrobe selections, especially when it comes to color. Chroma keying is an electronic process that replaces a specific color in a video image with an entirely different image. The most common example of chroma keying is the weather set of a TV news program.

On TV you "see" the meteorologist point to Cleveland on the weather map, but in reality he's pointing at a wall painted an even color of blue or green. The map is electronically superimposed over the colored wall.

What happens if the forecaster wears a blazer the same color as the background? It appears as if you can see right through his body; a map of Ohio replaces his torso. A nice special effect for a science fiction flick, but not particularly apt for a newscast.

Hats can present another big challenge. The brim of a hat throws a shadow that can easily obscure the face of the talent. Here, the best solution is to reposition the light to overcome the shadow.

Eyeglasses can also be a videography hazard. Studio lights can bounce off eyeglasses right into the camera—not a pretty picture. If your talent needs corrective eyewear to read cue cards or a teleprompter, there's not much you can do. Of course you can always suggest they memorize their lines.

What if your script calls for the use of eyewear? In this situation, the use of glasses is a deliberate wardrobe choice, not an optometric necessity. Maybe your talent plays the role of a scholarly professor; having her wear eyeglasses will increase her credibility. So have her wear the glasses at the beginning of the scene to establish the "scholarly look." As soon as possible, direct her to take off the glasses. You can also try using eyeglass frames without the lenses.

Sometimes everyday attire isn't the best choice for on-camera credibility. Even if a male scientist usually wears a suit and tie, most audiences will perceive him as more believable if he appears on camera dressed in a traditional lab coat. This positive use of stereotypes helps viewers quickly identify the roles performers play.

If you repeatedly shoot the same types of scenes, keep some clothing items on hand. The harried groom might forget his cummerbund or bow tie; you could save the day if you've got some available.

Remember, the best time to consider clothing choices is in pre-production. Before the shoot, take a long critical look at the entire set for any mismatched colors, patterns or other visual chaos.

Coordinate the color of the background, furnishings, clothing and props during pre-production meetings. Even makeup should complement all other elements of the scene.

A little planning here can save a lot of headaches later. For training productions, be sure to acquire the appropriate dress well in advance.

The telephone installer should appear in the official company uniform and the firefighter in protective garb. Use any specialized gear that's required for the job, like goggles, gloves, boots or breathing masks—as long as they don't obscure the talent completely from view.

Dress for Success

Remember those debates between presidential candidates Richard Nixon and John Kennedy back in 1960? This first "marriage" of political debate and television provides one of the best lessons in wardrobe and makeup.

Nixon refused to concern himself with matters of makeup and clothing. The result: his rumpled suit blended into the background of the studio scenery, giving him a disconcerting "bodiless" look, and his face displayed a lifeless complexion.

Kennedy, on the other hand, used the medium effectively by paying special attention to his makeup and wardrobe.

The rest, as they say, is video history. Make the right clothing and makeup choices on your next production, and you, too, may change the course of history.

Typical Makeup Kit Contents

- Makeup brushes
- Sea sponges (for water-based makeup), synthetic sponges (for grease-based makeup) and small "stipple" sponges (for detail work)
- Powder puff
- Water bowl

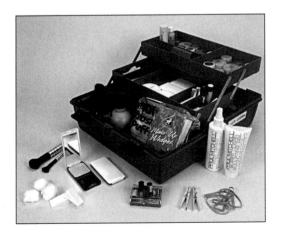

Figure 35.3 Contents of a makeup kit.

- Cotton balls
- Assorted shades of makeup base and mascara
- Assorted shades of eye shadow
- Assorted shades of lipstick
- Assorted shades of powder
- Pre-moistened towelettes
- Cold cream
- Tissues

Continuity Secrets

Videographers always worry about continuity; much of this worry concerns wardrobe and makeup. The best tips on maintaining continuity come from the trenches—professionals who have spent time on both sides of the camera.

Enter actor Jason Ricketts, a veteran of stage, corporate video, national TV commercials, network television and Hollywood feature films. You may have seen the blond, blue-eyed Ricketts on *LA Law, Simon and Simon,* the film *Innerspace* or the Quicken™ software commercials.

Ricketts learned first hand about the importance of scene-to-scene continuity when he worked as a photo double for Jameson Parker. Parker played A.J. Simon, the younger brother on the TV show Simon and Simon. The script called for Ricketts to do "hand inserts." Ricketts' hand, shot in place of Parker's, appeared during the final production, thanks to the editing process.

After examining Ricketts' arm, Parker requested his double's arm hair be thinned to more closely resemble his own. Parker was also justifiably vigilant about the manicure of Ricketts' nails.

Parker helped Ricketts understand how important it is to pay attention to continuity—the art of looking the same from one shot to the next.

"Even on a one-day shoot," says Ricketts, "the talent and crew need to be very aware of their appearance in terms of makeup and wardrobe. It's best to write down specifics before breaking for lunch or changing scenes."

Ask Ricketts to name a few of those specifics and he rattles off: "How many buttons are fastened on a shirt? Are pant legs rolled up? Was the talent wearing a belt, jewelry? What does the hair look like? How was the tie knotted? What about wrist watches, glasses and makeup?"

According to Ricketts, continuity is the mark of the professional.

"If you have a handle on continuity," insists Ricketts, "it just gives that much more professionalism to your entire production."

No one has a better handle on continuity than advertisers. Big budget TV advertising productions hold "wardrobe parades" where the talent models various outfits for the client, the director and ad agency representatives. During these pre-production sessions, everyone comes to an agreement on the same palette of colors.

"Everyone has a pretty good idea before going into the shoot what the wardrobe is going to look like," Ricketts says. "It's pretty organized. Because they often spend millions of dollars on these commercials, the producers want to reduce their surprises. They control everything they can."

But even the pros make adjustments on the day of the shoot.

"Sometimes on paper, or in theory, everything should look great—furniture, background, wardrobe, makeup," says Ricketts. "Usually everyone's on the same page, but sometimes fine tuning is necessary once all the elements of the scene come together for the shoot."

Ricketts recalls a recent experience on the set of a television movie involving his wardrobe and splattered salsa.

"The caterer brought Mexican food," remembers Ricketts. "I was putting some salsa on my enchilada when the sauce spilled onto my pants. I grabbed some soda water and a napkin and started scrubbing away. The stain came up and the shoot went on as scheduled."

This scenario is typical of the low budget production. "A big budget production would have several copies of the wardrobe on hand," notes Ricketts, "just in case there's a problem."

Other precautions also characterize the big budget productions.

"During the making of a major commercial," says Ricketts, "makeup artists remain on the set for the entire shoot. They hover over the talent, checking between each take for the slightest imperfections, paying particular attention when a closeup is about to be shot. As the budget decreases, one of the first things to go is the wardrobe, hair and makeup people."

Ricketts advises videographers to keep in mind makeup's most simplistic but most important function—to eliminate shine.

"You never want to look shiny," says Ricketts, who describes a grueling 20-hour shoot with a makeup person budgeted for only a half day. "We all looked great in the morning, but long after the makeup person left I knew we were getting shiny from perspiration under the studio lights. For continuity's sake, we touched each other up."

Whether it's on the set of *LA Law* or a simple training video for the local Red Cross, careful attention to matters of wardrobe and makeup provide a big pay-off a better, more professional production. Ricketts' sage advice has a ring of irony: "Do whatever you have to do to make it look like you've done nothing at all. The goal is to look *au-naturel*."

36
Time-Lapse Videography

Tim Cowan

If you've ever watched a scene on television or in the movies where clouds race overhead at a supernaturally fast rate, or where a flower appears to blossom in seconds, then you've witnessed a cinematic technique known as *time-lapse*. Often used to compress the apparent passage of time on film or video, time lapse consists of recording a few frames of film or tape, pausing for a certain amount of time, then recording a few more frames, and so on. When the sequence is played back, it appears that the action recorded is happening much faster than it does in real life.

In many ways, time lapse is similar to animation, using the technique of taking a succession of very brief shots to create the illusion of a steadily moving sequence. However, instead of fooling your mind's eye into thinking that a collection of still images are moving, time lapse compresses a lengthy event into a short span of time.

If you've ever wondered how you can perform this feat with your camcorder, wonder no more; this chapter will explain the process in detail, from setting up the shot to using the final footage in your existing video productions.

Before You Begin

There are several ways to achieve a time lapse effect with video, depending on how smooth you'd like the final product to be, what equipment you have and how much time and effort you're willing to put into it. However you choose to do it, there are a couple of constants you should be aware of.

For smooth recording, be sure your camcorder doesn't move at all during the sequence you're shooting. Otherwise, your footage will come out looking jumbled and unwatchable. This is best done by mounting your camcorder securely on your tripod and tightening the pan and tilt screws to make sure the camcorder doesn't move (known as a "locked down" tripod).

It's a good idea to switch the microphone off while recording a time lapse or animation sequence. If you don't, you'll end up with a quick snatch of many random noises on your soundtrack. Disabling the mike can be done fairly simply if your camcorder has an external microphone jack; simply stick an adapter that's not connected to anything into the socket and your camcorder mike will automatically switch itself off.

It's probably wise to switch over to manual focus while shooting a time-lapse or animation sequence. If you don't, your camcorder's autofocus will "hunt" for what it thinks to be the prominent feature in the shot to focus in on. If anything within range of the auto focus momentarily becomes more prominent than what you intended to shoot (i.e., a bird flying past while you're shooting the clouds), it could spoil the entire sequence.

If your camcorder can manually override its automatic light meter, you'll probably want to switch over to that as well. First make sure the manual level it's set on leaves enough light for your shot but not so much that it violently overexposes. It's also a good idea to make sure, before you even start shooting; that the light you're shooting with highlights what you're recording. After all, you wouldn't want your time-lapse sequence of a flower growing to be shot in silhouette, would you?

Most importantly, keep a close watch on your camcorder. If you're shooting out on the street, there's always the risk your camcorder could be stolen or damaged by someone passing by. Even if you're in your own backyard, there's a chance your camcorder could get jostled by a child or pet, or simply pitch over lens-first due to uneven weight distribution.

Calculating Time Lapse

Figuring out the length of time needed for either a time-lapse or animation sequence is pretty simple, once you know what you and your equipment are capable of. Since NTSC video always plays at 30 frames per second (fps), a single time lapse/animation shot of, say, 1/5th of a second will go for six frames. Five of those shots will equal thirty frames—one second of your completed video.

Let's say you want a 30-second time lapse sequence of clouds going by, and you've decided that one shot every half-minute will give you the desired effect. Assuming the five shots a second mentioned above, that means that 5 (shots per second) times .5 (minutes of real time) per shot times 30 (seconds of video time for the sequence being shot) equals 75 minutes of shooting time.

Being able to calculate the amount of shooting time needed has a number of uses. If you're shooting outdoors with battery power, for instance, it's vital to know just how much time you'll need to get the sequence you want, or if it's even possible given your batteries. Even if you don't need to calculate for battery time, it's useful to know just how long it will take in real time to get the sequence you want.

Shots Per Second

You may have noticed that I've gone to some trouble to distinguish between "frames per second" and "shots per second" when talking about time-lapse and animation videography. That's because, as I said above, NTSC video always plays at 30 frames per second, no matter what images you may have playing during that time. Ideally, the closer you can come to 30 shots per second the better, since your results will look a lot less choppy.

While a second doesn't seem like much time, it plays havoc with the smoothness of a time-lapse or animation effect. This is because both video and film rely on something called "persistence of vision" to convince you that a long string of still pictures is moving. Since your mind can't assimilate a series of fast-moving still images, it assimilates them instead as one continuous moving picture. The instant this collection of images slows down to the point where you can mentally register each single picture as such, the illusion's blown.

If you're using time lapse simply to show the passage of time (for instance, a shot of people setting up before a play or concert), then one shot per second will probably work perfectly fine. On the other hand, if you're hoping to do a sequence where a smooth flowing of the shots is essential (as in an animated sequence) then you're probably not going to be satisfied with anything less than 10 shots per second, and 15 or even 30 shots per second would be even better.

In-Camera Time Lapse

The easiest way to perform time lapse or animation is by using a camcorder with an interval timer function built in. Using these camcorders is very easy—you simply mount them on a locked down tripod, switch the interval timer function on and the camcorder does the rest. Depending on which model camcorder you have, interval timers can record as little as 1/10th of a second (3 frames) or as much as a full second of video each time.

If your camcorder doesn't have an interval timer feature, you can still do a reasonably effective in-camera time lapse or animation sequence—provided your camcorder has flying erase heads (most models manufactured during the last five years do) and you've got the patience for it. Set up your camcorder on a tripod, engage and then immediately disengage the record feature, wait for the amount of time you want between shots, and then repeat the process until you have the sequence length you want. To ensure that the camcorder doesn't shake when you hit record, you might want to use your camcorder's remote if it has one.

Depending on how well your camcorder responds to your pushing the

record button twice in rapid succession, and how fast you are on the trigger, you could get individual shots of as little as 10 frames (3 shots per second). While not as smooth as 30 or even 15 shots per second, it's still surprisingly effective.

One last technique that may or may not work for you is to utilize your camcorder's edit search feature if it has one. What you do is record a couple seconds of video, utilize edit search to get to the beginning few frames of the last shot, and then repeat until you've gotten what you want. Keep in mind, though, that this is an extremely labor-intensive process, and difficult to do if your camcorder's edit search function can't jog search.

Post-Production Time Lapse

If you want a bit more control over what your final product looks like, you might choose not to record time lapse in real time; instead, you can attempt it in post-production instead. The advantage to this is that you can experiment with the time between shots to find what works best far more easily than you can "out in the field." The disadvantage is that for a decent sequence, this requires a great deal of time and some fairly high-end editing equipment.

First, you'll need some footage that you can edit your time-lapse sequence from. Place your camcorder on a locked down tripod, just as you'd do if you were shooting an in-camera time- lapse sequence, and roll tape normally for as long as you've calculated you'll need for your sequence.

Remember the technique utilizing edit search on camcorders mentioned above? Well, it also works if you're using a VCR with edit search; jog shuttle and flying erase heads as a record deck. What you do is record a couple of seconds of video, engage the edit search, then utilize the jog wheel to creep back to a few frames right after the first frame of the last shot. It'll require a little practice to figure out just how far you'll need to advance the tape to leave just one frame, and a great deal of patience, but you can get 30 shots per second time lapse or animation this way.

If you're using an edit controller to run your playback and record decks, how close you can get your cuts to each other will vary depending on the quality of your setup. Controllers that utilize LANC or Panasonic 5-pin edit control protocols for playback control and infrared remote for recorder control probably won't be able to get much better than one or two shots per second. This is due to these edit controllers' inability to respond quickly enough for the extremely short cut-ins and cut-outs that time lapse and animation require.

Controllers that use LANC or Panasonic 5-pin protocols to control both player and recorder can get much closer—10 shots per second, if your playback and record decks are calibrated properly. By contrast, a professional editing setup utilizing time code and frame-accurate decks can give you true 30-shot per second accuracy—if you're willing to pay for them.

So What Are You Gonna Do With It?

Okay, I've told you a couple ways to do time-lapse or animation recording, and I've even covered why the more shots you record per second, the better it'll look. But what's it really good for?

I've already mentioned the most popular use of time lapse, compressing a lengthy event like clouds rolling by or people setting something up. So what else can you do with this technique?

Stop Motion Animation. This is the sort of animated sequence where an object that doesn't normally move, like clay or an action figure, appears to be in motion.

You've seen this a lot on television and in the movies, like those "California Raisins" commercials where clay models of raisins dance around and sing, or the original *King Kong*, where Kong was an articulated doll that appeared to be a fully moving giant gorilla.

To perform stop motion animation, you'll first need to create a space where you can set up your lights, camcorder, tripod and whatever you're going to animate. The advantage to this is that, once you've gotten everything set up, you can easily control the lighting and background so that all you need to worry about is animating your object.

Once you've gotten your object where you want it to start, turn on your camcorder and take your first shot. Remember that, for the animation to be smooth, you'll probably want to get at least 10 shots per second, and 15 or 30 would be even better. Then move your object slightly and take the next shot, and so on until you finish your sequence.

Keep in mind that for animation to look convincing, the audience has to get the impression that they're seeing the process of motion. For instance, if you're animating an action figure throwing a punch, you can't simply have your first shot with the figure's fist up, and the next with the punch fully thrown—you have to have the fist moving forward with each shot so it looks like it's getting there. To help you get some idea of how things move, it might be a good idea to watch a video of something you'd like to animate one frame at a time so you can see how it looks in real life before you try animating it.

Pixillation. A sort of variant version of both time lapse and stop-motion animation, pixillation is the process whereby a person seems to move in unusual ways. The music video *Sledgehammer* utilized pixillation so that Peter Gabriel appeared to slide around the room and even at one point seemed to be hovering above the ground.

To accomplish this, have your performer stand in one position while you take your first shot, then move forward slightly and stand still again while you take your next shot. If you continue this long enough, the person will appear to move across the room without walking. Make sure that whatever movements your performer makes from shot to shot aren't too radically different from each other, or the sequence won't look like a smooth movement.

Computerized Time Lapse

Performing time lapse or animation sequences utilizing a nonlinear editing system is surprisingly simple. Unlike videotape, digitized video can easily record one frame at a time. You simply lift the frame off your digitized video file, drop it onto the editing software's timeline, and look for the next frame you'd like to use. Once you're done, you save the results, turn your VCR on, and "print" your edited file to videotape.

Amazing as this nonlinear capability is, I think I'd better interject a few words of caution. First, the output quality of some older non-linear boards has often been described as "roughly VHS quality," which means that it isn't quite as good as VHS—you may notice some *artifacting*, which is what they call it when things that should be round seem to have little squares in them. Second, video files take up a lot of hard disk space-about 10-15 megabytes per minute of highly compressed video. Third, if you've got a lot of other boards or peripherals on your computer, there's a chance your computer will freeze up on you from time to time.

Of course, if you're one of the lucky ones with a miniDV camcorder, and a non-linear editor on your FireWire-equipped computer, you can create very high-quality time-lapse video.

You Could Always Cheat...

You might want to consider shooting a time lapse or animation sequence on an old Super-8 movie camera. Most Super-8 movie cameras have a single-frame option built in, which will guarantee a much smoother effect than you'd probably get with a camcorder or lower-end editing equipment. Some, like Minolta's better Super-8 cameras, even have an automatic interval timer.

To utilize a Super-8 movie camera for time-lapse photography, you once again set up your camera on a locked-down tripod, just like you would if you were using your camcorder. Take one frame using the single-frame option, wait for a specified period of time, and take another frame, continuing until you have enough footage for your needs. If your movie camera has the option of manual override of the automatic light meter, it's probably a good idea to utilize that so that your frames don't become darker or lighter abruptly. It's also a good idea to use Kodachrome 40 film instead of Ektachrome G, since the former is less grainy.

While it's harder than it used to be to process Super-8 movie film, Kodak can still do it. Your local camera store should have mailers for sale that cover the cost of processing and return postage in the purchase price, or check out Kodak's Web site. You simply stick the appropriate postage on the mailer, drop it in the mail, and wait for about two weeks. After you get the film back, you can either transfer it to video yourself if you still have your Super-8 movie projector and one of those film-to-tape transfer units, or you can have the job done professionally at your local camera or video store.

While this method requires both more money and a few more steps than any of the direct-to-video methods recommended in this article, it's worth considering if the effect is important to you; say, for a title sequence you intend to use more than once. It's obviously not to be used casually; still, you'll probably get smoother results this way.

All It Takes is Time

Once you've gotten the hang of time lapse and animation, you'll start to come up with a number of ideas on your own.

The great thing about these techniques is that they don't require a lot of expensive equipment that you wouldn't use for anything else—even high-end editing equipment, should you choose the post—production option, will be useful for all your videos. All time lapse or animation takes, really, is time—and the results will certainly be worth it.

37

Move Over, MTV:
A Guide to Making Music Videos

Norm Medoff

See video clips at www.videomaker.com/handbook.

On August 1, 1981, Music Television went on the air and changed the world of video forever. Now, years later, MTV—as Music Television is better known—has spawned several new music channels, and the making of music video has become an art form all its own.

On MTV networks, music videos are expensive Hollywood-style productions shot with multiple 35mm film cameras and featuring snazzy special effects, big name bands and exotic locations. But you, too, can produce great music videos—right in your own hometown.

All you need is a little money and a lot of creativity.

Keeping Costs Down

If you find it hard to believe that you can make a quality music video on a small budget, consider the following points.

Since you shoot music videos "film style," you can do it with only one camcorder. You just shoot many takes, which will provide different angles of the same action—as opposed to multi-camera shooting, which records several angles of the same action during one take.

Big-budget music videos often use fancy—read costly—transitions requiring digital video effects generators and expensive editors. But you don't have to use such transitions; you can produce an excellent music video using cuts to connect shots.

You don't have to pay a band to perform in your music video. No matter where you live, you should be able to find a local band interested in promotion. A music video gives a band valuable experience, both in making music videos and exposure to wider audiences. Your music video could also help the band of your choice secure concert dates or even a record contract.

Shooting live performances is not as difficult or expensive as you might think. Professionals employ a sophisticated and expensive camera dolly on a special track that allows the camera to move around performers; you can make do with an old wheelchair or shopping cart.

The good news is all you really need to make music video is a camcorder and the ability to perform insert edits. An insert edit allows you to add video to a sound track already on the tape. More on this later.

Getting Started

Before you do anything, determine how you'll exhibit your music video.

If you make it for private use only, you can select any music you want—no worrying about copyright and clearance. But if you plan to air your video on local cable or broadcast television, or even enter it in a contest, copyright becomes an issue. If you don't own the copyright on the music you use, you must obtain clearance.

This can be both time consuming and expensive. First, you have to contact the American Society of Composers, Authors and Publishers (ASCAP); the people there will direct you to the Harry Fox Agency in New York (212-370-5330). This agency will try to arrange clearance for you to use a particular song.

You can avoid this hassle altogether by using your own music or finding local talent to perform their own music. If you don't know any musicians, check out the local bars, which often feature up-and-coming bands and/or solo performers. These musicians can give you permission to use their music for your video.

Try to obtain recordings of potential performers. In most cities, you can buy CDs or cassette tapes featuring a sampling of local performers. Such recordings can help you choose both a group and a song for your music video.

HINT: when listening to the music, focus on the visual imagery that pops into your mind. Often the music you like does not conjure up any useful images. Don't be surprised if the song that inspires the best visual imagery is not the one you like most.

Pre-Production

Once you've selected your music and performers, you can begin the bulk of your work, pre-production. In this phase of the operation, you will do the following: devise a creative approach; write a treatment; make a storyboard; create a beat sheet; select a crew and record the song for lip sync purposes.

The creative approach. This is the fun part, where you generate the images for your video. There are three stylistic ways to proceed: illustrative, interpretive and performance. Illustrative video is the narrative approach that illustrates the story the song tells. Interpretive video uses the images you deem appropriate for the music, whether they relate to the lyrics or not. Performance video is just what the name implies—the musicians performing the song.

Many music videos combine all three styles of shots: illustrative, interpretive and performance. Resist the temptation to try one method and exclude the others. An all-performance video may rely too heavily on the musician's ability to be entertaining. An all-interpretive video may confuse viewers. An all-illustrative video may tax both the resources of your performers—can they really act out the story?—and your budget.

Some music dictates the style. For example, instrumentals tell no specific story, so the most effective approach is often an interpretive one where the images create a mood or evoke an emotional response. Country songs often tell a straightforward story, facilitating an illustrative approach. Rock music gives you the most options; many rock videos are all attitude—with little or no regard for viewer orientation

The treatment. Now that you've developed a creative approach, take the time to write a treatment or overview of your video. Simply describe in a few paragraphs what will happen in the video: the story, the characters, the mood. A treatment provides a convenient way of explaining the video to your performers and crew.

The storyboard. This step is crucial. The storyboard provides a visual record of the individual shots that will make up your video. Draw a series of rough sketches depicting the people, places and objects needed for these key shots. Put them in the order you intend to shoot them. Or use a Polaroid camera to snap the subjects and objects in the desired locations.

The beat sheet. Next, combine your storyboard with a beat sheet describing the music on paper. Match shots to appropriate sections of the music. Knowing what visuals you want and when you want them in your music video is the key to your pre-production planning. It's also the way the video professionals stay on target.

The crew. You may want to do all the work yourself. That's a lot of juggling: arranging lighting and special effects, directing talent and playing an audio version of the song while the performers lip sync. But you will be able to do a better job if you plan on a crew of at least two other people, one to act as a grip and audio assistant and one to shoot while you direct. If you don't know anybody, ask the band members if you can borrow their assistants or (known as *roadies*).

The recording. At this point, you should prepare a recording of the song you'll use as the sound track for your video. Not just any recording will do; you'll need an error-free version of the song on a safe and dependable recording medium. A CD is probably best, especially if you can take a CD boom box with you on location.

The next best sound source: a hi-fi version of the song—VHS hi-fi or Hi8 will do. This method creates some playback problems; you'll need an appropriate

video playback machine and sound system.

A reasonable compromise: a good copy of the song on audio cassette to play while on location through a good quality boom box or a portable sound system.

Quality is critical, because the performers must lip sync to this version of the song.

Be wary of cheap audio cassette players suffering from speed variations or muffled sound. If you use battery power, keep plenty of fresh batteries on hand.

Slow audio recordings can adversely affect the band's performance—not to mention cause horrendous editing problems later on.

Shoot It Right

Now it's time to shoot the raw footage for your music video. There are two ways to accomplish this task.

The in-camera shoot. If your camcorder has a flying erase head, you can insert the visuals onto a videotape that already has a copy of the song.

To get the song on the videotape, you can record the song at a live performance using a microphone mixer.

Or, record the audio using line output from a CD or audio cassette player into the line input of your camcorder or VCR. At the time of this recording, you will also record video, so try capping the camcorder and letting the picture appear black. This *blackbursting* of the tape records the video control track you'll use to perform the insert edits later on.

The next step: record video over the audio track with a video insert. You can record the entire music video shot by shot. This "editing in the camera" technique gets the job done without extra equipment or editing systems.

WARNING: this technique is not as simple as it seems; only use it if you can't edit. It may not work with every camcorder. If your unit doesn't give the performers a few seconds of audible preroll at the beginning of a video insert, they won't be in sync with the music when the recording starts.

If your camcorder does not give an audible preroll, try to find a VCR that does. Plug your camcorder's video output into the VCR and do the inserts there.

If that doesn't work, you can always insert interpretive video segments over those out-of-sync parts at a later time.

Editing in the camera will also mean you continually record onto one tape: your master. One serious mistake—like rewinding too far before you reshoot a given scene—and you have to start all over.

Shooting for the edit. If you have access to any kind of editing system, you should "shoot for the edit"—that is, plan your shots to provide plenty of material for the editing process.

Shoot scenes either in sequence or out of sequence, depending upon logistics. One easy way to organize your shoot: have the band perform the whole song in different locations or with different backgrounds and costumes. Tape three or four different versions of the song, and then mix and match shots from different versions. Shoot several takes of the lead singer singing the particularly dramatic parts of the lyric; be sure to get some extreme close-ups or unusual camera angles of the entire band. Keep close-up shots of the singers short; the lip sync may not be exact.

You can minimize time and travel for the band by cutting performance shots with location shots that help to emphasize the mood of the music, as in Figure 37.1

Experiment. Rent a fog machine to create a dream-like setting. (Ask your local theatrical or video production house where you can rent one.) Try fog filters for your camera lens. Place your lights in unusual places, like lighting from the side or below the performers rather than from above. Use colored gels over your lights or very narrow beam lights.

The shooting style of many music videos today incorporates vigorous camera movement and canted camera angles you may want to try yourself. Also common are combination shots taped while the camera is moving, i.e., zooming while dollying.

The usual rule of holding shots long enough for the audience to grasp what's going on also may not apply. Quick cuts—including cuts in the middle of pans, tilts and zooms—abound in music video. Most shots last two or three seconds.

Tracking your footage is the key to success when shooting for the edit. Keep a video footage log. By the end of the shoot, you should be able to consult your footage log to determine whether you've got the shots you need.

Don't leave the set/location until you're satisfied that you've shot it all. Check over your storyboard. Did you get every shot listed there? Review the key shots. Shoot them again if you're not satisfied.

Time to Edit

Once you have all the shots that you need, it's time to edit. If you have access to your own editor or one owned by a friend, go in with your storyboard, beat sheet, shot log and tapes and edit at your leisure.

If you plan to rent an editor or edit suite, the procedure is quite different. Here time is money, so plan all of your edits on paper before you enter the edit suite. Note: with the advance of nonlinear editing, you'll be able to experiment with more sophisticated editing techniques without paying an arm and a leg.

If you prepared a proper storyboard and beat sheet, you can review all of your raw footage, log it with your comments and your desired edit in and edit out points.

List the shots you want, the order you want them in and their exact location relative to the music. Such organization is particularly important for editing illustrative videos. Interpretive videos cut you much more slack; their looser flow allows

Figure 37.1 *Add cutaways to shots that match the feel of the song for a true music video look.*

you to put together seemingly unrelated shots without destroying the overall effect.

Be prepared to make 60 to 100 edits (or more!) for a four-minute rock or country video. Let the rhythm of the song suggest the pacing of your shot changes; shots and edits should match the tempo of the song. Try changing some shots on the beat of the music. You wouldn't want to do this on every beat, but it can prove effective, especially when it marks a noticeable change in tempo.

If your edit system includes a special effects generator or can perform some digital effects, experiment with transitions. Try some fancy wipes or digital effects like page peel or tumble.

Once your video is ready, make at least one copy of your edited master for viewing purposes. Keep the master in a safe place.

Break the Rules

Making music video is an art, not a science. Remember the conventions, but don't let them get in your way.

Go ahead and break framing, composition, exposure and transition rules—if it helps communicate your vision.

And by all means—have fun!

38

Practical Special Effects:
A Baker's Ten to Improve Your Video Visions

Bernard Wilkie

See video clips at www.videomaker.com/handbook.

Most videographers fall into one of two groups. First, the snapshotters—people who record scenes until their tapes are full, then view the disconnected events using the search button. Second, videographers and pros, who edit their tapes and employ all the techniques and processes necessary to obtain professional results.

It's not that people in the first group are unimaginative. They may not want to assemble their pictures or lay down soundtracks, but often they do wish to add variety to their videography. Thus this chapter can serve snapshotters as well as the more advanced. This list of special effects may also remind the experts that simple solutions can be the best.

Deep Water

Children mucking about in the pool occupy many tapes. However, all the action commonly occurs above water.

More exciting footage is possible with an underwater periscope, enabling videog-

raphy beneath the surface. Camcorders become underwater cameras without the need for expensive blimps or aqualung equipment.

The *periscope* is simply a rectangular box with a sheet of glass cemented into the bottom of one side. It contains two mirrors, one at the bottom, one at the top. Surface-coated mirrors provide better pictures, but even ordinary mirrors will produce good results.

Set up the periscope poolside and record events above or below water while holding the camera and peering through the viewfinder in the usual manner. Weight the device to overcome its natural buoyancy. Clamp it to something solid, like the pool steps. Those going to sea can mount the periscope on a boat or sink it into the water for shots of marine life.

Another trick to improve waterside close-ups involves the simple trick of placing a shallow tray filled with water and broken pieces of mirror below the picture. Ensure that the sun's reflections fall on or around your subjects. Then tickle the surface of the water with your fingers. Voila! The ripple effect says your subjects are waterside though they may really be nowhere near water at all.

Sound Skills

Sound is perhaps the greatest special effect of all.

The videographer seeking to create the atmosphere of a shopping mall need only shoot characters looking into a store window. Sound effects added later will provide the essential ingredients— children calling, skateboards whirring, the voices of people walking by, a distant police siren.

For the best location atmosphere, shoot scenes where you can capture the best and most interesting off-camera sounds. When on vacation, don't always aim for peace and quiet. If you want footage of your family on a foreign railway station platform, wait for the moment the train pulls out.

When creating a drama, always think sound. Sound can often say more than pictures. Imagine a scene with two people watching TV. Suddenly they react to the sound of squealing tires on the road outside. The sound of the crash that follows has them leaping towards the window. The skillful use of sound effects has fooled viewers into believing something horrific really did occur.

Bet you didn't know you can create a simulated echo using a length of garden hose and a funnel. Just stick the funnel in one end of the hose. Place both ends close to the microphone of a cassette recorder. Then speak. The effect is certainly weird.

Different lengths of hose produce different delay times. This contraption is useful for adding echo or creating monstrous outer space voices.

An even better echo results from linking funnel, hose and mike to one input channel of a stereo recorder while using the other channel and a second mike straight.

Mirror Dimension

Video pictures are two-dimensional, with height and width but no depth. This fact is used repeatedly to fool viewers.

The scene outside a window may be only a painted backcloth, but who can tell? A photograph of an object can look the same as the real object. Use photos when you can't acquire the real thing— a priceless museum exhibit, for example. Photographs, or even photocopies, can simulate multiple items such as meter dials or control panels.

The fifty-fifty semi-coated mirror, or beam-splitter, is a most useful piece of videography equipment. Used to superimpose one picture over another, it works because half the light passes through the glass while the other half reflects back from the surface.

Often employed to produce ghostly apparitions, it also has other, less spectral, uses. For instance, superimposing captions over pictures. With the mirror placed at a forty-five degree angle to the lens, illuminate the words as they appear. This technique can create opening titles, or overlay words or arrows on a demonstration video.

Superimpose graphics with a box housing the mirror and shielding it from stray light. Stand the rig in front of the camera. Light the caption from front or back. If lit from the front, place the lamps to either side, To remain in focus, the superimposed material must lie approximately the same distance from the camera as the main subject.

Used with a spotlight, plain mirrors stuck to a revolving drum provide a strobe effect. Two stuck back to back simulate the flashing lights of emergency vehicles.

Gun Fun

It's not unusual to see a TV actor, chest riddled with bullets, stagger and fall to the ground. If he seems to move awkwardly it is probably due less to his supposed wounds than to the fact he's trying not to trip over the operating wires running up his trouser leg.

This effect, using small explosive squibs secreted under clothing, is too complex and dangerous to discuss here. But there do exist safe alternatives, which, if used imaginatively, can appear as convincing as those in the movies.

You can use a bicycle pump for bullets in the chest. Record as a separate closeup, for it works only with a short tube close to the pump.

Suck some fake blood into the tube and seal the end with a small piece of tightly stretched party balloon. Hold the balloon in position with several turns of a rubber band. Placed under thin cloth, the pump will rupture the diaphragm, flipping the cloth realistically and producing a spatter of gore.

For continuity, shoot the garment on a stuffed sack. Then remove and place on the actor. A limp balloon containing blood and worn under a garment will produce a spreading stain when punctured by a spike attached to a ring worn by an actor. The flow will increase by keeping the hand in place and pressing hard.

A rat-trap set up behind scenery can punch out a piece of wall or knock a hole in a door. The hole must be pre-made, filled with appropriate material to disguise its true nature. Insert a captive peg from behind. When struck by the trap the peg ejects the filling, leaving a hole.

Rat-traps can also simulate a bullet hitting a mirror. Protect the front with a sheet of rigid plastic and cover the back with self-adhesive vinyl. This is essential to produce a really good shattering effect; without it the glass just breaks.

A bicycle pump with some talcum powder in the barrel will produce a convincing spurt of dust from rocks or concrete. Use energetically and apply a good ricochet effect on the soundtrack.

No bullet effect will impress without realistic sound. Conversely, a poor effect can often pass with a professional soundtrack liberally sprinkled with gunshots.

Smoke and Flames

This can be a touchy area; as always, Videomaker does not recommend you endeavor to create sequences involving fire, smoke or explosions without assistance from experts.

Movie and TV producers often rent an empty house or store when they need to create an outdoor fire sequence. To record in a studio is too impractical and too expensive.

However, property is property, so these big fire scenes are rigidly controlled to ensure they don't get out of hand.

In many cases fire can be simulated without actual flames. At night, backlit smoke rising from behind a building suggests it's on fire. Rooms powerfully lit, with smoke pouring from the windows, imply a house afire. Stretch a clear plastic sheet behind the window and pump smoke up underneath it. This ensures maximum effect at the window while preventing too much smoke from filling the room.

But the fact there's no flame doesn't guarantee total safety. Always ensure crew and artists have a clear exit to the outside. No one should have to stumble around in thick smoke and darkness.

Smoke, of course, is essential for all fire sequences. Much depends on the sort of smoke used. In moviemaking there are two types: pyrotechnic, and machine-made. Of the two, only the smoke machine is controllable. Pyrotechnics, once lit, will burn to a finish.

You can rent smoke machines; those who don't know where to look should contact a local theater or TV studio.

Smoke from reputable machines should cause no breathing problems, even when discharged indoors. Pyrotechnic smoke is appropriate only for exterior work or in places where it won't be inhaled.

Movie studios produce controlled flames by igniting propane. The gear usually consists of a fireproof and crushproof hose with a shut-off valve and pressure reducer. At the business end is a length of copper tube terminating in a sort of flattened funnel.

You can smatter small areas of flame around a set by using absorbent material treated with a dash of kerosene, burned on metal sheets or fireproof board. With the appropriate amount of smoke this will simulate the aftermath of an explosion.

House Mess

Many videographers must shoot in their own homes, which can cause problems when trying to capture scenes of dirt and degradation. Fortunately, it's usually possible to obtain materials easily cleaned up at the end of the day.

Freely spread sawdust, dry peat, coconut fiber, Fullers Earth, rubber dust and torn-up paper; all will disappear beneath broom or vacuum at the end of the shoot.

It's not easy creating convincing scenes of mess and filth. The camera has a habit of prettifying even the nastiest setups. It's therefore often necessary to exaggerate the dirty scenes.

For oil, food or paint spills, pour liquid latex onto a sheet of glass or metal. When set, spray paint the mess with any color. Peel off to provide a movable puddle; place where required.

Dead and dried vegetation often complement this sort of scene. Torn plastic sheeting sprayed nasty colors and wrapped around pipes, faucets and radiators also looks good.

To make metal appear rusty, wipe with a smidgen of petroleum jelly. Then blow cocoa atop the grease.

Cobwebs are great for dirty scenes, produced by spinning liquid latex in a special device called a cobweb gun. These guns are for rent, the fluid available from TV and theatrical supply houses. Spin webs over a collection of objects bunched close together for the best effect. Cobwebs won't straddle open spaces; string thin cotton across voids. Blow talcum powder onto cobwebs to make them visible. Don't apply to absorbent surfaces.

Caption Making

Electronic devices to produce lettering for videos are now available, either as separate equipment or as integral camera circuitry. Stick-on or rubdown letters come cheap and offer a variety of typefaces. Even magazines and newspapers will produce usable opening titles.

No one wants to engage in the laborious chore of cutting round letters with a stencil knife. But if you cut the letters or words as rectangles from white paper you can stick them to a white backing, the joins between painted over with white artist's paint or typing correction fluid. Photocopied, the joins disappear.

You can apply rubdown lettering or reusable vinyl stick-ons to a sheet of glass and place over various fancy papers or illustrations. You can also place the glass in front of three-dimensional objects like flowers or coins. Tabletop captions are simple to produce and offer more variety than stereotypical electronic images.

A tracing paper screen and a slide projector are also useful for graphic backgrounds. You need not project slides onto a flat screen; you can project them onto textured surfaces like rumpled cloth, rough plaster or piles of snow.

Interesting animations can result by fixing the lighting, the camera and the lettering to a common mount in which loose objects such as marbles, sea shells, sand, sugar or liquids are free to move around. When you tilt the rig, the camera perceives no movement of the backing; meanwhile, the loose objects react strangely, defying gravity and moving in a random and seemingly unpremeditated fashion.

President Matte

Miniatures are models placed in front of a set to extend the scenery or provide an effect unobtainable by other means. Mattes perform a similar task, but are simply flat paintings on glass.

This is an over-simplification, because there also exist creatures like "hanging miniatures" and "traveling mattes." But these characters would take us into deep technical water, and we're interested here only in the simple stuff.

Say the action occurs in the oval office of the White House. They won't let you tape there, and a re-creation would eat up your entire budget. So go for the special effect.

You may have to hire a table and some chairs—pretty safe stuff, because who, after all, recalls exact details of the White House?

Sensibly, you'll video your fake president against an easily-obtainable neutral wall. But at some point you'll have to show the room, or at least a convincing recreation.

One option is a *matte*. If you can procure a talented artist, take a large pane of glass, mount it on legs and position it between Mr. P and the camera. With constant reference through the camera viewfinder the artist can paint the oval office, leaving a hole in the middle to perceive Mr. P, his desk and the back wall.

Suppose you don't have such an artist. So get a photo of the scene, blow it up, cut out an area where Mr. P will sit, and, if your photo is black-and-white, add some color washes. There. The White House.

Miniature Quixote

Let's try a second example: the story of Don Quixote, the man who tried to kill windmills.

Unable to take your unit to Spain, you'll use plaster-of-paris, sawdust, sand, cardboard and other materials to construct a *baseboard model* of a sandy plain. On the horizon position a model wind-mill with motor-driven sails. Finally, mount the model on a rig supported from one side only. Put in a sloping floor in the studio and cover it with sawdust and sand. At the back paint a ground row to provide the horizon.

Finish up with three components: the sky backing, the main scene on the floor and, sticking in from one side with its supporting leg out of vision, a model of a sandy plain and the windmill. Make sure everything lines up and adjust the lighting to insure a complete blend.

When using a miniature, join the model to the set along natural boundaries—hedges, woods, roads—where the foreground, which we see as the background, will blend unnoticed.

Don Quixote will stand on the opposite side of the set from the windmill, pointing towards the background and yelling, "Kill! Kill!" He'll look at the distant windmill, which in fact sits right in front of him. Keep him stationary: if he strolls across the set he'll pass behind the model and expose the trick.

If you'd like to try a matte shot, paint on a board a simple sky and set it up where it will cover a busy background. Make sure the bottom of the board lines up with the top of a wall or some similar feature. Shoot it. The effect can be quite extraordinary.

Cardtoon Creations

Children exposed to television since birth accept everything on the screen without question, and quite often without interest. So why not give them the chance to participate as creative artists?

A cardtoon is the electronic counterpart of the puppet theater, where little figures move around on sticks. Recording the cardtoon technique on video gives youngsters the opportunity to design their own characters and write their own scripts. After the show they can sit back and view the results.

Unfortunately, video cameras don't work in the same way as the movie cameras that film Disney cartoons, so the action and movement must take place in real time. This is accomplished by cutting characters from stiff card and articulating them with tiny rivets and paper hinges.

Animation comes from fixing the various parts to hidden sticks or incorporating cardboard levers hidden behind parts of the background.

A simple example involves a picture of the sea, drawn as repeating lines of stylized ripples. Suddenly, up leaps a fish, attached to a cardboard lever and rotated between two layers of the wave pattern.

Still at sea, imagine a pirate ship sailing across, pushed or pulled from one side. Or a diver surfacing from beneath the waves.

All videography is a combination of long shots and close-ups, so we'll have to see the faces of the pirates onboard ship. You can make them speak using a simple up-and-down movement of the jaw. It's the eyes moving that most give the characters expression.

A few versions of the same heads—left and right profile, large and small—will result in a really satisfying cardtoon.

You can usually manipulate the parts by hand, though certain creatures may have to move faster than your digits can. In these cases you can employ thin dressmaking elastic as pulling springs.

Trick of the Light

Light, like sound, is too often taken for granted. It can suggest things which aren't really there; backlit smoke that looks like fire is but one example.

Take a look at movie scenes set in the countryside. The trees and the dappled sunlight through the leaves say we're in a wood, but the effect is truly produced by shining a lamp through holes cut in thin plywood sheets.

Look at the two people in the front seat of a studio automobile. We know the vehicle is moving because the background is receding—achieved via chromakey—but the scene would look dead if it weren't for the fact that shadows continually sweep across the faces of the actors. A spotlight, some plywood, flags on broom handles and some keen staff to wave them about and we achieve the effect of an auto in motion.

For night shots sweep a hand-held lamp from front to rear.

The prison scene looks a lot more sinister if a spotlight, trained on the floor, shines through a cutout silhouette of iron bars.

Many of these effects cause the autofocus to hunt. Switch to manual whenever this occurs.

Give Them a Try

Does any of this sound like fun? It is. Just remember, you'll never know unless you get out your camcorder and do it yourself. All it takes is a little imagination.

PART IV
Remembering the Journey

Post-Production Practices

If shooting video captures an experience, editing re-members, or re-composes it. Like a remembrance, or a dream, editing draws together various atoms of experience, bits of imagery and sound, into a new unified perception, a *gestalt*.

Like the memory of a beloved person or place, editing is selective. It doesn't recall the blemishes of poor exposures or the blunders of bouncing frames. But there's more to editing than simply removing the bad shots.

Editing, like shooting, is a creative act, not just a process of elimination. Done well, it gives rise to a new thing that is greater than the sum of its clips. It has a life, or at least a style, of its own. You have powerful tools at your disposal for making this new thing rise out of its elements. You have the use of timing and rhythm as does a composer of music. You have the use of juxtaposition, the placing of one clip behind another in such a way

as to imply the similarity—or of showing the dramatic differences—between the two. You have the use of cutaway shots that can visually enhance, or contradict, the statements of a narrator. You have the use of reaction shots, facial expressions showing the reactions of people, to the main action. You have the use of sound effects to amplify the importance of something on screen, or imply the presence of something that doesn't appear within the frame. And, of course, you have the use of music itself, which determines the emotional impact of your video perhaps more than any other element.

Then there are the dazzling special effects and transitions. You can turn your pictures backwards, upside down and black-and-white. You can make them look like they are made from mosaic tiles or solarize, posterize and otherwise victimize them. You can dissolve, wipe, peel or flip between them. You can make shot A wrap itself around a globe and fly off into space to reveal shot B. Yes, you do have special effects.

If you have just gotten a machine that delivers dozens of effects, do yourself, and your audience, a favor. Make a video that uses every single one of them: use every colorization and every three-dimensional animated flying transition. This will accomplish two things: 1) it will get them out of your system so you won't feel so compelled to use every single one in the videos that follow; 2) you can keep this glitzy hodgepodge of eye candy as a reference tape. In case you ever need to remember how a given effect looks when rendered, you can view it from the tape. Then, go forward and use effects tastefully, in ways that enhance the whole project.

This part of the *Handbook* will give you the benefit of many years of discovery in the world of film and video editing made by the many who have worked in this trade. Try these tips; they can help you take all those elements and make of them a whole new thing.

39
The Art of the Edit

Janis Lonnquist

See video clips at www.videomaker.com/handbook.

When Oliver Stone turned over the massive amount of raw footage that became *JFK*, editor Joe Hutshing knew it would be a challenge. "I wondered if it could even be watchable," Hutshing says. "It was so incredibly complicated. It was like looking at a schematic for a TV set and then imagining actually watching the TV."

From the mountain of raw footage, to the first five-hour cut, to the final three-hour-and-eight-minute editing masterpiece, Hutshing had to make decisions, consider choices and re-examine goals. This is editing.

Editing systems may range from sophisticated digital suites with all the bells and whistles to basic single-source systems consisting of a camera, TV and VCR. Still, the functions of editing remain the same:

- to connect shots into a sequence that tells a story or records an event,

- to correct and delete mistakes,

- to condense or expand time and

- to communicate an aesthetic.

Whether you're creating a Hollywood feature film or tightening a vacation video, the challenge is to take raw footage, and within the limitations of equipment and budget, transform it into something compelling and watchable.

Shooting With the Edit in Mind

Editing may be the final step of the production, but to make a truly successful video, you need to begin making editing choices in the concept stage. What will the overall look of the piece be? The mood? The pacing? Will you cut it to music? What kind of music?

There are several techniques that will help you plan. Prepare a shooting script, a storyboard or— if it is not a scripted production— an overview for your program. This will be the blueprint for your production.

A *shooting script* lists the action shot by shot, along with proposed camera angles and framing.

In a *storyboard*, actual sketches illustrate each scene. It's a good opportunity to see what will work before you shoot it.

An *overview* should include: the chronology of shots as they will appear in the video; approximate timing for each shot; and information about accompanying audio, graphics and titles for each scene.

Next, prepare a *shot sheet*. Make sure it includes every shot listed in your script or overview. Get several shots of each item on the list.

"You need a variety of shots," says Kevin Corcoran, vice president of Pacific Media Center in Santa Clara, California. "In a basketball game, for example, you get shots of the crowd, shots of the scoreboard, shots of the referee, shots of the environment. In action that's typically long and drawn out, you need to consolidate information. You need to have images to cut to in order to make it look smooth."

Even in scripted productions, Corcoran recommends getting a variety of shots.

"I always try to get a wide shot and a head and shoulders shot for each block of text," says Corcoran. "Who knows what you'll find when you go into an edit? There may be something that bothers you with continuity in the background of a wide shot. Now you have a place to go."

While Joe Hutshing had massive amounts of material to edit for *JFK*, Corcoran says the more common problem is too little material.

"Often there are large sections to be removed and no smooth way to cut," says Corcoran. "This is especially true when you're editing on a two-machine, cuts-only system. Ideally, you will have some other framing, another angle, a reaction or some other activity happening in the environment. If it's a person at a podium talking, you need an audience reaction shot or two or three. You must have cutaways to consolidate a half-hour speech without jump cuts."

You also invariably end up with footage you can't use, often due to the unexpected appearance of objects on tape that you never noticed during the shoot. Once, when editing the "dream house" segment of a TV program, I discovered a power supply right in the middle of the kitchen floor. Nobody saw it in the field and every sweeping pan— all wide shots— included the ugly box. Other than featuring a dream house with no kitchen, we had no option but to use the embarrassing piece of footage.

"There will always be things in shots you don't see when you're shooting," Corcoran says. "Things reflected in mirrors or windows, things in dark areas of the picture. It's important to change your framing to avoid having problems like this in the edit."

If you will edit your video to music, select the music in advance and time zooms and pans accordingly. If this isn't possible, shoot a slow, medium and fast version of each camera move. In general, shots should be five to 15 seconds in length. Know the pacing and shoot accordingly.

You'll enjoy a lot more options in your edit sessions if you aren't desperately "fixing it in post." Taking the appropriate technical precautions saves you from having to scrap otherwise good footage due to lighting, audio or other technical problems.

"In an event, things will only go wrong," Corcoran warns. "In weddings, for example, the light is nearly always bad. A camera light is essential, especially if you don't have gain control. And you'll need a lot of batteries for that light."

Good lighting greatly enhances the quality of your videos; invest in a lighting seminar if you need more information. As a rule, the brightest spot in your picture should be no more than 20 to 30 times brighter than the darkest spot or you'll be editing silhouettes.

You'll have trouble in your edit if you don't white balance several times during an event. This is particularly true during weddings, which may move from bright sunlight, to a dimly lit church, to fluorescent lights in a reception hall. If you don't white balance, the shots won't match— you may end up dissolving from a well-lit scene of groomsmen decorating the getaway car to a blue, blue reception.

Production Pains

Production can be exhausting, with long days of hard physical labor, but it's vital to stay alert.

On a particularly grueling corporate production a few years ago, a camera operator, who was also monitoring audio, removed his headset during a break and forgot to put it back on. Our talent, the president of the corporation, removed his lavaliere microphone to stretch, and sat down on it for the remainder of the production. Try to fix that in post.

Mikes can fall down, batteries can die, a cable can go bad. Without headphones, you may not know until it's too late.

"If you know from your headphones there's no hope for that microphone," Corcoran says, "You can unplug it and let the camera mike try. It's going to be better than what you'll get otherwise. Nothing can kill a production faster than bad audio. Wear your headphones all the time."

For most productions, steady images make the most sense. Always use a tripod. Hand-held looks, *well*, hand-held. There's a trend right now to overuse this technique, but avoid the *cinema verité* , or "shaky cam" look unless you're after a strobed look or the effect is actually motivated by something in the script.

Be sure to allow for preroll. When you switch a camera from the *stop* mode to *record*, it rolls back several seconds before it achieves "speed" and begins taping. Allow five seconds, 10 to be safe, before cuing the talent to begin speaking or executing your shot.

Unless your edit system is very precise (plus or minus two frames) you will have trouble editing to the word, so make sure

that you have two seconds or more of silence before your talent begins.

This is better than saying "action" to cue the talent: if the narration begins too quickly, you may end up losing two seconds of narration in edit to cut out your cue. Instead, count "five, four, three"... and cue talent after a silent count of two and one.

With high-end systems, you can encounter a similar problem. If the tape is checked and action begins too soon, you won't be able to back up over the break in control track to execute the edit.

To allow time for a good transition, instruct your talent to fix a gaze on the camera for two seconds before and several seconds after a narration. A quick, sideways glance for approval, a swallow or a lick of the lips before or after speaking may be difficult to edit out.

If you don't have control over the talent's timing and delivery— or example, when shooting a training session or wedding— your cutaways and reaction shots will be critical to mask cuts. Remember to shoot plenty.

In the Frame

Good framing and composition are vital in achieving aesthetically pleasing video that is cohesive and makes sense. A well-composed shot provides viewers with the information needed to follow the story. It reveals, through spatial relationships, the comparative importance of individuals and objects, and the effect they have upon each other. It focuses attention on details, sometimes subtly, even subliminally. Good composition can also disturb, excite and/or heighten tension if the script calls for it.

You can't fix poor framing and composition in post. A lack of head room will make your subject seem suspended from the top of the TV monitor. Framing a shot to cut at the subject's ankles, chin, hands or hem line is an uncomfortable look that doesn't allow "closure," a process in which the mind fills in the missing elements.

Remember the rule of thirds: place important elements in the top or bottom third of the screen. In a closeup, place the eyes at the one-third baseline. In an extreme closeup, the eyes are at baseline of the top third, the mouth is at the baseline of the bottom third, and, through closure, the chin and forehead are filled in.

Distracting or inappropriate backgrounds are nearly impossible to work around so pay attention to every detail when you shoot. In one production, a children's singing group performed a number in front of a blackboard. In the edit, I noticed one little girl standing directly in front of a large letter "M"— creating the look of two perfect, pointed ears. Again, saved by the B-roll.

Sometimes even balanced and thoughtfully composed shots don't cut together well. For example: if you're editing an interview or dialogue, cutting between head shots of the interviewer and guest, you need the heads angled slightly toward each other (to imply the interaction of the two) and off center, leaving "look space" or "nose room." Without look space, your interviewer will appear to address the edge of the TV screen. Centered, we have no sense of the spatial relationship of the two. They could be sitting back to back.

Similarly, maintain "lead room" for your subject to walk, run, bike or drive into.

Walk the Line

One production basic that can cause major consternation in the edit suite is "crossing the line."

Let's say you're shooting a parade passing in front of you, from left to right. A politician waves from a passing float, her back to you. You dash across the street and resume shooting, getting a great shot of her smiling face. When you go to edit, however, you'll find that you crossed the line: half of your parade marches left to right and the other half marches right to left. Cutting together footage from both sides of the line will create a bizarre montage where bands and floats and motorcades seem to run into one another.

Respecting the line is especially important in shots that track movement or where geography, such as movement toward a goal post, is critical to the viewer's understanding of the action.

Camera angles also play a role in the viewer's ability to interpret and believe the action. Let's say you want to show a child trying to coax a kitten from a tree. First we see the child looking up. We cut to the kitten cowering on a branch. We cut back to the child. The scene gains impact with the right camera angles. We see the child, framed left, looking up. Cut to a reverse angle shot looking down at the child, over the cat's shoulder, with the cat framed right. The camera angle duplicates the cat's line of vision. Cut to a low angle shot of the cat from the child's point of view. The edited sequence is fluid and believable.

There are two kinds of continuity you should monitor for successful editing. First: continuity of the environment. A made-for-TV movie has a scene in which a man speaks to his doctor. He wears a shirt with the collar turned up. Cut to the doctor. Cut back to the man, and his collar is flat. Cut to a two-shot and the collar turns up again. Productions on all levels are full of goofs like this one. To avoid adding blooper footage of your own, pay close attention to detail both in production and in the edit.

For the best possible editing situation, you also need to watch continuity of action. If your talent can give you numerous takes with identical blocking, you'll have lots of editing options. Cuts-only editing is at its best when you can achieve a multicam look by cutting to different framing on action. Look for the apex of the action— the full extension of the arm, the widest part of the yawn, the clink of glasses in a toast— and use that apex as the marker to cut to a new angle of the same action.

Motivate It

Transitions should occur only when motivated by something in the story.

A *cut* is the instantaneous switch from one shot to another. The most common transition device, it duplicates the way we see. (Just try panning or zooming with your eyes.)

A *dissolve* is the gradual replacement of one image by another. Use it to show a passage of time or create a mood.

A *wipe* is a special effect of one image pushing the other image off screen. With digital technology, the options are nearly endless. A wipe can erase, burn, fold, kick or flush the first image from the screen. Wipes signify the end of a segment and the complete transition to a new time, place or concept.

A *fade* is the gradual replacement of an image with black or vice versa, used primarily to begin or end a program or video segment.

Creative editing, using a variety of transitions, is still possible on a cuts-only system. If you can't fade in or dissolve, begin your shot out of focus and gradually make the image clear. A very fast pan— 15 frames or so of light, color and motion flying across the screen—is almost as effective as a dissolve. Allowing your subject to exit the shot ends a scene with the finality of a wipe. Cutting to a static shot, such as a close-up of a flower, a sign or a building, defines and separates scenes.

For greater insight, learn from the pros. Rent a well-done video and create an overview and shot sheet.

There are also seminars and many excellent books available on framing, composition and technique. For an in-depth study of media aesthetics, look for Herbert Zettl's *Sight, Sound and Motion*. Of course, editing is a practical as well as an aesthetic skill. On to the practicalities.

Editing Systems

Practically speaking, editing is simply copying selected video from the source tape to the edit master or record tape. A wide variety of systems and methods are available.

Single-Source Editing. You can perform single-source editing from your camcorder to your VCR. Your owner's manual will include complete directions; basically, you control the edit by pressing PLAY on your source deck (camera) and RECORD on the record deck (your VCR), pausing and releasing as you go. The transitions are cuts only.

This type of editing becomes frustrating quickly. As the editor you must locate edit points, manually set preroll, start the machines at the same time and react at precisely the right moment to control the edit. Frame accuracy is usually a problem. If you hit RECORD too soon, you suffer video noise between edits. Too late, and you lose frames on the edit master.

Expanded Single-Source Systems. The first investment single-source editors usually make is an edit controller. Most edit controllers allow you to shuttle to locate scenes, to mark in and out points, to read and display frame numbers either from a pulse-count or time code system such as SMPTE or RCTC.

These editors perform the preroll function automatically and start the machines together. Many systems give you: 1) the option of insert or assemble edit, 2) the ability to "trim"" add or subtract a few frames without resetting in and out points and 3) the ability to preview your edit. Some perform audio or video only edits and interface with a computer to store an Edit Decision List (EDL).

You can also expand single-source edit systems with an audio mixer, a switcher and character generator.

Multiple-Source Systems. These give editors the capability of A/B Roll Editing. The typical system consists of two or more source VCRs (A and B), which supply material to the video switcher or computerized editing control unit. There, the material is edited, combined with effects and sent to the record

VCR. Audio from the source decks is also mixed and sent to the record VCR.

Multiple-source systems allow an editor to connect two moving video sources with dissolves, wipes and other transitions.

In nonlinear systems, every frame is stored in digital form and is instantly available to the editor. Once you've designated an edit and transition on the computerized EDL or storyboard, the computer executes the edit instantly. You can grab a scene from anywhere in your source footage without waiting for a tape to cue. Experimentation becomes effortless.

As you move up to the more complex systems, do your homework. Read product reviews before you make the investment. Find out what peripherals you need for basic operations and efficient editing.

Investigate the availability of classes and users groups in your area. Is there a local production facility that rents a suite featuring the same system? You may need a back-up plan if your system goes down and you're facing a deadline.

Advanced Editing Systems. These systems feature Digital Video Effects (DVE), better compression, exciting animation, special effects, pro titles and more. They are revolutionizing editing, providing greater options, accuracy and speed.

The ramping of capabilities means a ramping of complexity; you'll need education and practice to get up to speed. The systems are relatively expensive and the technology is constantly changing. It isn't easy to know when to make the investment. Some videographers complain that editing functions have not been designed with editors in mind; they're waiting for upgrades to correct this. Others have found systems that meet their needs well, and are using them to produce amazing programs.

Again, do your homework. If you can, rent a suite and actually do an edit on a given system before you buy.

The Final Cut

It's pay off time. You planned ahead, you paid attention during production, and now you can relax.

Why? Because editing is going to be great fun. Enjoy.

40
Linear vs. Nonlinear Time Trials

Jim Stinson

When we're testing products like batteries or cables, it's easy to envy the editors of auto magazines who get to write stuff like:

The tires neighed and snorted as I slammed her through the slalom, but I twitched the wheel to rein them in.

While all we can do is sigh and murmur "vroom vroom" softly as we sit there swapping RCA plugs and dreaming of the fun we too could have if we tested products by hurling them through fabled road race courses.

What we can do, however, is use the metaphor of the racetrack to compare a typical linear editing system with a typical nonlinear system. Our Videomaker "test track" was a 10-minute video with 100 shots and numerous titles and effects. The course was designed to push two thoroughbred vehicles to their limits and then some—a face-off between old and new, analog and digital, linear and nonlinear.

The linear system is a fabled workhorse of advanced amateur and entry-level pro-fessional videographers, with upgraded torque and horsepower added by a state-of-the-art digital mixer. Its nonlinear rival is the new kid: macho, assertive, rippling the muscles of its Pentium II processor under the hood.

To rate the two, we'll push them into the pits and set them up for competition; then we'll compare them, *mano a mano*, for speed through the course, handling characteristics, driver comfort, and over-all quality.

So assume a look of breeding and old money, as if your other vehicle were a polo pony, while you stroll with us onto the test track.

Setup for Competition

To begin the trials, we stop watched two top-of-the-line pit crews as they set up the competitors.

At the beginning, the linear crew began arranging decks and black boxes, while the nonlinear crew had the computer's case open, preparing to plug a card or two into the motherboard's PCI-bus expansion slots. By the time they had the computer's cover back on, the linear crew had finished selecting their cables from an assortment of composite, S-video and FireWire options.

As the linear crew checked to make sure all of the edit control cables matched up between the edit controller and the decks, the frustrated nonlinear crew was surfing the Net for the latest capture-card drivers.

The hookup phase was no contest as the nonlinear crew cabled S-video and stereo audio lines to the PC and then right back again. A similar run to the single monitor and the crew jumped back, arms upraised, to show they had completed the first leg.

Meanwhile, the linear crew was running the feeds from two VCRs and a black box titling machine into the audio/video mixer and then out again to the record VCR, lashing up the four—count 'em—four required monitors (and trying to figure where the titler's preview output went).

But once the lines were lashed and the lug nuts torqued, the advantage switched to the linear side because their setup was complete and they headed for the starting line. In contrast, the nonlinear crew was still configuring the editing software, loading the capture card drivers, surfing the Internet for newer drivers, loading the newer drivers, configuring the software, surfing the Net again for technotes from the hardware and software vendors, reconfiguring the software...

To their credit, they were out there on the starting line, sweaty but triumphant, when the linear crew returned from a coffee break and the time trials were ready to start.

Speed Trials

Now it was time to push each beast, in turn, through a tough road course, negotiating the hills and curves of shot logging, data inputting, editing, revising, and outputting the finished product to tape.

In the first, shot-logging phase, the advantage shifted away from the linear machine as our driver juggled four source tapes, handwriting shot names and descriptions. The nonlinear system had the advantage of an automated logging system, which aided in the task of searching the tapes and automatically saving in and out points. After defining each scene on the nonlinear logging system, the software even placed a tiny image called a picon at the beginning of each clip, to identify it visually.

Next, the driver selected the shots he would use. On the linear system, he did this by placing check marks near the shot names on the paper shot log. On the nonlinear system, he clicked a check mark into a box near the selected shot names on the screen. When the stopwatch stopped, the systems were even. But when the digitizing commenced on the nonlinear, the balanced tilted because linear editing skips this chore, assembling the show directly from the source tapes. The linear system was immediately ready to move into the editing process, jumping into a distinct (but temporary) lead. With the nonlinear system, however, the driver would have to wait for the computer to finish batch digitizing the clips before editing could begin.

But at the editing backstretch, the nonlinear system started using its random access ability, and it cut in like an afterburner. With all the shots on the hard drive, the driver could grab them instantly, stick them on a non-screen timeline, manipulate them, preview the effect, try out transitions and superimpose titles, all at once.

The traditional linear setup didn't fare as well, with the driver searching tediously for every shot, making and re-making whole edits to revise the cut points, rehearsing and recording A/B-roll transitions.

When the driver had his shots lined up in the nonlinear system, he suddenly veered right off the track and headed cross-country. He was showing off nonlinear's audio capabilities: layering and balancing multiple tracks—in stereo yet—cross-fading music, effects, and narration along a path where the more limited linear machine just couldn't go.

Roaring back onto the test track, the driver turned the nonlinear system's stopwatch lead into what looked like a rout, because he was now screaming into the revision phase. On this part of the course, the old-fashioned linear system looked plain pathetic because if the driver didn't like a shot, he could only replace it frame-for-frame, or else re-assemble every blessed shot that followed it.

On the nonlinear system, the driver was pushing shots here and there, swapping footage instantly, trimming lengths, retiming transitions—even driving with one hand as if to show the competition how easy it was.

But at the last moment, the linear system bounced back. You see, the home stretch was output, where the arrogant computer-based system slowed to a glacial pace while the old-style edit system went zero to 100 in four seconds flat.

How come? Because it didn't need any newfangled "output." When the driver finally did get through his revisions, he was almost across the finish line. The upstart nonlinear system, on the other hand, had to "render" the project, i.e. build a single file containing all the clips, effects and transitions, indicated in the timeline. All the while the white knuckled driver hoped that the system wouldn't crash. (The driver never would forget the time when the final compositing had quit because the computer's power-saving feature cut in.)

So when all the dust had settled, there the two were at the finish line, with overall stopwatch readouts that looked nearly the same.

Handling Traits

And what was it like to drive them? Once again, each contender showed that they had both strong and weak points.

Though the linear system's mixer actually offered plenty of flips and wipes and other digital video effects, the nonlinear software let the test driver vary motion speed, color temperature, and every other image characteristic you could think of.

As for titling, both systems offer a wide range of backgrounds, typefaces and motion effects—more options than you could get through in a year.

In the win some, lose some department, the nonlinear system beat the pants off its older rival in sound management, as noted earlier.

On the other hand (as we keep saying), the storage abilities of the nonlinear system are laughable compared to the two-hour capacity of even a humble VHS tape. If you're going to go nonlinear, you'd better stick to very short movies, or get yourself a very large hard disk.

Comfort and Convenience

The traditional linear system was easier to learn because the equipment was simple and intuitive. You could say that the wheel, shift lever, and pedals were right where the driver instinctively reached for them.

On the other hand, the computer-based system faced our driver with a learning curve as steep as its acceleration graph: a screen packed with multiple windows and menus within menus within menus, all supported by a lengthy instruction manual that can most charitably be described as a nice try.

But once our driver had the controls down pat, the nonlinear software proved equally intuitive, especially the timeline metaphor that lets you see a model of your movie as you massage it. (Some systems use a storyboard metaphor, which also works well.)

And to be fair about it, the snonlinear system also takes longer to learn because it has so many more features, which means that there's much more you can do with it when it's time to sit down and begin editing.

The bottom line? Depends on your personal style and your types of programs. Computerized editing is a far more detached intellectual exercise than traditional cutting. If you feel more comfortable when you can get your mitts on actual knobs, sliders, cables and cassettes, then you may naturally relate better to the more traditional linear setup.

But if you routinely use computers, you've probably grown accustomed to mice, menu trees and all the little procedures that are now semi-standardized on PCs and Macs.

Overall Quality

So which car is better? Sorry, but there's no escaping that wishy-washy old "it depends." In general, the linear system will outrun the nonlinear on long projects requiring long clips, few cuts, few special effects and transitions and simple audio. The nonlinear system will blow the doors off of the linear on short projects with many cuts, transitions, effects and sound tracks.

A second-generation master tape produced by the linear system looks better than the master tape made from the nonlinear system (though a nonlinear system could easily beat it if you forked over several thousand dollars more than the cost of our test setup). That's because the nonlinear rig's output suffers from visual artifacts.

Third generation tapes (dubs made from either master) will of course suffer additional generation loss—unless the masters and dubs are made on DV decks connected by FireWire.

How about reliability? Here the nod goes to the traditional vehicle. If you'll permit a personal note, Videonics mixers and Panasonic VCRs have chugged away in my teaching lab five days a week for seven years without even a single breakdown. Desktop computers, on the other hand...well, you already know about their reliability, don't you?

Besides, linear editing commands a mature, slow-moving technology in which seven-year-old hardware is maybe 95% as capable as the very latest stuff. On the nonlinear side, products are improving so fast that seven-year-old hardware is practically worthless. (Try video postproduction on a 486 machine with a 200-megabyte IDE hard drive!)

Caveat: computer-based systems are fast becoming easier to install, more stable and easier to use. The results of this time trial could change almost daily, depending on the specific systems used.

And now, if you'll excuse me, I have to resume testing RCA cables.

41
Bullseye Editing: Stay on Target with Time Code

Jim Stinson

You can begin frame-accurate editing only with time code. You may be able to wipe and dissolve, mosaic and strobe, posterize, solarize, fly in and flip out. But without time code capability your most awesome effects are like walnut burl on a Yugo dash or frosting on a soda cracker.

A strong statement? Yes, and to understand it, you need to know what only time code editing can do for you and why its benefits are so essential to truly professional results.

Time Code Basics

Time code is a system for labeling each individual video frame with a unique address. Since there are 30 frames per second, a time code address looks like this example: 01:24:13:22, meaning one hour, 24 minutes, 13 seconds, 22 frames. (For some stupid reason, time code is sometimes displayed without the separating colons, and by the end of a ten-hour edit-

ing session, decoding undersized readouts that say 01241322 can make you crazy.) The great thing about time code is that a frame address is just like a street address: it never changes and so you can always find it. And of course, finding the exact frame is essential for precise editing.

Assorted Time Code Flavors

Currently time code comes in four basic flavors. *SMPTE Longitudinal* time code is a form of time code that can be recorded on any audio track. On a stereo VHS or S-VHS system, you can obtain rewritable time code by placing it on the linear (usually monaural) audio channel. Placed on a hi-fi audio track, the code is not rewritable, because you cannot replace a hi-fi audio track without replacing the picture as well. On the other hand, if you put the code on one of the linear tracks of the assembly VCR, you cannot make audio dubs to that track, because replacement audio signals are recorded only on the linear track(s). These would pave over the time code signal.

Vertical Interval Time Code (VITC) is recorded in the "vertical blanking interval" portion of the video signal. It must be recorded simultaneously with the video signal, and so is not a rewritable code.

Rewritable Consumer Time Code (RCTC) is rewritable, as its name declares. But RCTC is a proprietary system that is available only on certain 8- and hi-8mm cameras and decks and few edit controllers can read it.

8mm Time Code is an older system, again from Sony, that shares many characteristics with RCTC code and is also limited to 8mm cameras, mainly older industrial models. The two Sony systems are not compatible.

Control Track Code: though not a true time code system, this is the most commonly found control signal and is used in Control-L and Panasonic 5-pin edit control systems. Control track is limited for two reasons: it lays down an electronic marker only once every second (instead of 30 times per second) and it does not provide a unique frame address. Every pulse is identical. Some edit control systems can increase the accuracy of control track editing by interpolation. Using their own internal clocks, they calculate the positions of the 29 frames between each once-per-second signal.

How It Works

Regardless of their differences, all four types of time code work in essentially the same way.

A time code generator creates an electronic address for each video frame that is recorded on the tape at the same location as the frame. The generator may be inside the video camera or VCR, or it may be a separate black box connected to one or the other. It can even be a piece of computer software that is part of an edit control system.

With high-end professional equipment, the recording of SMPTE longitudinal time code can be done before, during, or after the audio and video are recorded. (This is also true of Sony's RCTC code. The R stands for "rewritable.") That's because 3/4 inch and Betacam formats place the code on a special dedicated track that is not affected by the audio or video erase heads.

To see how this upscale system works, pretend that you're making a corporate video about a new product. Here is how you might use time code in your production:

The time code generator in your Betacam SP recorder adds time code to your live-action footage as you shoot it. (Smart videographer that you are, you assign each tape cassette a different starting hour, so that you never duplicate addresses. For instance, the first frames of the first two tapes are 00:00:00:01 and 01:00:00:01.)

The Engineering Department has sent you some computer-animated footage of the product that has no time code. So you "post-stripe" the code on a tape that already contains a video signal.

In order to control the blank tape on which you assemble your edited program, you record time code on it before building your show. This is often called "blacking the tape," because a simple way to pre-stripe time code is to record an entire tape with the opaque cap on the camera lens.

By the time you're ready to edit, every frame of every tape has its own address. To see how handy these addresses are, let's look at how they improve your control over the editing process.

Time Code for Basic Editing

But first, remember what happens when you edit with the hour-minute-second counters on your consumer-level VCRs. For one thing, these counters are driven by signals laid down at one-second intervals on the tape's control track. That means they are accurate only down to the second, and a tolerance of one second in video is like a tolerance of one inch in a jet engine.

More importantly, the numbers are not fixed, but only relative. Suppose you are editing a birthday party from two original tapes. You begin by rewinding each tape to 00:00:00 and then logging your shots by the VCR time counter. So far so good. You lay down your first five shots from tape A, and by this point the time counter says, maybe 00:20:03.

Now you need an insert of the birthday cake from tape B, so you pop out tape A, insert tape B, and watch the counter reset to 00:00:00. You add the insert and then resume editing with tape A. But when you stick it back in, the VCR resets to 00:00:00 again. The only way to recapture the shot addresses on tape A is to rewind to the start, reset the counter, and begin all over again.

This gets old fast—very fast. Now consider the control that time code gives you.

First of all, time code lets you make a work print—a duplicate tape that you use in editing to save wear and tear on your precious original. On some systems you can make a "window dub", a work print with the time code visible in a rectangular window right on the screen. This makes it much easier to adjust shots by watching both picture and time code on the same screen at the same time.

Next, time code makes it easier to log your shots because each one has a unique starting and ending address. Some computerized edit systems include a database function that lets you log and describe shots automatically. Basically, you roll to the head of a shot and then mark it with a keystroke or mouse click. The software notes and stores the starting time code address.

Repeat the process with the end of the shot, type a brief description (like, "C.U.: Clyde gets pie in kisser") and continue with the next shot. You can see that shot logging is far less tedious this way.

With every shot logged, time code lets you locate shots much more quickly because each frame has a fixed address.

To return to the birthday example, if tapes A and B have time code, you can put either tape in a VCR at any point in its program and the readout will tell you exactly where you are.

Time code also makes matching action a piece of cake. Suppose you have a wide shot of an actor slamming a door behind him and you want to cut to a closeup of the doorknob just as the door bangs against its frame. Find the last wide shot frame before the bang and record it as an OUT point; find the first frame of bang in the closeup and record it as the next shot's IN point. Tell the edit controller to make the edit. That's it. Without time code, there's no way you can guarantee that accuracy.

Why is accuracy so important in matching action? Because the human eye/brain can detect mismatches of two frames or sometimes only one; and a glitch looks amateurish and distracts the viewer briefly from the program's content.

If you use A/B roll editing (to create effects like dissolves and wipes) time code is essential for precise control. To ensure that a dissolve takes place at exactly the right point in each program, you find the middle of the dissolve on each roll, note its address, and back up each tape by precisely the same number of frames. Then have your edit controller lock the two VCRs, roll them in sync with each other and the assembly VCR, and record the A-roll shot. When you begin dissolving out of the A-roll shot, you will be at the right starting point on the B-roll shot.

Last, but not least, time code permits auto assembly, if your edit controller has that capability. Here's how it works. As you have built your assembled program, the edit controller (often a desktop computer) has recorded the start and finish addresses of every single shot. (This record is called an Edit Decision List, or EDL.)

Now, suppose you screen your handiwork and note that a few scenes need to be longer or shorter—or maybe cut out completely. Instead of re-editing the whole tape, you simply go into your recorded EDL and adjust time code addresses as needed. Then put a fresh tape in your assembly deck and instruct the edit controller to rebuild the entire program. All you have to do is sit there and swap source tapes when the controller requests them.

Advanced Techniques

With the one-frame accuracy of time code, you can achieve effects that would otherwise be very chancy.

Take audio dubbing, for instance. If you've used this feature, you know that you can replace the audio that accompanies a piece of video with sound from another source. For example, suppose you've covered a scene with two cameras at once. The problem is that the video is superior on tape A but the better-quality sound is on tape B.

No problem. With time code you can synchronize the two tapes so perfectly that you could dub the A video with the B audio, matching lip movements perfectly.

Time code also enables you to edit two-camera shoots on the fly, as if you were switching them on location. Suppose you videotape a wedding with camera A locked down on a tripod for a wide shot while you tape a close shot with camera B.

Back at your editing console you can use time code to synchronize the two tapes, roll them together, and switch back and forth between them in real time as if you were directing a two- camera studio shoot.

You can use time code in the edit phase to recover from goofs during production. Here's just one example from the many I could choose: recently I made a training video that I'll call *Using Your Brand X Business Phone*, and the client was pleased, except for one tiny detail. When she saw a huge closeup of the telephone's HOLD key and red indicator light, she paused the tape.

"It's not flashing."

"What?"

"When your party's on hold, the red light flashes." she informed me ominously, "Where's the flashing red light?"

Thinking at warp nine, I devised an adroit solution: "No problem."

No problem? The flashing shot did not exist, and I couldn't afford to rehire a studio-quality crew, equipment, and lighting, just to shoot the blasted flasher. To resume production, even for one measly setup, would cost 2 Big Ones. It was evidently time for plan B.

My savior was time code. I did have a shot in which the indicator light started unlit, then turned on and glowed steadily. Using time code addresses, I identified six frames in which the light was off and six more when it was on. Then I told the edit control software to alternate these two tiny parts of the original shot, off—on—off—on—off for the six seconds I needed. The result: a flashing HOLD light synthesized from 30 shots, each one-fifth of a second long.

Limitations

After counting the blessings of time code at considerable length, I have to confess that I know of no 8mm or VHS time code system that does not involve trade-offs. For one thing, the 8mm and RCTC systems from Sony are not compatible with each other.

Both of these systems are proprietary, and can be used only with certain 8mm and hi-8 equipment. That means, for instance, that you cannot shoot hi-8 and then edit to an S-VHS master—unless you have an edit controller that can read one time code format and write another.

You must record VITC time code at the same time as the video and audio. You cannot pre or post-stripe VITC time code. To add time code to footage that doesn't already have it, you must copy the footage to another tape, laying down time code as you do so. That costs you one extra generation of tape quality.

SMPTE linear time code does not share the drawbacks of the other systems. It is rewritable, copyable, and transportable between tape formats. But, "Simp-tee" code must be recorded on an audio track—either one of the two stereo tracks or on the monaural track.

Many prosumer cameras and VCRs do not offer separate hi-fi and linear audio outputs, which are essential for splitting the audio from the time code. To get them, you must have your hardware modified by a third party vendor. This can cost hundreds of dollars and may affect your equipment warranties.

Finally, there is a limitation that people don't often consider: prosumer hardware. For frame-accurate editing to work, the mechanical systems of VCRs must operate extremely accurately, reliably, and above all, repeatably. And the machines must continue doing this over a long period of time. Many knowledgeable techies insist that prosumer VCRs advertised as editing decks, simply do not have the durability to do that. They are actually high-end consumer units, as their superfluous features such as one-touch recording suggest.

New units can perform frame-accurate editing, but only when the mechanism is unworn and in perfect adjustment. All makers of edit control systems point out that the accuracy of your edits is finally determined by the limits of your equipment.

The Future

Nevertheless, things are starting to look brighter. Sophisticated edit controllers are appearing. Some can handle several different types of time code. Time code generators and edit control systems are available from a number of companies.

Most importantly, desktop computers have become the dominant editing systems—especially since Windows and Macintosh platforms became more affordable. Computers handle time code with ease. Even more significantly, they cannot fully use their immense speed and power without it.

Meanwhile, a growing army of desktop video editors was impatiently demanding an address system equal to their computer control capabilities. They received that system, and so much more, with the arrival of the new digital video formats, such as miniDV, which has timecode integrated into its stream of audio and video data.

42

Seizing Cut Control:
What Editing Protocols Can (and Can't) Do for You

Jim Stinson

An editing protocol is a language that your video equipment uses to talk back and forth—a medium of communication and control. If you do even the simplest editing or simply start your VCR with a remote control, you are using an editing protocol.

The more you know about these protocols the better you can use them to simplify the editing process and produce more professional results.

Protocol Basics

To understand these electronic communication systems, you need to know how they work and what they can achieve. That way you can get the most out of the equipment you have now and plan an informed strategy for upgrading your system in the future.

Video editing protocols bear such poetic names as control-S and control-L, not to mention control-M, RS-232 and the ever-popular VISCA. There are many dif-

ferent protocols developed for different types and brands of equipment. It's important to see how they work and what they can do for you. But it's even more important to understand how different protocols support different levels of editing sophistication. For that reason, we've organized this survey by editing method rather than by editing protocol.

No matter how sophisticated these protocols are, you really have to know only a few things about them:

The simplest editing protocols communicate between the editor—you—and pieces of hardware—VCRs, camcorders, whatever. With slightly more upscale recorders, you can set up the assembly deck to communicate directly with the source deck. And with sophisticated outfits, an edit controller commands both the source and assembly decks.

These protocols communicate only two things: status information such as "source program is at 00 hours, 23 minutes, 19 seconds," and commands such as "switch from Record/Pause to Record."

Simple protocols communicate only commands, and only from the controller to the hardware. The decks cannot send back information regarding placement or programs. More complex protocols communicate both commands and status information, with commands traveling from controller to decks and information returning from decks to controller.

Now let's look at several different editing methods that use various protocols, beginning at the simplest level and working toward truly professional edit control systems.

Level 1: Eyeball and Zap

If you control your record and playback VCRs by wireless remote, you're using an editing protocol. Your zappers send coded pulses of infrared light to command your VCRs to Play, Record, Pause and Stop. It's only a one-way protocol because your VCRs can't talk back to their remotes to tell them how many minutes and seconds elapsed during the recording. But that's okay because you know exactly where both your source and record tapes are. In this very simple system, you are functioning as the edit controller.

For an even simpler example, suppose you're taping a show off the air and you come to a commercial break. Here's what happens, expressed in terms of an editing protocol:

The program—source sends you the status information visually. That is, it tells you it has reached a break by fading to black.

You then respond to this information by sending a pause command to the VCR to suspend the recording process.

While the commercials run, you glance at your VCR display to verify that it's in Record/Pause mode. That information tells you that your record tape stopped positioned where you wish to resume recording after the commercials.

The source—TV screen—gives you the status information that the program should shortly resume by fading to black again.

You now respond by sending the pause command to your VCR to resume recording.

Of course, you don't think about recording in this way; you just do it. But by viewing the process as a set of steps, you can see that video editing means sending commands to source and assembly hardware based on status information about where each one is in its program.

Level 2: Deck-to-Deck One-Way Control

If you edit video with your remote control, you are using a form of control-S editing protocol. The infrared pulses used by most modern remotes flow in a serial stream, coded according to this standard. The next step up is an assembly VCR, which uses the control-S protocol to command a source VCR directly. If you have two decks equipped with control-S plugs, you can have the assembly deck tell the source deck when to start rolling. It works like this:

You connect the control-S out jack on your record deck to the control-S in jack on your source deck. This means that commands flow one-way from assembly VCR to source VCR.

You set the controls on both decks for control-S editing (usually called the edit mode).

You find the end of your last scene on the record tape and enter Record/pause mode.

You find the start point of the next scene on the source tape and enter pause mode.

You start the editing process by pressing the Execute key—which is usually labeled Synchro Edit.

The record deck commands both decks to roll forward together, and at the same time shifts itself from Record/Pause to Record.

The result: a nice clean edit that begins where you want it to, give or take a few frames.

But notice what control-S doesn't do: it doesn't stop the recording where you'd like; you must do that yourself. Why? Because the source deck cannot communicate back to the record deck where it is in the scene, so the assembly deck cannot know when the scene ends. That is the big drawback of one-way communication.

To overcome this drawback, VCR manufacturers have developed two-way editing protocols.

Level 3: Two-Way Communication

The control-L protocol devised by Sony and the control-M system from Panasonic—often called Panasonic 5-pin—are two-way systems. They communicate both the commands and status information.

These protocols are used with external controllers. Controllers can be proprietary units, third-party black boxes or desktop computers. The controller connects to each VCR by a special cable.

With control-L or -M you can make more professional edits. That's because the VCRs continuously send their tape counter info back to the controller, so the controller always knows—generally within a second—how far each tape has rolled into its shot. This means that in addition to commanding both decks to start, the controller can also tell them when to stop.

This is very useful for edits such as inserts, where you want to replace only the video portion of a previously assembled scene. Suppose you've just dubbed a scene in which a football quarterback drops back, looks for a receiver and then throws a pass. The camera swings around and picks up the receiver in time to record the catch. Great shot, except that the middle part of it is a mess while the cameraperson is

fire-hosing the field in a desperate search for the pass receiver. To fix this, you want to replace the messy part with an insert of the coach's strained expression while the pass is in the air. With a control-L or control-M protocol, it's easy:

Select the record deck on the controller's pad and then find the start point of the insert and mark it by pressing the in set key on the edit controller.

Find the end point and mark it with the controller's out set key. The controller now knows exactly when to start and stop recording on the record tape.

Select the source deck on the edit controller, then mark the start of the coach's close-up with the in set key.

Select video insert mode on the controller (which will replace the picture, but not the sound).

Press the Execute key (sometimes labeled "edit start") to make the edit.

The controller will start and stop the tapes, controlling the insert record mode on the record deck automatically. When you check playback, the quarterback passes, cut to the tense face of the coach (while the roar of the crowd continues on the track), cut to the receiver just before he snatches the ball out of the air. Perfect!

If you tried the same thing with one-way, manual control-S, you'd end the coach's close-up a shade too late and pave over the receiver catching the ball. Then you'd have to go all the way back to the start of the wide shot, dub it a second time and try to insert the close-up again. You can see the obvious convenience of this two-way communication.

Level Four: Multi-Event Assembly

Here's another advantage of two-way communication. If the controller knows the locations of a scene, it can store that information in its memory as an editing event.

NOTE: An event can also be any kind of transition from one shot to another: cut, wipe, fade, dissolve, whatever.

The simplest controllers will "remember" only a few events. More up-scale models can handle up to 99 events, and some computer-based systems can recall 1,000 events or more. If 1,000 seems extravagant, consider that most two-hour feature films contain at least this many edit events.

Each event stored in the controller's memory includes the two types of familiar data: a command to start dubbing a new scene and status information telling where the assembly and source decks will be when the command comes. Together, all these stored events make an Edit Decision List, or EDL.

An edit decision list lets you re-edit a program automatically, making it practical to polish a rough cut of your program. Suppose, for example, that you have made a 30-event mini- production that is just dandy except for this problem: one lousy scene runs five seconds too long— and it's only the third scene in the show. If you were editing film you could simply snip out the offending five seconds and splice the film back together.

But you cannot physically splice video. Instead, you'd have to re-dub the scene to shorten it and then re-edit every single scene that came after it.

But with an edit decision list, you can find the offending scene on the list and shorten it five seconds by simply changing the numbers for its out point. Then you rewind your source and record tapes to zero, tell the edit controller to re-execute the stored edit decision list and stand back. The controller automatically rebuilds the whole program.

Enter the Computer

When you reach the level of complex, multi-event editing, you perform chores that require extensive computation and memory, and that of course, is where desktop computers shine. So it's not surprising that a variety of video editing hardware and software is available for bothMacintosh and Windows machines.

Computers can use control-L, -M, VISCA and even more sophisticated editing protocols. But before getting into that we need to clarify some points about desktop computers and video.

In video editing, computers can perform two very different types of tasks: signal processing and editing. Signal processing includes jobs such as generating computer-based images, creating effects and producing titles. It can also include video and audio switching and mixing. All of these chores demand that the audio and video program signals go through the computer, which processes them and then ships them back out again.

By contrast, using a computer for editing control does not require that the audio/video signals go through the machine. Instead, the computer deals only with the status information about where each deck is in its tape, the mode in which it operates (Play, Pause, Record) and the commands required to perform the edits.

The computer usually sends and receives this information through a jack called a serial port. The two most common types are the RS-232 and RS-422 ports.

These are communication protocols, not to be confused with the editing protocols we have been discussing. A communication protocol contains the laws governing the way the serial port sends and receives information, regardless of type. Like a traffic code, these "rules of the road" are the same for each vehicle, no matter what it carries or where it goes.

For example, a fax board, a mouse, a modem and a video editing system may all use identical RS-232 ports, even though they communicate very different types of information.

Serial computer ports permit more accurate editing, in part because they communicate information so quickly. A video frame lasts only 1/30 of a second. To address a frame accurately, the controller must send its commands in less than that brief amount of time. Control-S, -L and -M signals may take longer than 1/30 of a second because they carry the excess baggage of "housekeeping" information and because they repeat each message many times to ensure reception. Therefore, they are usually too slow to deliver frame-accurate editing.

But protocols running at the speed of serial communications can be more than speedy enough to do the job. This allows computer-based video editing protocols to achieve true frame-accurate editing if—and that's a big if—they can identify individual frames.

Level Five: Frame-Accurate Editing

All professional video editing allows you to specify the precise frame with which to start an editing event. Frame-accurate editing is absolutely essential for matching action from shot to shot.

For example, suppose your actor, visible to the waist in a medium shot, sets down a coffee cup. You want to cut to an insert of the cup as it clicks on its saucer.

If you've tried to do something like this with amateur video equipment you know how dependent you are on dumb luck for a match. That's because the human eye is so sensitive that it can detect even a slight mismatch between the action and the sound.

To match action from shot to shot, you must be able to mark the exact frame in each shot where the cup touches down. But to mark a frame you have to identify it. There are two ways to do this: by counting frames as they speed by or by giving each frame a unique address.

Some edit controllers, like the Viewport from Selectra Corporation, count frames by closely monitoring signals inside the editing decks. By doing this, they can achieve 1 or 2-frame accuracy. The trouble is, so finely tuned is each unit to a particular VCR model that it

will work only with that model. The more common method of addressing individual frames is time code.

Basically, time code provides a unique and permanent address for every single video frame, expressed in hours, minutes, seconds and frames. For example, the twelfth frame past one hour, seven minutes, and thirteen seconds has this address: 01:07:13:12. Unlike a VCR footage counter or an hour/minute/second display, which simply count seconds or inches of tape from the zero point you set, a time code frame address does not change, no matter where the tape is when you start playing it.

So with the communication speed of a serial port, the computational power of a computer and the accuracy of time code, truly professional video editing is possible. But before you rush out and invest in a system with time code, you should be aware of these facts:

8mm and VHS video systems were not originally designed to include time code, so all time code systems are improvisations that demand sacrifices of one sort or another.

There are many types of time code, all of which are incompatible and some of which are not transferable.

Some versions of VHS and S-VHS time code do preempt one sound track and may require modification of the sound input/output jacks on the edit VCRs—at a potential cost of $200 per deck or more.

Digital formats, such as miniDV or Digital8 have time code included in their datastream, and their IEEE-1394 ("FireWire" or "iLink") connections allow precise control. Although still a bit more expensive than analog prosumer gear, the improvement in accuracy and image quality is worth the extra money.

Types of Time Code

Time code comes in several mutually incompatible versions:

SMPTE Longitudinal Time Code. A form of time code that you can record on any audio track. SMPTE, pronounced "SIMP-tee," stands for the Society of Motion Picture and Television Engineers, which establishes film and video standards. Usually referred to simply as "SMPTE time code." Used in VHS or S-VHS systems with both linear and hi-fi stereo tracks.

SMPTE Vertical Interval Time Code (VITC). A form of time code recorded in an unused portion of the video signal: the vertical interval. Generally impractical for VHS or 8mm applications.

Rewritable Consumer Time Code (RCTC). A system developed by Sony and available on consumer-level 8mm camcorders and decks.

8mm Time Code. Available on a few Sony industrial VCRs and camcorders, not compatible with RCTC.

Practical Implications

So there's a quick survey of video editing approaches and the protocols that serve them. How can you use this? For now the obvious answer is to determine the most sophisticated control system that your equipment can handle and then use it.

And in planning for the future, remember:

If you have a computer, look hard at computer-based editing systems. After all, you've paid for a lot of computing and communicating power, so why not use it? Editing systems consisting of computer cards and software are cheaper than comparable stand-alone hardware-software packages.

Consider all your equipment needs. Some protocols were designed for 8mm and others for VHS, but editing considerations are not the only reason for selecting one format over another.

Find out what each system will sacrifice. In the small-format video world, trade-offs and compromises are inescapable. In selecting an editing system, be sure you can live without those capabilities it can't provide.

Do your homework. The field of prosumer video editing is brimming with new systems and products. So survey your options carefully before you choose a system.

Finally, whichever way you go, don't be cheap. Invest in the most accurate and versatile editing system you can manage.

Because precise video editing can make the difference between true programs and mere footage.

43
Titles:
From the Simple to the Sophisticated

Larry Burke-Weiner

See video clips at www.videomaker.com/handbook.

The situation: a family reunion. They'd all rather be home wrestling a garden weasel than sitting through the hundredth retelling of how Grandma and Grandpa met over a red snapper.

The problem: you're producing a video of this event that has to be entertaining and cheap. Martial arts are out of the question. Your only hope of rescuing this snoozefest is good titles.

Fade in audio over black frame: Grandma going on about the Edsel and how fluoride ruined the autoworker's union. Fade in white serif letters on black: *What the Heck Does That Mean? Productions Presents.*

Fade out title. Then, when Grandma begins to speak: *It was a Great Big Red Snapper.* The title appears just as Grandma says it. Hold on title for a few beats, then fade to black.

Fade in slowly on Uncle Sol nodding slowly as Grandma speaks. Grandma asks if he's heard this one before. Sol says, "Oh, but I *love* that story. Please, continue."

Fade in white sans serif subtitle: *I can't believe she's gonna tell this story again.*

Grandma begins to tell it again. Sol continues nodding. Fade in white subtitle: *I should've spiked the Tang.*

Fade to black. Fade in white cake with *Welcome Levines!* in red icing. A hand comes into frame and slices a big chunk out of the right corner of the cake. The piece is removed to reveal *REUNION 2000* in red rub-on lettering.

The rest is literally a cakewalk.

As this example shows, innovative titling is an effective and relatively inexpensive way of taking raw (and rather dull) footage and spicing it up.

The choice whether to use an electronic character generator (an electronic keyboard that generates type on the screen), your home computer, press-on lettering, cake icing or sidewalk chalk is up to you—and your wallet. In any case, with a little planning and ingenuity, you can create titles that communicate tone, setting and background information in a visually appealing way.

So what's out there for the videographer who wants to explore titlemaking but doesn't want a degree in electronics or a second mortgage? Let's go shopping.

Titling, Guerrilla Style

As with most aspects of videography, there are many ways to approach the problem of getting titles onto the screen.

If you want interesting title work but don't want to shell out for Digital Video Effects (DVEs) or a character generator, your best bet is the local mall.

As many gaffers will tell you, a good lighting equipment store is your local Walmart, K-Mart, office supply store or hobby shop. This maxim also holds true for finding supplies to create titles.

A trip to the local Ben Franklin can yield colored art board, index cards, a plethora of ink pens, glues, rub-on lettering, felt letters, stencils, clear plastic gels, colored chalk and even plastic google eyes for a doll.

Tacky sounding? Bob Dylan didn't think so when he stood in front of a camera holding cue cards with lyrics scrawled on them for his *Subterranean Homesick Blues* video. Try tossing index cards onto a desk with the credits written on them, or printing the titles on a sheet of paper as the tape rolls. The makers of a Jack Kerouac documentary inked the titles onto a roll of toilet paper, then unraveled them slowly past the lens. Art board with press-on letters—say, white type on black—can be as effective as the titles on Woody Allen's films.

Placing titles on Plexiglas (because it's cheaper than glass and more durable) and shooting through it is attractive, particularly if the background has rhythmic movement, such as the ocean or trees blowing. Try a rack focus (going from one end of the focus ratio to the other) to bring the titles suddenly into the frame.

In-camera titles—those you create with your camera's titler, if it has one—are an easy and cost-free option. You can use them to effectively present time and place, or to convey simple ideas without much pizzazz. However, the placement of the titles, typestyle, color and amount of type tend to be very limited.

For a wide selection of type styles, color and special effects, and a little extra expense ($100-$1,000), use a stand-alone electronic character generator (CG).

With a CG, you can create titles with drop shadows, borders and differently spaced lines that move in a variety of directions, from traditional scrolling (rolling up the screen) to slow fades (where the words appear and disappear in an almost hazy, soft focus).

CGs come in many shapes and sizes, with a wide range of quality, features and price points. One guide to help you with your selection is the nanosecond rate (the speed at which the monitor's electronic beam changes brightness). The sharper the image, the lower the rating in nanoseconds. Thirty-five is the baseline for broadcast-quality titles, so if you're planning on doing broadcast work, use this as your touchstone when selecting a CG.

Let's Be PC About It

For a more extensive selection of font and effect options, a home computer (PC) is the answer.

With computers, you enter a whole new and complicated world of titling. Terms such as alpha channels, anti-aliasing and keying are suddenly thrust into your vocabulary. Font decisions, color choices, RAM and megabytes all dictate the creative palette from which you can create your titles.

Software such as Adobe Premiere, Ulead's Media Studio Pro and Crystal Graphics' 3D Impact! Pro are wonderful packages that can produce amazing titles as well as textures and objects. They also ask a lot from your computer's memory configuration, as well as additional hardware add-ons such as a graphics accelerator and/or a video playback board.

But once you've got the necessary hardware, the tools of PC graphic design (together with a few cheap tricks) can be the answer to creating great looking titles on the home computer. There are a lot of great type creation software packages you can use. Some, such as Macromedia's Fontagrapher, require a knowledge of typography, while others, such as QuarkXpress and Adobe Illustrator, ask only that you be able to type.

For image manipulation, texture creation and special effects, you can use a powerful 3D paint program such as Fractal Design Painter or Adobe Photoshop. With the help of software plug-ins (additional third-party programs that add features to the original package), these amazing programs give you the tools to render any texture or style you can imagine—without hassling with genlocks (the synchronizing of two video signals) or twenty patch cords.

One plug-in that works with a number of software packages, Kai's Power Tools 6.0, has a seemingly endless supply of textures, fractal designs (colorful designs that form repetitive patterns—very cool stuff), gradients (the blending of colors in a pattern) and special effects such as spherizing. It also comes with a program called Quickshow that lets you fade in and out titles in any sequence you want for as little or long as you want.

Once you have your titles, you can position them on your storyboard, directly within your editing system, or save them to a file for later (or repeated) use. If your computer has the capability of producing analog NTSC video, you could run your titles out to analog videotape, or, with an IEEE-1394 port, out the FireWire port to DV.

But none of these tools are of much value if you are not producing titles that your audience can read.

Tools Don't Make The Title

Font style, color, composition, motion, pacing and background all affect your titles' readability. So when you're making these decisions, make thoughtful choices. Ask yourself what would be a visually appealing choice that will compliment the tone of your video production.

In all of your choices, try to match the style of the titles with the style of your video. For example, letters stenciled on an AK-47 for a wedding video could be just the right touch if the newlyweds happened to be NRA chapter presidents.

Here's an example from Hollywood: since the subject matter of the feature film *Sneakers* is computer cryptography (creating and breaking security codes), the opening begins with nonsensical titles in a bold white font; the words appear as though someone is typing them on a computer screen. A few seconds pass before the letters rearrange themselves to form a real name (as if breaking a code). Thus "Fort Red Border" becomes "Robert Redford."

A recent trend in print typography is to challenge the reader's deciphering capabilities by using illegible fonts obscured by photos and miscellaneous art (*Wired* magazine, for example). In some instances, this approach works. However, for video, since viewers cannot ponder a title unless they hit pause (which defeats the whole purpose of video), illegible and obscured type will only incite them to do something more constructive than watching a video with illegible titles.

Don't let this be your mistake. When faced with forty or fifty thousand font styles, the temptation is to overindulge. Fonts are like ice cream flavors; there are millions available, but trying them all at once will make you and the viewer sick. Using ten different font styles for one title conveys a multitude of negative messages such as "confusion," "disorder," "ugly," and "rank amateur with a CG." But if that's the mood you want to create, by all means, overindulge.

Fonts breakdown into two styles: serif and sans serif. This book is printed in a Serif font. This is a Sans Serif font. The difference is in the end strokes of each letter (Sans is French for "without," as in "without a curling end stroke.") Serif fonts are easy on the eye and usually convey warmth. Sans serif type is bold and direct—great for subtitles and in-your-face titles. They can also be cold and unforgiving.

Once you've decided which font is most appropriate for the style of your video, be consistent with its usage. If one title is Garamond Condensed, all the titles should be Garamond Condensed. Sub-titles should never change font style unless you are using the style change as a humorous or dramatic tool.

In addition to selecting a font style complimentary to your video, you should choose the colors for your titles with the same amount of care. Day-glo colors can connote a feeling of garish energy or pop culture. They may suggest aggressiveness or playfulness. Muted pastels hearken to a more pastoral, gentle tone. They suggest subtlety.

Besides the psychological connotations, keep in mind the readability of your color selections. Red or other bright, saturated colors tend to bleed over the type and create a fuzzy look, so avoid them. On the other hand, dark gray fonts will get lost on a black background. Your colors should blend, but make sure the viewer can see them clearly. A blue sky with white fonts and a drop shadow to make the white pop out is a nice blend.

While your titles should be distinguishable from the background, they should not compete with the background. If the background is busy, keep the titles a solid color. If the background moves, keep the type out of the line of movement.

When selecting background images, be creative. If you're doing a production on National Parks, use tree bark as background. If it's a wedding, why not use lace or a shot of the bride's bouquet? Experiment with shadows on a sidewalk or clouds. You live in a world made up of textures and patterns, all free for your use.

Just as you are careful in font, color and background selections, consider the visual message you will convey with the composition of the titles within the frame. In the feature film *Speed*, the titles appear in an angle inside an elevator shaft. Each title wipes clean every time we descend past a steel girder, only to be replaced by a new one. The viewer senses motion and action—as the film's name implies—and has a visual cue to a plot element.

Consider your own titles. Will they appear over moving images? You can center them if the images are pure location with plenty of screen readability, like the Grand Canyon or a freeway.

Titles can appear in a fixed location for stability of composition or in different spots for a more playful sequence.

The rate at which titles appear affects the tone you set for your production. Titles that slowly fade in and out may carry the message "this is a serious/scary/mysterious production." Sudden cuts can give the feeling of urgency or even intentional sloppiness. But no matter how quickly or slowly your titles appear, be sure to leave them on screen long enough for the viewer to see them. A general rule is that a viewer should be able to read a title three times slowly before it leaves the screen.

The same applies to blocks of text. If it scrolls, use the Star Wars rule: take your time.

All of these considerations are useless without knowing your safe title area. This is the area where titles won't be cut off the screen. Use the "eyeball" principle: if it looks good on your viewfinder, screen or computer monitor, bring the titles in another three-quarters of an inch (roughly ten percent) all the way around.

One good thing about titling for video is that the aspect ratio (the dimensions of your screen—TVs are generally 3:4) is usually the same for you as it is for your viewers. This also comes in handy when you generate titles on a computer screen—it usually shares the same aspect ratio (or close to it) as a TV screen.

Annoying Titling Travesties

You've seen them. You've probably done them. Now you're going to hear all about them: titling travesties (subtitle: titles that close eyes and doors).

Mistakes fall into many categories, and each has its own special way of annoying the viewer. Let's begin with my favorite: ignoring the safe title area. The title of your production is *Weddings Are My Life*, but all we see is *dings Are My Li*. I'm annoyed just thinking about it. Remember that you're likely to lose anywhere from 5 to 15 percent of your image around the screen—known as the home cut-off. Allow for the extra space, particularly if you're running titles close to the edges or using subtitles.

Next to nonsensical titles, the other major no-no guaranteed to annoy is the pacing of vital information. It's no big deal to the viewer if the production credits scream past or blip for a few seconds, but we the viewers better know what the name of the production is, where it's taking place (if listed) and what the circumstances are (if you have a scrolling intro). If it's on the screen, it better stay there long enough for our brains to make sense of it. Read it three times, slowly. If you can, chances are your viewers can, too.

However, beware the title that sits on the screen for half an hour. There is no better cue for your viewers to wisecrack about the duration of the title hold and how that must be a precursor for the rest of the video. Ouch.

Fonts. I'd like to say enough said, but I realize the temptation is strong. Don't play mix and match with your fonts. Bouncing around styles and colors will only confuse. Also, be aware of your style choice as it relates to the content. While you're at it— stay away from busy background textures that cause a *moiré* pattern (an image that appears to throb) on your screen.

If you're going to use objects arranged to spell words, make sure the viewer is able to identify the word at a glance and the object. Why the object? Because we the viewers will assume the object has bearing on the content (that's why it's used for titles), but if we can't make it out, we'll miss the meaning (and be annoyed that we're missing something).

These are the major mistakes with titling. Of course, there are others, such as illegible handwriting, misspellings and color bleeds, but then I'd just get too annoyed talking about them all.

Titles can be as fun to make as the production footage, particularly if you get to make fun of your Grandmother's story about meeting Grandpa. In which case, you might want to make the subtitles transparent 4-point type—to stay in good with the family.

44
Adventures in Sound Editing: Or How Audio Post-Production Can Make Your Videos Sound Larger Than Life

Armand Ensanian

See video clips at www.videomaker.com/handbook.

Imagine the tape you made of your kids' last campout. It's got some crickets on the sound track. At night, didn't those crickets seem to chirp louder as campers grew quieter and the sky grew darker?

When Bobby started telling Billie Jo the story of the cricket that ate sisters, didn't those chirpings seem even louder to her? With some equalization, reverb and creative mixing of your original sound track, you could let your viewers hear those monster crickets the way Billie Jo heard them.

Sound editing can turn commonplace video events into adventures that seem larger than life. With some simple electronic equipment—some of which you may already have on your stereo system—you can polish up any raw sound track. Here's how.

Know your Sound

To make good video, you need to understand light; likewise, to make good audio, you will need to understand sound.

Webster defines sound as "mechanical vibrations traveling through the air or other elastic medium." How many times these vibrations occur in a second is the *frequency* of the sound. A tuning fork vibrating back and forth 1,000 times per second generates sound with a frequency of 1000 cycles per second or Hertz (Hz). If it vibrated 200 times per second, we'd hear a frequency of 200 Hz.

Variations and combinations of frequencies account for all the sounds we hear. At best, the human ear can recognize frequencies between 32 and 22,000 Hz. We can perceive frequencies below 32Hz, but as vibration, not sound. This range deteriorates with age or abuse. It is not uncommon for senior adults to top out at 9,000Hz, while losing all low-frequency sounds below about 150Hz.

People with ear damage will adjust sound to match their deficiency. For example, a sound engineer at a rock club may set up the PA to produce ear piercing highs to compensate for such a loss of his high frequency sensitivity. So if you plan to use the services of a studio, make sure your sound engineer can hear.

The interaction of two or more frequencies creates a third sound called *harmonic*. Harmonics, also known as overtones, give sound its life, allowing the human ear to distinguish one voice from another. Poor recordings reduce or eliminate harmonics, turning sound into an unintelligible mess.

Good recordings recognize the harmonics of a sound track, and enhance them—with the help of some snazzy audio postproduction devices.

The Mixer

The single most important tool for audio production is the *mixer*.

A mixer's number of input channels determines how many different signals—read sounds—you can work with simultaneously. You need at least two for mono recordings and four for stereo. More inputs allow you to control more sound sources.

Remember that the tape speed and audio recording format will determine the frequency response of the recording medium. A VHS tape recorded on the linear track at EP speed may yield a frequency response no higher than 7,000 Hz. So stick with hi-fi audio if possible.

The first step is to hook up the right cables to the proper inputs and outputs. Do not scrimp on cables, because bad connections can create a lot of buzz and noise. A good mixing console will allow an input signal to be either mike or line level. This compensates for low-level microphone signals and high-level line inputs, such as those arriving from VCRs, cassettes or CD players. Output will be mono or stereo, depending on your VCR.

Mixers have sliders, or *faders*, that control volume for each channel. A master fader controls overall volume. Mixers will also feature all or some of the following: attenuation, equalization, cue sends, pan pots, solo switch, monitor control, Volume Unit (VU) meters or Liquid Element Displays (LEDs) and echo/reverb.

Attenuation cuts down high input signals to prevent overloading.

Equalizers allow precise tone adjustment of selected frequency ranges.

Cue sends can send the signal to the video tape, headphone or monitor speaker.

Pan pots control the spatial position between right and left stereo channels.

The *solo switch* "listens" in on any individual channel without interfering with the recording process.

Monitor controls adjust headphone or speaker volume. We'll look at echo and reverb later.

The Equalizer

An *equalizer* (EQ) is the most common piece of signal processing equipment. It allows you to divide the entire (human) 32—22,000 Hz audio spectrum into separate bands you can adjust independently. With an EQ, you can raise or lower the levels of these bands to change the tonal characteristics of a sound, reduce noise and even create certain audio effects.

Many mixers have a simple EQ built-in, allowing you to adjust high and low frequencies independently. This is fine for broad tonal changes, but for more control, you'll need a graphic equalizer. Though a handful of mixers offer this level of tonal control, most graphic EQs are external units.

Whereas simple EQs use knobs—one for treble, one for bass—graphic equalizers have vertical sliders that show the amount of correction you applied to each band of frequencies. Thus you can tell, at a glance, what the graphic EQ is doing to the audible spectrum. Hence the term graphic.

Think of the graphic EQ as a vastly expanded bass/treble dial. The more sliders, the greater the selective control.

Say you have some music on your soundtrack that sounds flat and unexciting. Try boosting the 100 Hz and 10,000Hz sliders, while lowering the 1,000 Hz to 500 Hz sliders. The graphic pattern on the equalizer will look like a suspension bridge—what you'll hear is rock concert sound.

If you're working with classical music, you may need to elevate the middle frequencies to bring out specific instruments. And Aunt Trudy's squeaky voice may require you to slide down the high and upper midrange frequencies.

Hooking up a stand-alone EQ is easy—simply connect it between the source output and mixer input. Or, if you want to alter your whole sound mix, place the EQ between your mixer and record deck.

On the Level

In the world of audio post-production, the strength or "level" of an audio signal is as important as the way it sounds. We have our ears to tell us that a sound is tonally correct; for clean recordings we need something to tell us about the signal's level.

VU meters and LEDs monitor the levels so you can prevent overloading and distorting the signal. VU meters use a unique scale calibrated in decibels (dB), which measure actual signal strength.

VU meters boast an ascending scale of values that start with negative numbers. Instead of being at the bottom, zero is near the top of the range. Why? Because zero dB indicates a full-strength signal. This way it's easy to remember that if a signal strays much above zero, distortion may result.

Sometimes, simply setting a record level based on the VU meter isn't enough.

Some sounds jump wildly from loud to soft, making them distorted one second and virtually inaudible the next. The difference between these very loud and soft sounds is the *dynamic range.*

Changes in dynamics, at least within reason are good. It's the dynamic range of a recording or vocalist that coveys realism

and power. Streisand's vocal style is dynamic, my grandmother's isn't.

But when recording sounds, too much dynamic range can be a problem. Thankfully, there's help available from a nifty device called a *limiter.* A limiter reduces instant signal peaks, such as a loud scream or feedback from an electric guitar. A limiter allows you to set a maximum signal level; the unit holds all sound under that limit.

A *compressor* works somewhat like a limiter, though its effects are less dramatic. Radio stations use compressors to maintain high signal levels without fearing distortion. The small sacrifice in realism may be worth it if you plan on taping loud live concerts or monster truck rallies. Good production mixers have limiters built in, for controlling sounds at the time of recording.

Reverb

Sound loses energy as it moves through the air. Sound strength drops 6 dB every time you double the distance between a mike (or ear) and the sound source. This is why good microphone technique includes moving the mike close to the sound source during live shooting or narration dubbing.

Sound waves bound off walls and hard objects. Echoes are bouncing waves that follow more than 1/20 second after the original sound; reverb is repeated reflecting waves that sound almost continuous.

Post-production electronics can simulate both these effects. Echo and reverb units can run the gamut from simple tape loops, playing back what you record immediately, to sophisticated digital reverbs simulating rooms of all shapes and sizes.

Reverb units bring fullness to sound. With a reverb you can make a three-piece band playing in the basement sound like a major concert hall event. Reverb emulates the sound of a big hall. But watch out—too much reverb will make things muddy.

Reverb also adds authority to a narrator's voice. Rock and roll's ultimate DJ, Cousin Brucie, uses a lot of reverb.

Reverb, particularly when applied to drums, can make a band sound bigger than life on tape. Now you know how rock bands maintain the ambiance of the concert hall on recorded media.

Once it's on tape, you can't eliminate reverb easily from a recording. That's why recording engineers add it later for maximum control. Use sound-absorbing dampeners—rugs, blankets, egg cartons—in small studios whenever recording to capture a clean dry sound. Then employ electronics in post-production to simulate real room resonance.

Digital delays are similar to reverbs. They electronically delay the input signal for a selected amount of time. You can produce echoes of extremely short duration. Use digital delays to double a vocal, and you can make it sound as though two people are singing together. And, if you take one output channel from the delay and run it through an EQ, you may even make it sound like two different people. The possibilities are infinite.

Phase shifters delay the incoming signal slightly, causing the delayed signal to partially cancel the original. This causes whooshing sound effects—good for planes, trains and other speedy objects.

Going for the Take

Okay, you've got your mixer and your equalizer. Now you must decide which sounds you need for the mixing session.

Much pre-recorded material is available, but you may want to experience the thrill of creating you own sound effects (SFX) like they did in the old radio shows. Crumpled cellophane, for example, makes good rain or eggs frying. But keep in mind you only have two hands to work with and will need them for the mixing board. If you can, have an assistant with you during a mix.

Say you're audio mixing a wedding tape. You can "sweeten" up all your location audio through use of both technique and equipment. You may need to clean up and equalize the live audio vocals a bit.

You'll also want to work on the sound from your outdoor segments—selectively equalizing background noises or reduce them manually during spoken passages. That waterfall, for example, near the wedding party blocked out some of the vocal interaction. An EQ can help, as well as a touch of reverb on the vocalists' voices.

When possible, record musical interludes preceding the ceremony—such as a soloist's number—on cassette rather than relying on a room mike during the shoot itself. If the bride doesn't object, the artist may have a professionally recorded tape of the same material you can use to replace the live track. After all, we are not focusing our attention on the soloist unless there are a lot of close-ups requiring lip sync.

In almost any kind of production, consider using background or "wallpaper" music track for continuity. It will fill in those silent gaps often associated with live footage. One cheap trick; try an inexpensive keyboard with built-in rhythm sounds as background. Use the individual slider on the mixer to boost the volume of the background gently during these silent periods.

If you're relying on pre-recorded music, find selections that don't clash with the theme of your video. For a wedding, don't use anything overly aggressive or dynamic; instrumentals are a safe bet. You may wish to sprinkle in some sound effects like ambience or laughter. Stock music of applause and laughter may follow special introductions at the reception.

By now you will have run out of hands. Starting the CD player just in time while cross fading from the live track to an over-dub makes this a job for an octopus. Pro studios use computer sequencers and remote controls to help. It is best to try a few dry runs before actually recording onto the final tape. While practicing, send the mixer output to a cassette recorder so that you can listen later. It is very difficult to be objective while working the mixer.

With most projects, you'll find it challenging at best to audio edit the entire length in one pass. Use cuts and scene breaks in the video for audio transitions and segues. Have an assistant keep records of tape count and time passed as editing cues. Have them act as audio directors, coaching you through the moves.

Complicated editing may require sound recording the output to a tape recording before laying it on the videotape. A minimum of two output channels will be sent to corresponding tracks on a tape recorder. You may use the individual left and right channels of a stereo cassette recorder for two-track mono recording. This is ideal for adding narratives that may require numerous takes. You can then mix the two-track master directly onto the video. You'll have control over each individual channel.

Successive generations do add noise, but what a small price to pay for such flexibility. You can also add noise suppression or filters for the final mix-down. Four, eight, twelve, sixteen and twenty-four track audio recorders are available at recording studios for complicated mixes of numerous audio elements.

It takes a lot of practice to learn proper audio mixing technique. Even a small video switcher/audio mixer demands a lot of attention. The results, however, are light-years ahead of what you produce in-camera.

45

In The Audible Mood:
Sound Effects and Music, Evocative, Legal and Inexpensive

Armand Ensanian

Imagine *Popeye* without *toot toot*, *Casablanca* without *As Time Goes By*, *Jaws* without *bum BUM bum BUM bum BUM*. The soundtrack is the very lifeblood of a video, setting the mood and enlivening each scene of your work.

The trick is to find the right set of sounds to accompany your video. The options are many, from creating sound effects yourself—a la old-time radio—to buying mood music from a music library. In this chapter, we'll explore these options, and discuss the legal and financial ramifications of your choices.

The Copyright Challenge

The availability and affordability of a simple special effects generator (SEG) with built-in audio mixing has prompted many videographers to try adding music to video during editing.

It's harmless fun, attaching a favorite song to a tender moment between mother and child, a hot rock tune to fast-paced footage at the track or some Benny Goodman to Grandma and Grandpa's 50th anniversary party tape. After all, you're not planning to make thousands of copies for distribution. What are the chances that the composer, publisher or lyricist of the songs will ever see this tape, anyway? Slim at best; no one's going to bust you for borrowing a tune or two for personal use.

The trouble begins when you turn pro or even semi-pro.

Every serious videographer will eventually land a real-world assignment: a wedding video, local commercial or contest entry video. Whatever the application, you cannot use someone else's work without permission in any video offered for sale, profit and/or distribution. Music is like photography, sculpture or any art you can hold in your hands. Yet the fact that you can't hold it in your hands makes music ripe for theft by otherwise law-abiding citizens.

You needn't stoop to theft. You can buy great sound effects and music for your videos from a number of sources. The first is the most expensive. It involves buying the rights, for one-time use, of a pop song performed by a noted artist.

Your clients will request this sort of thing often. After all, most clients can only relate to what they know and hear on the radio. You'll have to explain that you simply can't use the pop song without getting permission from the artist's representatives first.

This means the artist's publisher or licensing agency, such as the American Society of Composers, Authors and Publishers (ASCAP). You've seen the name on almost every record, cassette tape or CD you own. ASCAP operates like a big collection agency, collecting fees for the use of their artists' material. It's a fair system that keeps artists from getting ripped off—but it's an expensive one for videographers.

The price you pay for one-time use rights of a pop song depends on how you distribute the tape. Radio stations, TV stations and local bars with a jukebox pay agencies like ASCAP a blanket fee.

This fee may range from thousands of dollars for a radio station to a few hundred dollars a year for that jukebox. (For more info, call ASCAP in New York 212-595-3050.)

Unless you expect your video to make a big profit, paying for rights to use a hit tune may be unwise. There are cheaper ways to go.

One way is to create your own music. If you've the talent and the time, this is a worthwhile option. You can use your desktop video equipment to make music; MIDI interfaces to Macs and PCs provide the musically inclined with unlimited creative potential.

Software and sound cards can produce digital sound that rivals the best CDs. Some even come with hundreds of digital sound samples, ranging from pianos to xylophones.

Despite all this automation, you'll still need to have a musical ear, something that's not included. If you don't have an ear for music, find local musicians to participate in your production. They'll have both the equipment and the ear you need. You'll find, however, that the more people involved, the longer the process. Just make sure all parties sign an agreement transferring the rights to use the material over to you. Handshake deals don't carry the weight they once did.

The simplest solution, and perhaps the best bang for the buck, is to buy the music from a music library. The quality of the material will be as good as or better than you'd expect. You have heard music library soundtracks all your life. Commercials, opening scenes to sport events, TV news shows and presentation videos all use music made to augment video productions. In fact, most of the music is so much like what you hear on the radio that it's often easy to persuade clients to use music library tracks. Moreover, much of what's available was written and composed by award-winning musicians; the quality is first rate.

Libraries

You'd be surprised at how well music libraries categorize music ... broadcast promotional, broadcast show theme, corporate imaging, corporate icon build, credit roll, documentary, events such as weddings, birthdays and anniversaries, industrial presentations, news and information, movie soundtrack, retail presentation, retail promotional, sound-alike, sports/action and underscore—to name a few.

All of us have a good sense of what these many categories sound like. Just watch some TV or look at a promo tape at the local travel agency. Still unsure?

Music libraries will gladly send you a tape or disk of material for approval for a specific tune; some even offer sample CDs.

There are many ways to buy music from music libraries. The traditional method charges per needle drop. This means you pay only for a specific selection of music used from their records, tapes or CDs. The term comes from the days of the phono-graph, but applies to tapes and CDs as well. Most large music libraries will help you find exactly what you need. Computer databases allow quick searches from tens of thousands of titles.

The dollars add up quickly here, but you may obtain a very distinct tune, for a one-time production, that's not available from any other source. This is ideal for videog-raphers who don't produce a lot of videos, but who need a unique style of music when they do.

Music library charges for needle drops depend mostly on the extent of use. For example, the charge for broadcasting the same music on the same video will vary, depending on whether it's broadcast in Smalltown, USA or Metropolis. Determined by the number of viewers, nee-dle-drop fees may range from $150 for a network TV broadcast opening title to only $50 for a local TV program. The only other cost: $20 for leasing the CD that contains the material. You make your choice from the CD, report the piece of music you will use and pay for it.

There are other methods of dealing with music libraries. You may buy a produc-tion-blanket where you pay only for the music for a given production. Typically you select a CD with a variety of musical themes from a library, and use as much of it as you wish, provided you restrict that use to that production. Costs depend on the size of the project and distribution, but average in the hundreds of dollars.

Annual-blanket fees allow you to buy licensing rights to a group of CDs for unlimited use during a given year. Popular with high-volume videographers, radio and TV stations, fees from reputable large firms are around $1,000 for a group of over a dozen or two different theme CDs for the year. You can also sign multi-year contracts with CD upgrades.

The best value: a buy-out library. Here you pay a one-time price for a CD, or set of CDs, that contains the music you need. You may use the CD as often as you wish, for as long as you live. You can better appreciate this deal when you know that you can purchase buy-out libraries for as little as $5 per sample tape from small independent producers advertising in the back of magazines.

But cheap can also mean poor quality. Listen to sample disks and tapes before you buy.

Sound Subscription

You may want to subscribe to a library service. Here you receive a CD every month or two, and have the option to keep it for a one-time fee. Beware, though: you may soon find yourself way over-stocked, and stuck with soundtracks you'll never use.

It's like that videotape collection you started a few years back; despite the vari-ety you always seem to stick with a few favorites. A variation on the subscription theme: lease the entire contents of a music library for a term, such as a year, and receive new CDs every month or two to add to that library.

One last note of caution: music libraries are licensed to individuals or production companies. Borrowing a library CD from a friend, and then putting your name on the finished product is a direct violation of copyright.

Stay honest. It'll pay off in the long run.

46
Dig 'Em Out, Dust 'Em Off

Jim Stinson

Sooner or later, you'll come upon that drawer awash in color print packets, that carton bulging with slide boxes, that crate stuffed with round rolls of movie film and you will feel regret and a sort of mild guilt. All those precious records. All that effort and money. And you never look at them.

Why? Because there has never been a quick, convenient way to do so. Sure, you can arrange prints in albums, but for slides or movies you have to haul out the projector, wrestle with the screen, load the trays or thread the film and round up a patient if unenthusiastic audience.

So you don't, and there your family pictures sit, year in and year out, in mute but eloquent reproach.

But now that you have a camcorder, you have a way to turn your photographic archives into video programs that are as exciting as they are easy to show. You can transfer your movies, slides and prints to tape. And in doing this you can enjoy the fun and satisfaction that truly creative work delivers, because you aren't just

copying materials, you're transforming them into video programs.

So this chapter offers a bundle of tips to help you copy photo materials to video. We'll start with suggestions that apply to all video copying, whether the source material is movies, slides or prints. Then we'll cover transferring movies and slides.

The chapter that follows this one concludes with a look at the challenge of videotaping photographic prints and other artwork.

General Tips

Here, then, are tips that apply to all types of video copying.

Copying with a Monitor. First, always use an external monitor to set up and check your work. It will give you a far more accurate image than you can obtain from any camcorder viewfinder, and in copying it's essential that you frame your originals precisely. You can run a cable from most camcorders to a monitor through composite or S-video jacks.

Adjust the monitor to show color, brightness and contrast as accurately as possible.

If you can't generate color bars and set up your monitor on a vectorscope, use a good quality store-bought video as a reference standard. Once you figure out how the tape should look, you'll be able to use it to dial in almost any monitor.

I set up my monitors with a commercial exercise tape. It has high video quality, carefully lit flesh tones for many skin colors and bright but realistic colors.

Rely on your monitor as you check and adjust focus, composition and color balance.

Controlling Color. Color balance is critical in copying. Usually you want to match the colors of the original, but sometimes you need to change the original colors to correct them or create special effects. To help control color, make sure you set the camera's white balance correctly. Most modern movie and slide projectors—as well as the lights on pro copy stands—use lamps with a color temperature of 3,200K (degrees Kelvin). That is exactly the color temperature of your camera's "indoor" white balance setting, so set your camera to indoor—or "incandescent". If the color balance seems off when viewed on your well-adjusted external monitor, warm the light or cool it with photographic color compensation (CC) filters obtainable at photo stores.

For more color control, run the image through a color processor before recording it. To do this, don't use the "corder" part of your camcorder. Instead, just run the video signal out of the camera section to a color processor. From there, run the signal to a VCR.

Then, using your camcorder as if it were simply a camera, record the processed image directly on the VCR.

Steady as She Goes. Since a still image shows no movement of its own, it reveals the least little bit of camera shake; so be sure to set up your camcorder on a sturdy tripod. Here's a cheap trick if you need to

set your camcorder securely on a table or other flat surface. Obtain a bolt threaded to match the unit's tripod socket and use it to secure your camera to a simple plywood base plate. Recess the bolt into a single piece of plywood, or put small feet on either side of the bolt.

If your camera has a remote control, by all means use it. You'll find many camera *on/off* buttons on the handgrip, so pressing them will no doubt jiggle the camera. Using the *record/pause* button on the remote guarantees jitter-free images. On some camcorders, you can even control the zoom, add titles and engage autofocus from the remote. Perform photo transfers, and the infrared remote becomes your best friend.

Keeping Things Quiet. Your final program will probably have sound—narration, music or both—but when you're recording you don't want to pick up projector noise or other ambient sound. To prevent this, disable your camcorder's microphone by inserting a plug into the external mike jack. The trick: use a plug that's not connected to anything. I find that a mini-to-RCA converter plug works fine with my camcorder.

On the other hand, you may wish to use the video soundtrack as a notepad to record data about your pictures as you transfer them. ("That's Aunt Florrie and the film can says Lake Runamuck, Summer, 1956.") When you assemble this raw footage into your final program, you can use these vocal notes to help create your final narration.

These tips cover all types of copying. Now let's look at copying movies and slides, beginning with some suggestions that apply to both media.

Copying Projected Images

First of all, decide whether you want to use conversion hardware or simply record the projected image off a wall or screen. You've seen film/slide transfer systems—arrangements of screens and mirrors that let you set up projector and camcorder at right angles and videotape the projected image.

While these can be quite useful, they sometimes limit your flexibility in selecting which portion of the image you want to record. It may be better and cheaper for you to project the original images onto a white surface and aim your camcorder directly at that surface. To make high-quality copies with this front-projection system:

- Make sure that projector and camcorder are perfectly level and that their lenses are at the same height.

- Place projector and camcorder as close together as possible, at equal angles to the screen. This will keep the image borders rectangular and the pictures undistorted.

- Use a smooth, white screen—never a beaded movie screen. I find that the white back of a poster printed on glossy stock works fine. Almost any high-quality white paper will work. Don't be afraid to try many types of papers—after all, they're cheap.

- Place the screen as far away as you can, while still getting an acceptably bright image (use your external monitor to check). The farther the screen from the projector and the camera, the shallower the angle between them—and the smaller the chance of picture distortion. Put it too far away, and the camcorder may try to compensate for the dimmer image by cranking up its gain. The result: a grainy image.

By setting up your copy operation like this you can make high-quality transfers and retain better control over the images. If those images are 8mm or 16mm films, you'll need to make a few extra adjustments.

Copying Movies

Movies are simple to copy to video because the screen proportions of the two media are the same: 4 to 3. Simply adjust your camcorder lens so that the projected image fills the frame and you're in business.

But movies are also hard to copy because of differing projection speeds. Your camcorder, of course, records thirty frames (images) per second. Film, on the other hand, runs at 24, 18 or 16 frames per second. The trouble is, these are only nominal speeds, and none of them matches the 30 fps of video. As a result, an unpleasant flickering effect often marks video transfers.

Fortunately, most film projectors have speed controls that you can use to vary the actual number of frames per second. The best way to use the speed control is to change the film projection speed gradually while eyeballing the effect on your monitor. When you've found the best setting for your video outfit, leave the projector at that speed and transfer your footage.

You can also stop many projectors so that they project a single frame as a still image; you can transfer these frozen moments to video. The problem is, once the movie film stops in the projector gate, it's protected from the hot projector lamp by a thick glass heat filter that degrades the quality of the image. If you have a good four-head video system you may get better results by transferring the movies in real time, displaying a selected frame as a video still and then re-copying that.

A final tip: because of the magnification required of tiny 8mm film images, they often show dirt, fuzz and fingerprints. Clean movie film by pulling it gently through a fine, lint-free cloth saturated with film cleaner bought at a photo store (or with carbon tetrachloride).

Never use a water-based solution; film emulsion is industrial Jell-O and you can guess what happens when it's soaked in water. Also, be sure to clean the gunk out of the projector's film gate—that's the pair of metal plates holding the film as it's projected.

Copying Slides

Whether you use rear-screen or front-screen projection, copying slides is much like copying movies. But copying photographic transparencies can be trickier because their shape does not match that of a video screen in two crucial ways. First, slides orient vertically—portrait—as well as horizontally—landscape. Secondly, even horizontal slides will not fit a TV screen because the ratio of their sides is three to two instead of four to three (see Figure 46.1). So, in transferring slides, you must compensate.

Timing Your Slide Transfers

A slide will sit there on the screen until you advance the projector tray, so how long should you roll tape on each photo? I find that between 5 and 20 seconds per image is ample, depending on how much there is to see. Clearly, you don't need as long to look over a simple road sign as you do the ceiling of the Sistine Chapel. If you doubt that 20 seconds is longer than anyone wants to stare at even the most visually interesting slide; try it, you'll see.

Pre-Programmed Slide Shows

Many dedicated slide photographers show their programs on two or more projectors, dissolving between them with a programmer unit controlled by pulses on an audio-cassette. The cassette also carries a synchronized sound track.

By far the easiest way to transfer slides to video is by creating a complete program in this manner, plugging the cassette player audio into the camera's line-in jack and recording the program in real time. Another advantage: your complete program is first generation video! The big disadvantage: you can't compose your video frame for each slide.

But whichever way you choose, you'll end up with video footage that you can edit into a program you'll be proud to show.

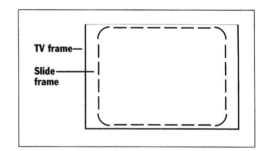

Figure 46.1 *Copying slides can be tricky due to their shape.*

47
Easy Copy

Jim Stinson

Video is probably the most convenient and effective way to display precious movies, slides and photographic records. The preceding chapter showed you how to transfer movies and slides to video. Now we'll look at the fascinating process of copying photographic prints and other forms of flat art.

To transfer flat art to video, you'll need a copy stand, copy lights and suitable backgrounds on which to place your subjects. You can put these elements together cheaply and easily.

Your Basic Copy Stand

A *copy stand* simply holds the camcorder, lights, and artwork (see Figure 47.1). Still-camera copy stands, available at better photography stores, aren't always suitable for video, however. Many video lenses won't focus close enough at the camera-to-subject distance imposed by copy stands intended for 35mm cameras. You'll often get better results by assembling your own video copying stand.

In building a stand, consider how to orient it. You can use your tripod as camera mount for a horizontal stand, but holding the artwork in place vertically can be a nuisance (see Figure 47.2).

On the other hand, a vertical stand requires you to build a special mount for your camcorder, since it's usually impractical to shoot straight down from a tripod (see Figure 47.3).

I find it's best to compromise. A rig set up at about 30 degrees from horizontal lets you use your tripod, but still prevents photos from sliding off their backgrounds (see Figure 47.4).

Elemental Light

Now that you have a place to shoot your photos, you need to light them. Rig your lights on either side of your artwork support, at 90 degrees from each other and 45 degrees from the artwork and camcorder. Since the angle of incidence equals the angle of refraction (as you recall from

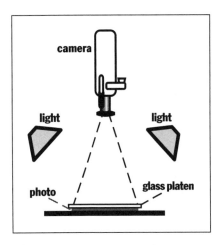

Figure 47.1 *A copy stand.*

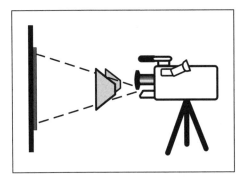

Figure 47.2 *A horizontal stand using a tripod.*

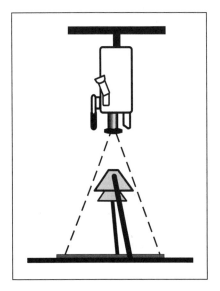

Figure 47.3 *A vertical stand requires you to build a special mount for your camcorder.*

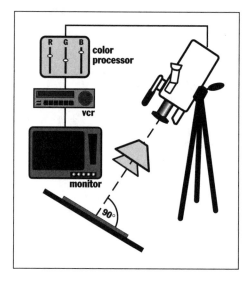

Figure 47.4 *A rig set at about 30° works well.*

high school physics), lights placed 45 degrees off the lens axis will not reflect into the lens (see Figure 47.5).

What kinds of lights? If you copy nothing but black and white originals, a pair of fluorescent shop lights work great and cost very little. You can't use them for color, though. They may seem to match daylight white balance, but they really don't—and the mismatch will be visible when you copy color prints.

For color, use halogen (quartz) lights, whose 3,200-degree Kelvin color temperature exactly matches your camcorder's "indoor" or "incandescent" white-balance setting. (Never use standard household bulbs. Their lower color temperatures create a slight orange cast.)

For inexpensive halogen lighting, buy a couple of clamp-on work lights and fit them with halogen PAR floodlights.

Or modify a halogen work light—the kind with two heads on an adjustable floor stand—for a better and more versatile lighting setup.

If you're at all handy, you can modify one of these work lights for use as an all-purpose video light. Or dismount its heads and rig them as individual copy lights. Simply replace their electrical cords with plugged power cords, add a junction box to the lamp yoke, and

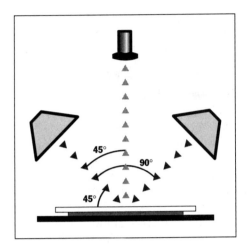

Figure 47.5 *Lights placed at 45° off lens axis won't reflect into the lens.*

remove the lamps from their yoke to use them with your copy stand (see Figure 47.6).

If you don't care to mess with electrical modifications, you can buy these same halogen heads as individual work lights that squat on low feet.

With camcorder and lighting in place, you need a way to support—and display—your photos or artwork.

A Little Background

For simply copying drugstore color prints, you may not need a background for your artwork. Instead, simply set your rig so each print fills the frame. But to transfer images that don't fit the 4-to-3 video proportions, you need a background to fill the rest of the frame around them.

For an "invisible" background, you can place your photos on black cloth or cardboard. But this is rarely successful since bright copy lights invariably bring out some backing texture, resulting in tattletale gray on your screen. It's better to back your artwork with a dark, pleasing color like burgundy or rich blue.

Textured backgrounds (like burlap, monk's cloth, or velvet) contribute even more. The 45-degree lighting accents the background texture, adding dimensionality to the shot. For variety, try inexpensive dining table place mats. They come in a wide range of colors, patterns, and textures—and they're just the right size.

Perfectly flat prints and artwork are rare, so you may need to press them against their backing with a *platen*—a simple pane of glass. Professionals use "optically flat" glass free of imperfections, but you'll do well enough by checking the glass sheet for ripples and distortions before you select it.

Be careful—glass is notoriously reflective and hard to light.

Overcoming Problems

To light materials for copying, you must cope with reflections and excess heat from the lights.

To defeat reflections from a glass cover or from glossy color prints, make sure to set your lights at the recommended 45-degree angle. If that doesn't work, try a polarizing filter on the camera lens.

To completely kill reflections, some professional copiers use polarizing gels on the lamps as well.

Heat from your lamps can cause problems. Halogen bulbs put out enough of it to damage delicate artwork if positioned too close. Fortunately, two 500-watt halogen lamps throw plenty of light on the

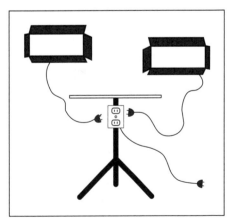

Figure 47.6 *A work light modified for use as an all-purpose video light.*

subject from four or five feet. (Be careful how you handle hot lamp housings. Professional gaffers wear gloves.)

But you need to keep the lights as close as you prudently can, to increase "depth of field"—the zone behind and (especially, in copying) in front of what you focus on that's still acceptably sharp.

In copying, you're usually working with your zoom lens set at its longest focal length and focused at its shortest distance. Both reduce depth of field to the minimum. Since the size of the lens opening also affects depth of field, you need lots of light to achieve the smallest possible aperture.

As an experiment, I aimed a camcorder at a copy surface and held a ruler at right angles to that surface. I studied the resulting image on a monitor to see how much of the ruler remained sharp as its markings got farther from the plane of focus. In ambient room light the depth of field extended less than 1/4-inch from the plane of focus. But when I turned on two 500-watt halogen lights the zone of sharpness jumped to over an inch.

You need as much depth of field as possible for two reasons. First, despite your best efforts, your camcorder may not be absolutely at right angles to your artwork. Part of the art will be soft if the depth of field is too shallow.

Second, you need a bit of focus leeway. In copying, you often have to focus by moving the camcorder instead of by adjusting the lens. You generally have your lens set at its closest focus, and can't adjust it any further. To fine-tune the focus you have to move the camcorder slightly instead.

Why not get around the problem with your lens's macro setting? Many camcorder lenses offer macro capability for close work, and yours may be fine for copying.

I tested the same camcorder with it's 8-80mm zoom and 4-foot minimum camera-to-subject distance and found that at 80 mm, the minimum field was 2.25 by 3 inches—small enough to copy wallet-size pictures.

When I switched to macro (which operates only at the wide-angle end of the zoom range), the minimum size was almost the same, but the lens hood was now only half an inch from the artwork! Needless to say, this makes lighting the art impossible.

In shooting photographic prints a color processor patched between camcorder and recording VCR can be indispensable.

You can color-correct age-faded photos. Typically, elderly color prints that have been exposed to light show a sickly magenta cast. Adjusting color balance can reduce this somewhat. And color adjustments can improve the originals: pump up the orange glow of a sunset or enhance the blue-white of snow. Add a slight sepia tone to black and white prints for an old-time effect (but don't overdo it—it can look hokey). Or set the colors to zero to guarantee black and white originals don't take on a color cast when viewed on a color monitor.

The Copy Session

Changing camcorder distance, focus, lighting, backgrounds, and such can be tedious, so the best procedure is to gang all similar artwork together. Sort photos by size, and then by portrait (vertical) or landscape (horizontal) orientation.

Sort by proportion, too: prints from 35 mm originals are usually 3 to 2; instamatic or Polaroid prints have different proportions. The more originals you can copy before you have to change the setup, the faster the process will go.

One nice thing about copying is that your finished programs can be first generation video. To do this, I create a complete music track, dub it to an assembly tape, and then load the tape into a VCR.

Next, I set up my camcorder to copy photos and cable it to the VCR. Using the video insert mode, I record the camera signal, timing it to the music as I would any editing element.

For stereo hi-fi sound and/or narration, copy the finished tape, relaying the music and adding voice-over narration. The final program tape will still be only second generation.

What's in Store?

The exciting future of video copying is already here! You need only a desktop computer, a card that accepts and outputs NTSC video, and a software package for retouching images.

You import a photo from your copy camcorder to your computer, which digitizes the video signal from NTSC. You take all the time you need to revise and improve the picture to your taste, reconvert it to NTSC and export it to your VCR.

You can remove Junior's acne; improve color, brightness and contrast; erase cracks and spots from that priceless family daguerreotype. Software packages will combine photos, mix photos and artwork, and "morph" a person into somebody (or some thing) else!

All the hardware and software you need for a "Digital Darkroom" is currently available and—with the rapid decline in prices—widely affordable. But even without computerized retouching, you can use your videography skills to pull those wonderful photos out of dark drawers and dusty albums and put them up on the screen where everyone can enjoy them.

PART V
Tell the Story Better

Traditional Distribution Channels

Your video has not been completed until it has been seen. Yet, with all the advances in tools for creating video, tools for getting video out to viewers have lagged behind. Nevertheless, there are ways to put your work before the eyeballs of an admiring audience.

Currently, distribution comes in two flavors: traditional methods and "new media" methods. Traditional methods range from VHS tape duplication and mailing to leasing time on cable television. These methods tend to be expensive, difficult to negotiate or both. On the positive side, they can deliver relatively high quality video and audio to your audience. They "tell the story better" than the new media channels.

Part V of the *Handbook* will explore those traditional media that can be accessed with a reasonable amount of work at modest expense. These include cable TV by way of public and leased access methods, and broadcast TV through UHF and low power TV stations. We've also thrown in some tips on planning VHS tape distribution and promoting your products and video services.

"New media" are currently the opposite of the traditional media. That is, they are generally easier and less expensive to access, but currently deliver lower quality audio and video to your audience. In other words, they enable us to "tell the story sooner" rather than "better." This is fast changing, but we're getting ahead of our story. We'll visit the new media in Part VI, but let's explore the traditional media first.

48
Commercial Distribution: Mapping Your Way to Financial Success

William Ronat

See video clips at www.videomaker.com/handbook.

Making video means working with motion, sound, color, words, composition, light—all combined in a maximum creative effort. That's fun.

Many have produced enough video to get good at it. This means putting in long hours planning, shooting and editing programs. That's work.

When you work for a company, you expect a paycheck when you finish your work. That's reasonable.

When you work for yourself, however, you can expect to work just as hard to sell your product as you did to make it. That's life.

Fun and Profit

To sell your product, several things have to happen. People have to know your product exists. They have to decide that the product is something they want. Then they have to pull their money out of their pockets and hand it to you.

This exchange an take the form of tickets at a theater, credit card information over the phone, money from the advertisers running commercials on your show or checks from all those networks buying your product. It all boils down to this: you get paid.

That's what you want. You may travel many paths to this destination. You can sell directly to the consumer through magazine ads or direct mail; you can work through a distributor; you can sell your program to broadcast TV: or you can buy airtime from a cable company, and then sell commercials during your show.

The path you choose (see Figure 48.1) depends on your product and, to some extent, how deep your pockets are. Being in business involves taking risks, which means you sometimes have to shell out some cash before it starts to flow back to you.

Is It Good?

The first step in the process is to take a long, very critical look at your product.

Is this a show that other people will want to watch? Would they pay money to

Figure 48.1

spokesperson's delivery, you could be in trouble; one of these flaws alone could flag your project for rejection.

Let's say that (based on your long and successful track record) your product is intellectually compelling and technically flawless (congratulations). Now you're ready to distribute that perfect video. Let's start with the cheapest way to do it— by yourself.

Birth of a Salesman

This method seems fair and natural to most people. You worked hard on your video; it's only right that you should enjoy all the profit from this effort. Remember, however, that this also means assuming all the risk and fronting all the money for the advertising, postage, dubbing costs, shipping and so on. Also figure in time and effort for answering inquiries about your product, as well as packaging and addressing these packages when you do make a sale.

Let's say your glorious video details the proper maintenance of inboard marine engines. Is this program any good? You think so. Your friends think so. Even your enemies admit to liking it. But these opinions won't do you much good when it comes time to sell. What you need is a review.

Check all the magazines covering your subject (*Inboard Boating Illustrated, Marine Engine World* and so on) and read them. Do they review books or videos of interest to their readers? If they do, write a professional cover letter to the editor of each, describing your show in a straightforward manner. (If they don't, send out letters (see Figure 48.2) anyway; your product may be good enough to set a few precedents.) Don't hype your product at this point; journalists don't like that approach. Also send along VHS dubs of your program for the editors to pass on to their reviewers.

Don't expect replies, unless you also enclose a SASE (Self-Addressed Stamped

watch it? Would you? Ask your friends to watch it and tell you what they think. Now do the same with your enemies. That should give you a nice range of viewpoints.

Is the target audience large enough to justify the production costs? You can produce the greatest video ever on *Growing Beans in Sandy Soil,* but if this subject doesn't appeal to enough people, then you may not get your investment back— even if every potential member of your target audience buys your product. You have to believe strongly in your product— just make sure you can justify that belief.

Your next consideration: production value. Producing on consumer-level equipment will work for some uses, but if you plan to sell your show to a broadcast or major cable network, you may want to work with professional gear. Also, use the best crew and talent that you can afford. If your show is perfect, *except* for the lighting, *or* the audio, *or* the on-camera

Figure 48.2

Envelope) in each letter. If you want your dub back, then send a SASE with enough postage to cover its weight.

Getting a review is not a quick process. The reviewer has to view your show, write up the review and get the review to the magazine. Depending on the publication, it may be two to three months from the point the review reaches the editor's desk until it actually gets into print.

If the review is favorable, your show has just taken a big step toward legitimacy. A major publication (*Inboard Boating Illustrated,* no less) has complimented your show in print. This tells your potential buyers that your program is for real, that you did not just fall off of a turnip truck and that the video is worth spending their money to see.

Another way to lend legitimacy to your tape is to have an expert introduce the material. You can also put the expert's picture on the cover of the tape's package. Potential buyers see their favorite inboard engine expert on the tape and say to themselves, "Hey, I trust Joe Inboard. This tape must be good." Of course, Joe Inboard will probably ask you to pay for his image, or he may even ask for a piece of the action (such as a percentage of the profits).

If *Inboard Boating Illustrated* uses a rating system (four little boats equal excel-

lent, and three little boats equal good), then you have a perfect element for your next step in selling your product-advertising in the magazine. In your display ads, you'll feature "FOUR BOATS-*Inboard Boating Illustrated*" as prominently as possible.

The cost of your ad will vary according to the number of people who read the magazine, how much space your ad takes up and how many colors appear in the ad. As you might expect, the price goes up as readers, colors and space requirements go up. If you run an ad run more than once (which you almost always have to do to have any impact), the amount per issue goes down. Magazines often offer breaks for running an ad three times (3x) or six times (6x).

Fulfillment

Once the ad runs, the orders start to pour in.

But how do you get them? Do customers order by credit card over the phone (any time of the day or night)? Or do they send you a check? Do you wait for checks to clear before you send customers their tapes? What if the check bounces? Which—if any—credit cards will you accept? What about a money-back guarantee?

You can avoid some of these headaches by working out a deal with a fulfillment house. I once used a service from a company that made dubs of my show and kept them on hand. The company provided me with its 800 number, which I used in my ads. My customers placed their orders with this company. The company performed a number of services for me:

- recording the pertinent customer information,

- accepting credit cards,

- waiting for checks to clear and

- sending out the videos, using a preprinted slip sleeve that I provided.

At the end of each month, the company sent me a statement telling me how many units sold during that time, along with a check-minus the fees they charged for their fulfillment services.

This is a good, convenient service; but it does mean additional expense. Also, fulfillment services typically require the assurance that they make a minimum amount of money per month, which means selling a minimum number of your videos each month. If the total falls below this, you may pay a penalty.

Direct Mail

Another method of reaching potential customers is direct mail. You get direct mail from advertisers at home all the time; you probably think of it as junk mail. But it's only junk if the ad is trying to sell you something you don't want. That's why you must make sure that the people you mail your ad to are the people who want your product.

There are companies that sell lists of business names, preformatted on sticky labels for easy use. These lists break down according to type of business, number of employees, region of operation and so on. Be selective, and you can buy the right list for your target market.

Also, *Inboard Boating Illustrated* probably sells its subscriber list to advertisers. Consider buying subscriber lists from appropriate publications.

Once you buy the right list, find a local company that handles large mailings. This way, you won't have to stuff any envelopes. If you've ever licked more than twenty stamps at a sitting, you'll know this service is worth the expense.

When you determine the price of your program, remember to figure in the advertising costs. For example, if you buy a list of 10,000 names from a magazine, print up 10,000 ads and stuff 10,000 stamped envelopes for your direct mailing, you've shelled out some serious money. The direct mail industry considers a one percent response rate "good;" that's 100 orders from a mailing of 10,000. So you could lose money if you don't charge enough for your product.

Libraries & Video Rental Stores

After you've fully exploited the inboard boating market, exploit libraries.

There are public libraries, college libraries, high school libraries and more. Most have videotape departments stocked with a variety of videos; especially popular are how-to programs.

Check out the magazines covering this market: *Booklist* (you won't find it at the newsstand, it goes out directly to librarians); *School Library Journal; Library Journal;* and *Wilson Library Bulletin.* Each of these publications reviews videotapes, so getting a review is a good place to start.

The local video rental store (see Figure 48.3) is another option. One method would be to walk in, ask to see the owner and try to sell the show then and there. But this would be much like trying to teach a pig to sing. It wastes your time and annoys the pig.

Rental stores buy their programs almost exclusively from major distributors that publish catalogs every week. One is the Major Video Concept catalog

(800-365-0150), which boasts lots of four-color ads for Hollywood features. One page advertises foreign films and other videos.

Distributors

When you distribute a product, such as a videotape, there is a certain amount of infrastructure that has to be in place. You must let the customer know that the product is available, you must be able to take orders and fulfill them. This is true whether you have one product in your line or a thousand. Obviously, this infrastructure is less expensive per product if you have a thousand products. This is why there are companies called distributors.

In the early years of the motion picture industry, there were no major film studios. The people who owned the theaters needed product. Viewers didn't ask for much, they would watch a man petting a dog or a horse running down a road and be happy. But they always wanted more.

Finally, the people who owned lots of theaters and thus needed the most product decided to make their own movies. That's how the major studios were born. Distribution was the key.

For video, there are hundreds of small specialty distributors. One distributor might serve a market of gardeners. Every few months the distributor sends a catalog to these gardeners featuring all the books and videos on pruning and planting. The gardener can pick out several and order all of them at the same time. Convenient.

You can usually find a list of distributors at your local library. Books listing all the video products available for the current year often include a list of distributors for ordering purposes.

The names may not tell you much about the distributors. So before you send out copies of your program to distributors, call the companies and chat with the owners about your product. Even if these people show no interest in your video, they may recommend distributors who might.

Figure 48.3

When a distributor decides to handle your show, you'll negotiate a contract. You won't sell the show, but rather license it. This means that while you still own the material, the distributor will receive a percentage of the retail price of product sold. This could be a healthy percentage, like 75 percent. Or it could be less, depending on how good a negotiator you are and how much the distributor wants the product.

When negotiating remember that the distributor is paying for advertising, fulfillment, storage and so on; you assume none of the risk. These services are worth something.

The Big Screen

You've always dreamed of seeing your work on the big screen. This means selling your work to film distributors—not an easy sell.

Major film festivals are a good place to show off your work to distributors. Events such as the Berlin Film Festival and the Sundance Film Festival are places to see and be seen by "the players." Before you can enter your show, however, check the entrance requirements. Pay particular attention to format. If you shot on video, you will have to transfer the master to 16mm or 35mm. Expect to pay a couple of thousand dollars for this process.

There are many video festivals as well. Entering and winning prizes for quality and great content can't hurt your chances of interesting a distributor in your show. And the price of a dub for your entry will be less expensive than a 35mm film print.

Seeing your work "on the air," either on a broadcast or a cable station, is always a thrill. Again, there are many avenues you can take to reach this goal.

Vid News is Good News

Many local news shows encourage videographers to be on the lookout for newsworthy events. (see Figure 48.4) Even in large markets, a news director has only a limited number of crews to cover the station's area.

Given the improved quality of consumer gear, stations are more likely to get footage of dramatic events as they happen if a quick-thinking videographer happens to be on the scene.

The value of the footage depends on its newsworthiness; most stations pay $50 to $100 for short news pieces. You can expect a lot more if you get the Loch Ness monster, Bigfoot or some natural or man-made disaster on tape.

If you live in an area which has a low-power television (LPTV) station or a small cable company, you might be able to get on the air by forging a partnership. If the station or company doesn't have a production arm, make yourself useful to them by offering to:

1. cover city council meetings,

2. produce promotional spots, and

3. shoot public service work.

Could you do this out of the goodness of your heart? You could, or you could offer to do all this work in exchange for airtime on the channel, say an hour a week.

On the Air

Now you own an hour of airtime; how do you profit by it? There are a number of

Figure 48.4

ways to boost your bottom line. If you live in an area that attracts lots of tourists, create a show that reviews restaurants, profiles interesting people and recommends local hot spots. Sell 30-second spots on the show to some of these same businesses. Then produce the 30-second spots for them, charging for production services.

Or take your hour of airtime and sell it to a syndicator. The syndicator will then fill the time with an infomercial or an entertainment program, commercials included.

Or you could run a telethon and ask for donations from viewers in order to keep your telethon on the air. (If you succeed at this one, be sure to tell me about it.)

Leased Access & Satellite Time

If you have a great programming idea and live in an area where the cable company carries more than 36 channels, you may be able to lease your own time. The Cable Act of 1992 requires cable operators to set aside a certain percentage of their channels for lease by independent programmers.

This sounds good; but some cable companies hesitate to sell time to independents, and others charge too much when they do. Contact your local cable operator (see Figure 48.5) and ask about its leased access policy; be ready to fight for the time that is legally yours.

If national exposure appeals to you, try a satellite TV network such as Channel America. Channel America is a 24-hour satellite channel affiliated with 110 full and low-power TV stations and cable affiliates in the U.S., reaching some 25 million households.

On-air hours are open to programming in the following categories: Talk, Events, Emerging Sports, College Network, Nostalgia, Travel, Lifestyle, Talent Showcase, Outdoors, Hunting, Fishing, Financial Opportunities, Your Money, Music Across America, New Age, Hobbies,

Crafts and Collecting, International and Health and Fitness.

Airtime on Channel America isn't free, however. And the satellite channel won't pay you for your programming. There are ways around these financial obstacles, however. Say you've reformatted your inboard marine engines video; it's now a half-hour show with breaks for commercials. Approach national inboard engine manufacturers; ask them to buy the time on Channel America for your program, in exchange for exclusive commercial time. Of course, they would also pay you for the production.

Destination Distribution

Distribution paths are many. Some involve risking your money; some involve spending your time; and others involve sharing your profits with others.

The key is to find the ones that work for you and your video. Persevere and you can sell your show.

All it takes is a little effort and a little luck... and a great product.

Figure 48.5

49
Getting Cable Time

Mark Bosko

You've shot hundreds of hours of video. You know the difference between trucking and panning. You can set up a three-point lighting arrangement. You understand white balance, focal range and depth of field. You have mastered desktop editing and you've even been paid a couple of times to capture an event on tape. For the most part, you are an accomplished, semi-professional video producer.

So what's missing? Simply put: an audience. To produce a video is one thing. But to have it seen by others is a whole 'nother ball game.

Getting your video aired on television has historically been difficult. But that is changing. With the advent of the 500-plus channel cable television system wired into homes across America, there's going to be a lot of air time to fill. And who better to do that job than those already experienced in low-budget video production? More choices for the viewer means that specialized and niche programming will become more popular than ever. Cable channels needing low-cost programming are going to turn to alternative, non-commercial sources. Cable operators will be open to new ideas and new technologies to fill these needs.

Though this brave new world of opportunity for videographers is still on the horizon, there are currently several ways to get your feet wet in the world of mass distribution. You can prepare for the fertile future by finding that elusive viewing audience now.

Public Access

In 1984, the Cable Communications Policy Act stated that local governments could require cable operators to provide one or more channels for public use. This is how public access television was born. Parodied in movies and often thought of as a dumping ground for amateurish video production, public access still provides videographers with the opportunity to broadcast their work to a potentially large audience. Public access is available free to anyone—regardless of background or agenda—who has something to say to local viewers.

Public, educational and governmental (PEG) access television channels on cable TV serve a wide range of community organizations. These include churches, Lions and Rotary Clubs, local political organizations and high schools and colleges, to name a few. Through PEG access centers, thousands of community groups and individuals produce an incredible volume of new local programming each week.

Since all cable operators are not required to provide access (it is a function of negotiations with the communities they serve), finding a channel to program your work is not always easy. Your local cable operator is the best place to start. Inquire as to the availability of its access channels. If you're lucky, one already exists in your community.

Because public access is a locally governed entity, operational rules, equipment availability and technical assistance will vary. Some studios feature membership fees, lighting rigs, high-end gear and fully-trained crews, while others may have feeder decks that merely transfer your self-produced tapes to the system. Regardless, it's a low-cost method of getting your videos broadcast to local TV audiences.

The Alliance for Community Media, a nonprofit, national membership organization founded in 1976, is one of the best places to get information on public access television. The group represents the interests of over 1,000 PEG access organizations and public access Internet centers across the country. You can contact this group at Alliance for Community Media, 666 11th St, NW, Suite 740, Washington, DC 20001-4542, (202) 393-2650. Or visit them on the Web at www.alliancecm.org for more information.

As for programming content, once you clear all broadcast standards and practices, your options are wide open. A recent review of access shows in the United States found the following as common topics: restaurant reviews, talk-shows, local news, stand-up comedy, dating, job hunting, classifieds, automobiles, local band showcase, fishing, electronic gaming, Internet instruction, independent video showcase and a program that gauged local nightspots by their attraction to singles. Limited only by your imagination, public access provides videographers the simplest manner to get productions on the air.

The educational and government access channels offer only limited opportunities for videographers. Government access is information and programming concerning civic affairs (which is usually seen as text scrolling on the screen or city council meetings), while educational access is often used by organizations that air telecourses as part of continuing education programs. Both are local origination channels, which mean those who oversee the channels also decide what airs. Those same folks are also responsible for production and likely have an in-house staff. Contract work that pays well is sometimes available. For videographers with advanced gear and skills, it's worth checking out.

Leased Access

As with public access, videographers can thank the U.S. Government for yet another distribution avenue. The leased access provision of the Cable Act of 1992 requires cable system operators, that meet certain criteria, to make a portion of their capacity available to all comers for a reasonable fee. Through the law, Congress placed a vital means of distribution within the reach of most independent videographers. Combined with today's low-cost, high-quality video production technologies, leased access makes it possible for just about anybody to have a professional-looking and accessible television show.

Unbelievably though, few videographers take advantage of leased access. Why? Two reasons: First, a general lack of awareness on the part of videographers and second, the reluctance of many cable system operators to allow unaffiliated programmers onto their systems. This is understandable when you consider that cable systems must sacrifice potentially high-profit channel capacity for local productions that may interest only a small portion of the subscribers. Regardless, you are now aware of leased access and the law is the law, so there should be nothing holding you back.

To get started with leased access, call any local cable system that operates within the parameters that we mentioned earlier. Check on channel availability, open airtimes and most importantly, rates. Leased access time, usually sold in half-hour units, generally costs anywhere from $25 to $300 for a 30-minute block depending on where you live, and the time block you'd like to purchase.

If the thought of cablecasting your message to thousands of viewers for less than the cost of a fine meal sounds exciting —it gets better. Not only does leased access availability give videographers a showcase for their work; it also allows this activity in a commercial atmosphere. Unlike public access and the other government television involvement, leased access allows for commercial material within the programming. In other words, you can sell advertising time just like a real broadcaster.

Depending on your initiative, salesmanship and financial situation, you may view leased access as a moneymaking opportunity or a necessary out-of-pocket cost. Either way, it's a chance to develop and distribute videos to a once-unavailable audience.

On With The Show

With public distribution of your video skills finally within reach, it's time to work on ideas possessing audience appeal. Maybe you can recycle some past

video project that you did into a broadcast program. Recital or play videos could become a local talent showcase series. Be sure to obtain the proper copyright releases on any copyrighted material including plays, music, etc. If you shoot weddings, maybe a weekly wedding show, sponsored by area bridal shops, will gain interest. Or, if you've shot a bundle of seemingly useless "around-the-town" type stuff—parades, parks, cityscapes, and the like—local merchants and government may look favorably on a "home town tourist's guide" much like those run in big city hotels. For those developing new shows, again, the sky's the limit as long as you observe applicable broadcasting ethics.

Low Power TV

Finally, if public and leased access channels are unavailable, don't forget that other options exist to distribute your work. Low Power Television Stations (LPTV) sometimes operate much like leased access stations. Airtime may be available for purchase from the station, with the producer retaining commercial time during the show. Individual LPTV stations and some multi-station networks are scattered throughout the nation; they exist on the upper bands of the broadcast signal. Beware: LPTV may be an endangered species. With its reapportionment of broadcast spectrum for Digital Television transmission, the FCC may eliminate low power stations in the coming years.

Another avenue to distribution is reality TV. The popularity of these shows means they are always looking for footage. Scan the content of the programs currently airing to determine if you have or can shoot footage that fits their needs. Along the same lines, local news programs also provide a channel of distribution. Any footage of natural disasters or one-of-a-kind events, that you have exclusive footage of, may find its way onto the evening news. With some planning, research and a little ingenuity, there's no reason why your video project could not be "on the air." After all, getting your video broadcast to an audience outside of your living room is the final goal of any distribution effort. Additionally, it can be a gratifying and even profitable experience for a semi-professional video producer.

50
Paths to Broadcast Television

Mark Bosko

Let me relate to you the story of a fellow videographer who "made good." John started like most of us, goofing around with his parents' film equipment. Though just a child, the creative art of cinematography really clicked for him. He created one film after another to the delight of his family and friends.

Though the passion for this "art" festered inside John, he became frustrated with the what seemed like a wasting of his time. The whole purpose of producing a movie was for an audience's enjoyment. After five years of basement screenings, family and friends hardly qualified as a legitimate audience anymore. John knew there just had to be a better way, but didn't latch onto it quite yet. Much later, in college, John enrolled in the school's teleproduction class. He knew a little bit about TV production, but was still mainly a "film" guy. It was here that he discovered what would later "rule" his world—videotape.

Even with excellent marks, John still yearned for that elusive "audience."

Luckily, he was outspoken about this need, and a professor took notice. The professor, as it happens, was on the board of the local public television station. He was primarily responsible for development of new local programming.

Thinking of John's desires, and some of the super productions the students were showing, the teacher came up with a "Young Filmmaker's" showcase program. The show would give aspiring film and videographers (now John's medium of choice due to cost and time considerations) a place to present their programming to a potentially large viewership.

Happy ending: John got his audience, the station got quality programming, and viewers got some alternative shows to watch.

Not all stories involving broadcast TV are so inspiring, and sometimes PBS networks are the hardest nuts to crack. But the example above does point out the many opportunities that exist for videographers looking for distribution of their productions within these "hallowed halls." For some reason or another, broadcast television stations have the image of an "insider's club." You've got to know someone or already work there in a lower capacity to get an in. If you weren't a part of the community's filmmaking "elite," your chances for broadcast were nil. Maybe ten or so years ago that was true, but today, this is simply not the case. Especially for dedicated, experienced videographers.

One of the reasons for this "opening" may be attributable to the increase in number and types of broadcast outlets on the map. Before everyone in America had cable, there existed a strong distinction between broadcast and cable fare. Broadcast programming was free, contained a some locally produced programming (created at the station, not by independents), and carried the network shows as they "came down the line." Cable, on the other hand, carried new movies and other, non-traditional television material. It was also perceived (rightly so) as being very costly.

As the years passed and more media moguls developed, the number of cable stations quadrupled, the cost for the service plummeted and the demand by consumers who "wanted their MTV" skyrocketed. This led to the confusing mix of cable and broadcast stations that currently exists on your channel selector. This influx of new entertainment choices spurred a huge void in the supply of programming able to fulfill the scheduling needs of the stations.

Another reason broadcast stations unlocked their doors to outsiders was the fact that now they wanted to compete with the trendy and popular cable networks. The old, stodgy rules of operation were changing, and any new face that had something to add to the party was welcome.

And, in recent years, a new broadcast outlet, Low Power Television (LPTV) became popular. The limited signal put out by these stations reaches a relatively small, geographically close audience. The LPTV stations tend to be carriers of downloadable national satellite programming, mixed with an unusually high amount (for broadcast) of local fare.

While all of this is certainly encouraging (if not educational) for the future of aspiring videographers, what real opportunities exist now, in the present, for those of you who can't wait a lifetime for their dreams to be fulfilled? What does broadcast television offer you in the form of distribution?

VHF

You've got a job (that you like) and possess some fine-tuned production abilities. You own a little equipment, no Industrial Light and Magic, but a respectable "studio" on your own right. You've made some industrials, a whole slew of weddings, even an instructional tape on gardening for your spouse. You got some good ideas for programming that you think will go over big with the locals, but how do you get it on the tube?

Time saving tip number one: skip the VHF channels in your broadcasting area. VHF slots, usually reserved for network affiliates, offer the independent videographer little in the form of finding an audience. The stations are network controlled, meaning they have mega-bucks at their resources. They're not rude, but why would they want to mess with your $1.98 Talent Show when they can program a re-rerun of *Who's the Boss*? It just doesn't make sense for the big boys to play with you.

About the only exposure you may achieve through a VHF outlet is sale of news-type footage. And this comes from personal experience. My town was literally burning down. A huge fire started in the historical district, and I happened to be at the right place at the right time with my camcorder. I got some great shots before any of the large news crews showed up. They were aware of my presence and asked to buy the footage for inclusion in the coverage of the story. I was only too glad to succumb to their wishes. But don't plan on getting rich from selling news footage. These deep-pocketed network guys could only scrape together $50 for the whole 30-minute tape. And then I didn't even get an on-air credit!

UHF

If you've seen the Weird Al movie of the same name, then you are aware of the possibilities available for independent programming to air on these channels. While it's not quite as zany as the Weird Al film, opportunities do exist (especially in smaller markets) for videographers to find an audience.

Many UHF stations are becoming network affiliates (FOX has conquered quite a few), so the chances with these stations are slimming. If you live in a large TV city (one with more than four or five UHF channels), then you should be able to locate a willing outlet. In the Cleveland-area, a late-night television host on a UHF channel hosts a viewer's film's series. It's a great show comprised of shorts (one shot-on-video feature has played) broken up with interviews of the videographers. There is no pay involved, but the audience is pretty big, and loyal. The program is also popular with local advertisers who recognize the local customer base tuning in.

If there is a late-night gig in your city, why not hit up the host with this idea? It makes his or her job infinitely easier, and becomes attractive to sales personnel at the station.

Sunday morning talk and "city" shows also seem popular with UHF channels. Easily produced, these programs focus on community events and personalities. Often, the production may center on one specific area, and the show is a submission from a freelance videographer.

Fairly new to the broadcast arena, Low Power Television Stations are basically UFH stations, only with less signal ampli-

fication. These stations function much like their big cousins, only with a greater concentration of the local goods.

Cashing in

Knowing that there's some distribution avenues available in UHF and LPTV is good news, but, you'd like some compensation for your efforts, right? Well, just like the VHF networks, these stations pull in the reins when it comes time to pay. In fact, you may be the one paying them to show your program.

WAI-TV, part of a three-channel LPTV network in Cleveland and Akron, offers air-time for sale. Going for $250 and up per half-hour (depending what day and time you buy) the channel is a natural for independents. "We have space for sale just like every other television network. It just happens that ours is available to the independent," says Bill Klaus, owner of the station. Klaus makes it clear that the reason an independent can buy time from his network is because it is affordable. "Sure, someone with a home-grown production could go to their local VHF station and buy a half-hour of time to broadcast the show, but they'd probably have to mortgage their house to do it. My network makes it affordable, and we often barter time as well so the videographer can actually make a buck."

Bartering, as Klaus mentioned, is another favored option of UHF and LPTV programmers. What this means is that you retain some of the commercial time allotted within your programming block. As an example, let's say you buy a half-hour of air time for $200. That's the flat rate. With that price, you are the owner of all 8 minutes of commercial space. You can deal with the station, letting them keep 4 minutes of ad time, dropping your payment to $100. And, it also works in reverse. If they want to buy your show (yes—that actually happens sometimes), they may offer you the commercial time in exchange for any payment. This way they don't have a cash outlay, but fill their schedule. You, on the other hand, have found a profitable distribution outlet.

The large, network-affiliated stations will not likely be interested in your bargain broadcasting, but many independent stations will take a look. Local interest, interview shows, documentaries, community affairs and sporting events are all good ideas to present to any small broadcaster in your neighborhood.

When you go door knocking, bring an attractive demo and a professional presentation package outlining your programming ideas. This packet should present your work in its best possible light. While you may think you are the only indie out there (or at least the only one in the neighborhood), the fact is that programming managers deal with many proposals from many people. "It's probably hard to believe, but I get at least a proposal a week from independent video producers," states Klaus. "Most of the production ideas have no substance. If I got one backed by a demo, or put together in a professional manner, I might pay more attention to them. But too many of them look like a half-hearted effort to get a show on the air."

"I don't mind working with independents," Klaus continues. "In fact I like it, but they just have to be more professional in their approach. If a videographer wants to make some money by getting his programming on the air, instead of spending it to buy time, he should prepare the idea as completely as possible. That would be the type of producer I would look to work with."

There are no set rules here. Just remember that it's the fact that people are able to view your production through their television sets for free that's important, not the amount of bills you have wadded in your pocket.

The multitude of small and low-power broadcasting outlets has created a void of original, low-cost scheduling alternatives. There are only so many stations that can broadcast The Andy Griffith Show at any one time, and it's that fact that opens up the audiences to you.

51
Access Your Community

Art Aiello

Videographers face a number of obstacles in the pursuit of producing the perfect video. Untapped niches are becoming harder to find. Technology is changing overnight. And for those hoping to pursue the craft full-time, there's the issue of putting bread on the table.

For many producers, the easiest route to securing equipment access, finding a potential audience and keeping a full stomach lies in working for someone else. The tradeoff is that you give up control over content. It's not your vision up there on the monitor. It's someone else's.

There is hope, however. More and more producers are finding public access television an effective and inexpensive way to produce the kinds of shows they want to put on the air. In fact, many public access television centers will allow you to take any idea you like and turn it into video.

They'll often loan you the equipment, teach you how to use it and give you air time on a local cable channel. So what do you need to do? Often all you need is the right demographic profile (i.e., you must live in the geographic area served by the access center) and a small investment in training. Best of all, you can produce for public access television and keep your day job, too.

The Basics of Community Access Television

We have the Cable Communications Policy Act of 1984 to thank for giving us public access television. This act stated that local governments, in their capacities to grant franchises, could make cable operators provide one or more channels for public use as part of their franchise agreements. It does not say, however, that every cable operator has to provide channels for community access. Community access is strictly a function of negotiation between cable operators and the governments in the communities that they serve.

Fortunately, many communities have been prudent enough to ensure access for their citizens. The Alliance for Community Media, a Washington, D.C. based non-profit organization that represents the interests of community programming nationwide, has identified almost 1,000 access centers across the country. They acknowledge, however, that the number exceeds that amount.

"We are aware that there are more than 1000 centers," says Barry Forbes, Executive Director of the Alliance. "And although we're a little unsure about South Dakota, and we've had some difficulty locating centers in Wyoming, there should be at least one access center in every state."

The Cable Act of 1984 provides for three different kinds of community access-public, educational and governmental (PEG). You will inevitably find the most opportunities through public access, although I will discuss educational and governmental access later.

Public access is for everyone to use, regardless of your background or your agenda. These channels are at the disposal of any person or organization who has something to say to a local audience.

Getting on the Air

Because public access exists on a local level, the rules for using access centers vary from community to community.

"Never say things like 'most access centers,'" Forbes says, "because there is no such thing. Every community has different rules for use of their facilities." Nonetheless, there is common ground shared among access centers. According to Karen Toering, Operations Manager for the Milwaukee Access Telecommunications Authority (MATA), the first step is to apply for membership with your local access center.

"To become a member of MATA, you must either be a resident of the city of Milwaukee," Toering says, "or you must be a non-profit agency serving Milwaukee. Then you have to pay the annual membership fee." Individual membership in MATA costs either $10 or $25 per year. In either case, you are eligible to produce shows, vote in member elections, and receive the MATA newsletter. The $25 membership, however, gets you discounts off workshop and dubbing fees, and you get equipment insurance coverage.

Toering explained that you then must take an orientation course and three training courses before you can start producing shows. The training courses cost $15 each and explain the basics of video production. This includes legal issues, MATA rules and regulations and nuts-and-bolts stuff like portable field production and editing.

"Every member is required to take these courses because they are specific to MATA," Toering says. "We realize that many experienced producers may already know how to operate our equipment. But our facility is open to everyone, and everyone must take the training regardless of experience." Not every access center charges a fee for these courses. The West Allis Community Communications Corporation (WACCC) is the access center in the suburban Milwaukee neighborhood of West Allis where I live. WACCC charges a $20 annual membership fee, yet charges nothing for the several training programs it requires its members to complete.

Once you successfully complete the training, you'll receive an ID that grants you equipment and facility access. What happens next is up to you.

No Beta Equipment Here

You're probably thinking that the equipment available to members must be outdated and old-fashioned. Probably just some old cameras and portable VHS decks—hand-me-downs from the local community college. Just enough stuff to make you feel like you're producing a real video, right?

Not true. Many access centers have a healthy inventory of prosumer equipment. It might not be Beta-SP, but it's probably not Beta I, either. In many centers, you will find S-VHS or Hi8 camcorders or docking recorders for field production. And there will be field monitors and lights and lavaliere mikes, too.

In the studio (yes, some centers have one) you will look up and most likely see a lighting grid. And—and are those three studio cameras I see? Wow! They even have tally lights! And a teleprompter! Does my $25 a year include the teleprompter? You're kidding! Is that the control room in there? A phone patch, too? No way!

Participation in public access may be inexpensive, but it is no small potatoes. According to the Alliance, access centers churn out more than 20,000 hours of original programming each week. And they give members the tools to do it. Whether it is field production, studio work, or A/B-roll editing, your local access center will provide you with equipment that is more than adequate to produce a very attractive show.

Locating an Access Center

In spite of the apparent abundance of access centers throughout the country, locating one near you is not always easy. Don't get too excited until you know for sure whether or not your community has an access center.

You might try contacting your local government first and asking for information regarding public access centers in your area. Your local cable operator would most likely have that sort of information as well.

If you have a little more money to kick around and are curious about the scope of public access nationwide, the Alliance publishes a directory of all the access centers they are aware of.

Surfing the Net

For those of you who have more than a passing acquaintance with cyberspace, many access centers have or are developing Internet addresses and websites on the World Wide Web. According to Kari Peterson, Executive Director of Davis Community Television in Davis, California, the Alliance is currently probing the powers of the information superhighway.

Peterson explained how visitors to the infobahn can find information about public access. Use a search tool," she says. "For example, you might look for topics under `access' and `television.' The information you would get that way might lead you to more specialized information that, hopefully, will tell you how to get involved with an access center near you."

Peterson said that many Web sites are being developed with direct links to other sites. "You might, for example, find a Web site for the southeast region of the United States," she says. "That site could, in turn, take you to the Anytown Access Center Web site."

A good starting place is the Alliance's Web site. It has lots of general information and a long list of links. You can find them at www.alliancecm.org.

Public Access vs. PBS

Many people confuse public access television with a host of other distribution options. One of these is public television. Public television, better known as PBS, is not the same thing as public access television. For one, they receive their funding from different sources. Many community access centers get funding primarily through franchise fees paid to the local government by the cable operator. Public television, on the contrary, receives funding largely through private fund-raising, state appropriation, corporate underwriting, and Federal grants.

More importantly, it is the program directors at public television stations who decide what goes on the air, and they draw from a large pool of shows produced by high-paid videographers and filmmakers. In public access television, *you* decide what goes on the air.

Likewise, do not confuse public access television with leased access. Leased access requires that you purchase time from a cable operator in exchange for the opportunity to run your own commercial messages. Leased access is another way to get on the air. And though it may cost more than public access, it allows you to make money.

Government and Educational Access

Even though public access will be your most viable avenue, you may find limited opportunities in government and educational access.

Government access is just that—information and programming concerning civic affairs. Most of the content deals with local government meetings. And people are tuning in.

"A 1992 survey in our area indicated that 30 percent of local cable subscribers were watching government access television," says Hap Haasch, Cable Administrator for the City of Ann Arbor, Michigan. "This stuff is very important to our viewers. We don't dare screw up our coverage of city council meetings or school board meetings. If we do, the phone starts ringing off the hook."

Educational access is used by those organizations wishing to provide access of a strictly educational nature. You will find many community colleges and continuing education programs making use of this kind of access. Educational access stations often air telecourses as part of continuing education programs. Some public schools produce shows that benefit high school students as well.

Opportunities for you to produce in these two arenas are not as common for one important reason. Many government and education stations are local origination channels. That means that those who oversee these channels also decide what gets put on the air. They are also

responsible for much of the production, as well. As such, they will more than likely have an in-house production staff.

"The educational community wouldn't participate unless they could be assured of a certain level of production quality," says Ann Flynn, Executive Director of the Tampa Educational Cable Consortium, better known as the Education Channel, in Tampa, Florida. "We allow volunteers to produce and occasionally write scripts. But the actual shooting and editing falls to our paid staff."

Flynn adds that the Education Channel is not typical of all educational access centers throughout the country. "We strive to be a PBS clone," she says. That means they go beyond the telecourses and air a wide variety of educational programs.

More advanced videographers can take advantage of this situation, however. "We work with five or six contract producers who we pay to develop programs for our center," Flynn said. " I would say that those kinds of opportunities are available to all video producers."

The same goes for government access. "The real inroad to broadcasting on government access is the contract work," Haasch says.

So What Do I Put on the Air?

The sky's the limit when you decide what kind of a show to produce. A quick glance at a monthly program schedule for MATA reveals sports talk shows, live religious call-in shows, job hunting forums, and senior citizen programming, just to name a few.

"We have a couple of producers who put on a really wild weekly show," Toering says. "They ask viewers to call in and report any news they know about. When the caller is through, they try to guess the caller's age. That's the extent of it."

Whatever you decide to produce, remember the ethical responsibilities that come with such freedom of expression. Most access centers will not censor any programming as long as it doesn't violate the law. But don't think that you can stir up a hornet's nest of controversy and then hide behind the shield of the access center. They tend to refer complaints about programming to the producers responsible.

In other words, you will take the heat. Public access may not be half-time-at-the-Super Bowl television, but it is television. And as such, it has the power to influence people. That power is available to you, whoever you may be. If you understand that power and use it responsibly, public access television may just prove to be the production opportunity of your dreams.

52
Promotion Strategies: Fame and Fortune on a Budget

Mark Bosko

If you don't tell people about your video, they won't even know it exists. The more aware the public is of your work, the greater fame and fortune you'll eventually achieve.

You can promote your work in many ways—from buying expensive full-color magazine advertising to sending out a simple press release. Full-blown promotional plans practically guarantee increased video sales, but they're often too expensive for the first-time videographer.

But there are less expensive ways to promote your work. In this chapter, we'll survey a selection of promotion strategies that will cost you little more than the price of pen and paper—and some hard work.

The Press Release

The press release is the most widely used and abused promotional technique. This one-page synopsis tells the media what you want them to know about your video. Media outlets such as newspapers, magazines and broadcasters receive hundreds of these daily; to make sure yours receives proper attention you need to a) submit it in regulation form and b) make it stand out from all the others.

Press releases follow a standard format. Deviate from this format and no one will read it. This sounds harsh but it's true. If you want the press to read it:

Keep it short. You don't need 10 pages to communicate your message. One page is best, two is the maximum. If it is longer, re-write it; include just the basics.

Title it. Without a headline, they won't know what the release is about. Trust me: they won't take the time to read it to find out. The best headlines are short and to the point.

Say who sent it. The upper left-hand corner of the page should set out the following information: your company's name, address, telephone number and—most important—a contact to call for further information. You're the best contact;

if you can't do it make sure you choose a contact well versed in all aspects of your video.

Provide a release date. Write "For Immediate Release" on the press release; this tells the press that they can use the information revealed in your release right now. If you don't want the information released until a specific date, then provide this "embargo date" in place of the usual "For Immediate Release" (i.e., "For Release October 10, 2003").

Use the standard form. Type the release, double-spaced with ample margin areas. Check for any errors in spelling, punctuation, grammar or content; this kind of mistake screams amateurism.

Now you think your press release is going to look like all the others—and if the format is correct, it probably will. But Joe Reporter down at the Daily Globe doesn't care about fancy formats; he's looking for interesting content. Write your release so it not only answers the stock journalism questions—who, what, where, when, how and why—but also leaves the reader wanting to know more. Appeal to the natural curiosity of the reporter—without getting cute—and no doubt your video will see some press.

Mail, fax, transmit electronically or hand deliver your release to every possible media outlet. The wider the distribution, the better your chances of getting press.

Radio and TV Interviews

Turn on the radio. Flip through the stations. Listen. You hear a lot of talk, don't you? Radio stations live on music and talk; they have plenty of music, but they need the talk. That's where you come in.

More than 80 percent of the 11,000 radio stations in the United States air some sort of interview program; getting on one of these interview shows is easier than you might think. Best of all, it's free.

Call the stations you know that air interview programs and ask how they book their guests. Send the person in charge of booking a press release, some data sheets naming the subject, cast, crew, locations and length of your video and a cover letter. The cover letter is important; think of it as a sales letter selling you and your video. This letter should persuade the booking manager that your video is an ideal choice for the program—due to its exploitative elements, controversial theme, local interest or whatever "hook" will prove irresistible to the station.

With any luck, someone will give you a call to find out more about your project and determine if you would make a suitable guest for the show. When you get the call, be sure to answer all queries with confidence and grace; you want to prove you're a coherent, interesting individual who won't freeze up during the program.

Interviews are great promotional vehicle, but they do offer one distinct disadvantage: lack of control. You don't have the benefit of complete pre-planning. You can't predict what the interviewer will ask you or what part or parts of the interview will air. With live, call-in formats, you face the additional challenge of fielding questions from the listening public, who may or may not approve of you and your video. Two suggestions for handling radio interviews:

Restrict contact with the media to yourself or one or two other people associated with your production. You should coach these people—talent, director, producer—on appropriate responses to possible questions. You certainly don't want to hear your cameraman giving out details that contradict your press release.

Listen to the radio program before appearing on the show. Observe how the DJ deals with guests. By listening, you'll discover if the host and callers are friendly or abusive, the show is live or taped and whether the focus is straight news or fluff. Prepare your answers based on your observations.

Now that you've successfully conquered radio, turn your attention to the other half of the broadcast spectrum: television.

Compared with radio, the market for TV guests is small; the chances of landing an interview are smaller still. To boost those chances, approach the TV station about an interview in the same way you approach the radio stations, but do more follow up. Don't wait for someone to call you; make those return calls yourself. TV people are always busy, and they believe that if you want to book yourself on a show you should do all the work.

You do have one advantage: you're promoting a video. Videos are visual and naturally lend themselves to the medium of television. Include a trailer of your best scenes—those most visually compelling—with your press release. This way the person in charge will know you have something unique to offer.

Most local television interview programs are not overwhelmed by low-budget video producers trying to land spots as guests on a regular basis. You're one of a kind; play up the glitz and glamour you bring to your hometown.

Tell the show's producers about any publicity stunts you plan and ask that they cover it to air with your interview. Keep the subject exciting and visual, and you shouldn't have any problems.

Screeners/Trailers

"We want to see the video."

That's what you can expect to hear hundreds of times while promoting your project. The press wants to see a *screener*. A screener is a full-length, promotional copy of your tape, provided free of charge to media personnel upon request.

They need to see first-hand what your video is all about. Screeners allow reporters to check out such considerations as budget, acting, effects and pro-

duction values. Now your video must live up to the expectations you've created for it. Did you exaggerate too much in your press releases?

The press uses screeners most often for reviewing purposes; media outlets occasionally request them as well, primarily to check out authenticity. Nobody wants to devote space to a "phantom" video, especially national magazines.

Providing these screeners for everyone who asks becomes an expensive proposition. There is, however, a low-cost alternative: the *trailer*.

Short compilations of your best scenes, trailers can accomplish all a full-length screener can—at less than a third of the cost.

A general rule of thumb: use screeners for press outlets that want to review your program and trailers for those who just want to "take a look." Some suggestions:

Use a disclaimer on screeners. When duplicating screeners to send to media outlets, superimpose or key the words "For Promotional Use Only" over the video during its entire duration. Not that the press is dishonest, but if your video lands in the wrong hands, nothing will stop those hands from selling the video as their own. The practice of using a disclaimer offends no one; it's always better to be safe than sorry. This tip comes from a video newsletter in California investigating a small cable station making illegal dupes to sell in Mexico. So protect your property!

Duplicate screeners and trailers on B-grade tape. Most of these videos will be viewed only once or twice, making a high-grade tape unnecessary. The press outlets can handle a little dropout.

Use a copyright. Place a copyright notice prominently on all screeners and trailers. Put a notice physically on the tape and within the program itself.

Limit trailers to 5 minutes or less. This is adequate time to show your tape's highlights. Keep it "lean and mean."

Make your trailer available in broadcast formats. Some broadcast press outlets may want to include your trailer as part of the story on your project. Don't miss out on this free advertising by not having the proper tape available. Most stations can work with three-quarter-inch format. It's cheap and widely available.

Include any televised press on distributors' screeners. Tag a mini-trailer of any televised news stories about your project onto the beginning of the screener. How impressed will that distributor be when he sees your story as it appeared on *Entertainment Tonight?* A lot more impressed than he'll be when you just tell him about it. Be sure to check with news agencies concerning legalities of duplicating such stories.

Press Kit

It's now time to compile all of the press you've received thanks to your promotion strategies and organize it into a press kit.

The most useful weapon in your promotional arsenal, a press kit represents the culmination of all your efforts. Its purpose: to show the attention your project has received, proving that your video is worthy of further coverage such as newspaper space and TV airtime.

Some people say to include only the big "headline" stories. I say include everything—from that one-paragraph blurb in your local new paper to the full-page story in *Variety.*

Press is press. The more you can show a media outlet, the easier it is to get more coverage.

Some tips on putting together a decent press kit:

- Use a high-quality bond for reproduction of the originals. The heavier weight paper lends a classy look.

- Check the clarity of copies, especially when articles include photographs.

- Place articles in descending order of importance, starting with the most prestigious—usually national media.

- Articles buried in the editorial section should be shown with the cover or masthead of the publication; copy it and place it on the reproduction with the article.

- Allow for only one article per page, unless the articles are extremely short and from the same publication.

- Use the proper tape formats when including televised or radio coverage.

- Put all the print elements along with a cover letter into a slick folder.

The press release P.R. strategy alone should garner enough press for a substantial press kit. And if you've employed the other promotional strategies as well, you may find you cannot include everything—your kit would be so thick, you'd go broke on postage. So choose only the best of your material for your press kit.

Publicity Stunts

Publicity stunts can be a great low-cost promotional technique, attracting both media and public attention to your video project.

Used with great success by the film industry, publicity stunts have traditionally accompanied the release of new movies. The "golden age" of publicity stunts was the 1950s, when the following tactics drew big crowds:

Nurses in theater lobbies, placed there by smart promotional men asking viewers to sign medical releases, in the case of heart attacks brought on by the shocking subject matter they were about to see. A particular favorite of science fiction and horror film promoters.

Bogus pickets, carried by "protesters" hired by a film company's publicity department to demonstrate against a film's sex and violence quotients.

Film "banning," which implied that a movie's subject matter was so offensive it should not be shown in certain areas.

Your own publicity stunts don't have to be so melodramatic, however. Try setting up a live magic act in the video stores stocking your *Magic Made Simple* video. Or, celebrate the release of your *Keep Our Town Clean* video with a litter collection contest for local school kids; offer the winners some free production time. Any stunt you can think of that involves the public and creates interest in your video makes for good publicity.

To ensure success: keep the press informed and keep it legal. Check out all local laws that may apply to your particular stunt.

9 Ways to Cut Promotion Costs

Mailing and distributing press kits and screeners can prove expensive. Still, there are ways to economize:

Don't use envelopes when mailing press releases. Tri-fold the paper and staple the bottom.

For big mailings use pre-printed postcards—they're cheaper to mail and cheaper to produce.

Order return address stickers bearing your company name from one of the many mail-order catalogs that offer such merchandise. They look good and cost substantially less than printed versions.

Shop for supplies and copy services at a large office supply store. Many of these places duplicate the same document for as little as two cents a page.

Mail screeners fourth class. Fourth class costs less than half the first-class rate and takes only two to four more days to deliver.

Save on postage by using air bubble envelopes instead of cardboard VHS tape mailers for screeners.

Put together a trailer instead of mailing out full-length screeners; this reduces postage and duplication costs.

Save big on phone bills by using toll-free phone numbers when calling distributors, TV/radio stations and publica-

tions. You'll find them listed in the toll-free directory at your local library.

When making long-distance toll calls, place them when it's most cost effective; keep time zone changes in mind. Contact your long-distance carrier for specifics.

The Bitter Fruits of Publicity

Execute a proper and thorough low-budget video promotional campaign, and your life will drastically change.

The good news is people will see your work. The bad news is more complicated.

First, you're apt to lose much of your leisure time to your publicity efforts. Sure, it is great to come home, pop open a beer and settle in to watch Divorce Court. But you're not going to get on any magazine covers that way. Not that you must devote every waking minute to the promotion of your video—we all need some time to relax—it's just that sooner or later the marketing machine you create will take on a life of its own. Instead of playing cards with the guys or hitting the mall, you will probably find yourself fielding phone calls and writing letters. If the process threatens to consume you, write "free time" right into your work schedule.

Promoting your video does not have to alienate you from friends and family, though this often proves the case. So go shoot hoops with friends or spend some quiet time with your spouse when you can.

The second source of grief that accompanies promotion efforts: reviews. Reviews are a necessary part of the marketing process; you'll need to develop a "thick skin" to survive the nastier negative criticism. Remember, you are sending out your video to literally hundreds of outlets, hoping to generate publicity and resources for a press kit. Among all these people watching your tape there will undoubtedly be some who don't like your work, for whatever reasons.

Who cares? What do critics do, anyway? They sit in front of a monitor all day, pointing out faults in something they never had the guts to try to do themselves. These people make their living by proclaiming what—in their own minds—is good and what is bad.

At least this is what you must tell yourself when bad reviews come in. You will, on the other hand, admire the intelligence and good taste of the reviewer who raves about your show.

There's one sort of criticism to which you should pay special attention—that of your fellow videographers. Send your video around to other producers who have "made it" in your field. Suggestions and insight from such individuals are very valuable, often saving you time and money on your next production.

A final thought: prepare yourself for the fame that will haunt you after your name begins to appear in the media. No longer will you be able to venture out into the world a nobody. You'll become a local celebrity and if your video is a big hit, national fame will follow. Standing in the limelight is fun, but it can be dangerous.

If you're flitting around like some self-important media butterfly and your project takes an unexpected turn for the worse, your fall from grace will hurt all the more.

Promoting your video should be a fun and exciting experience. Be sure it stays that way!

Tell 'Em and Sell 'Em Again

With the right promotion strategies, fame and fortune can be yours. The key is persistence. It takes time to create and execute a promotion plan, but it's worth it.

If you believe in your video's success, as you surely do, nothing can stop you.

53
The Demo Tape

Mark Bosko

With so many of today's videographers relying on their video skills and equipment for income, marketplace competition is keener than ever. To succeed, these courageous entrepreneurs (or hopefuls) need all the help they can get.

A good demo tape should be your number one marketing tool. There's nothing like it to showcase (and sell) your videography talent. A well-done demo attracts new clients, creates good public relations, and can even lure competent employees.

Unfortunately, a good demo is not that easy to make. In this chapter we explore elements of the demo tape—its reason for being, its creation, its uses. Once you see what a demo can do, you'll wonder why you never got around to making one before.

Why You Need One

Say Uncle Bob, the dentist, needs a marketing video. He wants to feature basic information on his facility—friendly staff, low prices, after-work appointment hours. He'll show the tape around to factories and large corporations. The vast numbers of employees within these companies permit him to offer attractive discount plans.

In production terms, the video sounds easy. Some interior shooting. A couple of staff interviews. You'll finish it off with narration and graphics. You're a member of the family, so getting the job's no problem, right?

But during your meeting with Bob, he asks to see something you've done. A representation of past work. Some evidence you're competent to make a video to his liking.

Uncle Bob wants a *demo*. You have one, don't you?

If you're like many small video companies and independent producers, the answer is probably no. But it takes more than smooth talk to convince clients—even Uncle Bob—that they can trust you with their money. Videos aren't tangible things. Until a camera comes out of the bag, they're just talk and writing.

Investing in someone's videography skills without having viewed his work is like buying a car based on nothing but a sales pitch.

Videos often record those once-in-a-lifetime events. A potential client must be certain you'll get it right the first time. He can't stage his daughter's wedding again because you forgot a microphone. He needs a good look at your "credentials."

The demo also is a simple way to attract new business. It shows off the power of the medium. It gets your foot in the door.

As a fund-raising tool, a demo can't be beat. Whether you hope to make a low-budget feature, a social issues documentary or an instructional tape, it takes more than expendable income to pay for a vision.

To paraphrase, "Demos talk. Bragging walks." Investors must see proof of your abilities. No amount of pipe-dream description will get you the cash you need.

Low-budget producers often shoot a couple of scenes of the planned work, and present this "demo" to potential investors.

J.R. Bookwalter of Akron, Ohio, is the definitive real-life example. He admired the work of Hollywood producer Sam Raimi (*Darkman*, *The Evil Dead*), and pegged him as a possible backer. Raimi screened the novice filmmaker's previous efforts. He was so impressed by the badly exposed Super 8 "demos" he agreed to finance a low-budget film.

To the tune of $125,000.

"Raimi told me that of all the proposals he'd received at that point, only mine was accompanied by a representation of my experience," Bookwalter says. "I'm sure if I hadn't screened my films, the deal would never have gone through."

This isn't a common scenario, but it proves the demo's potential.

Just remember: No demo has more impact than a bad demo, while a great demo pays the bills.

Creating a truly effective demo tape takes more than some assemble edits and a blank tape. Careful planning is the first step.

Consider the Content

You want to show off only your very best work in your demo tape. Scan all your videos, noting outstanding shots, imaginative camera-work and good production values. You want to show a broad spectrum of abilities.

If a particular vacation video looks good, include it. Earmark for use any wedding footage that came out better than normal. Sporting events, community functions and film-to-video transfers all provide raw material for a demo.

Don't be impatient. Getting the best possible footage may mean scanning three entire weddings to find that gorgeous sunset kiss sequence. Any extra effort invested at this point will only make the demo that much more powerful.

Let's say your video services have just become available. Let's also say that you really haven't had any legitimate (paying) jobs yet. Sure, you've goofed around with camcorders for a couple of years. But until now, videography wasn't something you'd considered a career choice. How can you put a demo together without footage?

By creating what you need.

For example, to target the wedding and event video market, you'll need footage of a wedding or two. Check nuptial schedules of area churches and get permission from some couples-to-be to shoot some footage. You don't have to cover the entire wedding. Just get a few shots good enough to convince a prospective client you can handle the job.

Nobody wants to be your first client. If you include wedding footage in your demo you'll appear to have experience in this area.

One caveat. Be certain you really can adequately produce a wedding video. Acquiring a few stray shots and actually shooting and editing a cohesive and attractive ceremony are two very different things. You don't want to misrepresent yourself.

Which leads us to another option for acquiring demo footage: Shoot it for free.

Don't cringe. I realize making money is the whole point. But we all should pay a few dues. Free production work is one way to do this.

Hundreds of organizations gladly accept the donation of video work. Any nonprofit entity (your local food bank? SPCA?) is a good place to start. Call. Explain your situation. Make your offer of free service. Beyond getting demo material, this philanthropic practice increases your working experience. And it's not bad for your reputation, either.

Inform local press of your charitable video "donations." This is great free advertising. Doesn't it feel good to help out others?

A Manual of Style

How you edit your demo can have as much impact as its contents.

First, set an appropriate length for your tape. Your projected audience pretty much determines this. A 3-minute demo isn't really long enough to warrant an award of cash.

Nor would you want to solicit a commercial account with a half-hour production. The client wants a 30-second spot, not a TV series.

Rule of thumb: Keep it short. For the general production market, 5 to 10 minutes is about right. It's not so long the viewer gets bored but not so short a potential client will question your experience. There's ample time to display your best work, professionally and courteously.

Applying the term "courteous" to a demo tape may seem odd, but your customers lead busy lives. They have better things to do than sit through your 30-minute extravaganza. Like anyone else, they want to get their information as quickly as possible. (You can always include supplemental materials with tapes you send to major funders.)

You've decided on a 5-minute demo. Now you're ready to edit footage, right?

Wrong. We're not done planning: Determine the order and style for presenting your experience before you start cutting. And it's a good idea to create a detailed script. Map out the order of each segment of footage.

Now consider style. Who's your target market? How can you reach them most effectively?

If they're serious business people, try a straightforward presentation—interspersing your footage with defining graphics and augment it with a clean voiceover. The key here is quick-paced editing with a clear demonstration of your abilities.

Perhaps you plan to approach several markets using a single demo tape. Intersperse interviews shot expressly for the demo—remarks from enthusiastic past clients—with cuts of your footage.

Taping these interviews means a little extra work, but it's well worth it. The boast of a satisfied customer impresses potential clients more than any claim you can make.

The testimonial is popular for all facets of advertising—just check out the commercials during network prime time. Using this technique in a small-town framework pays off especially well.

Business people and ordinary citizens see neighbors and friends—familiar faces—on the tape. If a prospect's competition or friend up the street is using you, chances are good you've found a new client.

The truly motivated may want to host their demos. The hosted demo is an innovative approach most smaller production companies seldom take advantage of.

If you're not a smooth talker, find someone who is. Create a script and have the emcee introduce each clip or segment. You can structure this many ways, for a serious, comedic or down-home feel.

Take care, though, if you're going for laughs. Your sense of humor might not be that of the general public. It's less risky to be serious.

Consider your host's setting, wardrobe and narrative. All these play a big part in the presentation's effectiveness.

Experiment. You may want to combine styles. Try a hosted demo that includes interviews with satisfied clients. Voiceover client comments while rolling footage from a particular job. Incorporate shots of your equipment in the demo.

Interview yourself—talk about customer satisfaction, your state-of-the-art gear, your sincere goal of creating the best possible product.

Shameless self-promotion adds a personal touch, and it's as popular as the testimonial. Again, just take a look at any network TV program for abundant proof.

Technical Concerns

Planning and assembling a slick demo does you no good if the tape itself is defective.

This sounds obvious, but it's a legitimate consideration. When putting your production together, work with the lowest generation tape available. If you use a wedding shot, pull it from the original footage.

This applies to any material included in the tape.

If raw footage isn't available, be sure your editing set-up allows for the cleanest possible dub. Remember—you'll dub the demo again for client copies. If you use a second-generation shot, it will be fourth-generation when a client views it.

Seen much good-looking fourth-generation footage lately?

Keep a quality-control check on footage, graphics and narration recorded specifically for the demo. Your amazing shots won't impress if they're book-ended by amateurish on-screen intros from an inexperienced host. Be sure graphics and voiceovers complement the rest of the tape. It may seem funny to have your voice talent talk like Elmer Fudd, but does this really show you in a professional light?

Distribute your demo on VHS. Other video formats are gaining popularity, but VHS is still most widely used. Businesses, organizations and individuals welcome this tape size. If someone requests a different format, make it available. Conversion services are plentiful and cheap. It's worth the expense if you get the job.

Packaging affords you another chance to get a jump on the competition. True, you can't judge a book by its cover. But it doesn't hurt to impress when there s a chance to. Most demos I see come in plain cardboard or plastic sleeves. *Boring*. Why not take advantage of those nifty full-sleeve insert shells? Create a cover or design with photos from your business, high-tech images or even your logo. The 8-1/2 by 11-inch layout facilitates low-cost color copies. Slip your custom cover in the sleeve and you've got a package that really stands out.

Or you can print professional face or spine labels on colored stock. Consider dubbing onto tapes with colored shells. It costs a bit more, but it separates you from the masses.

Mail your demos in bubble-lined envelopes—they cost less to mail than cardboard shippers, and they allow you to insert additional materials. Always include a cover letter, a short note informing the recipient of your intentions.

If you're sending the tape in response to a request be sure and point this out. Enclose any press you've received. Letters of recommendation are good, too. Praise from past clients impresses potential ones.

Show It Around

There are a surprising number of additional ways to get your tape to roll where it counts most:

Event videographers should keep a tape available for loan at photography studios. Drop a couple at local bridal and tux businesses.

Film-to-tape transfer specialists might leave a copy at film shops.

Attend county fairs, business expos or video industry trade shows with demo in hand. These functions are tailor-made to sell your business. You'll meet business owners and potential clients face-to-face.

Present your demo to church, school and community groups. These organizations always have some sort of function in the works. Allow them to witness the advantages of having videotape recording of their event.

Following Up

People can be lazy. Your demo may pique their interest, but you're dreaming if you think it's enough to inspire every viewer to make the call. It often takes some additional sales effort to get the job.

Back to tooth-man Bob. As he considers a marketing plan, he gets your demo tape in the mail. Until now, direct mail, print ads and weekly shoppers had seemed the way to go. They require little effort on his part, pricing is reasonable, and audience delivery is good.

But now he's struck by the impact of live, talking images. Potential patients can "tour" his high-tech facility. Nurses and staff can show their friendly faces. Bob himself can make an earnest plea for healthy teeth. Done well, the demo may intrigue Bob enough to give you a call.

Or you could call him. Dispel the high-cost myth associated with video production. Explain how the video will attract clients just as it attracted him to your work. Let Bob know he can be as involved as he wants in making the tape.

Putting together an effective, professional-quality demo is not something you do one afternoon out of boredom. It takes patience, planning, creativity and hard work. You want to show yourself at your best.

To do just that, keep a few key ideas in mind.

Include only your best work. Don't be impatient when scanning your videos for footage. You want to show a broad spectrum of ability. If necessary, create the footage you need.

Set an appropriate length. Rule of Thumb: Keep it short. Five to 10 minutes is about right.

Plan order and style before editing your footage. Keep in mind your target market. When possible, create a detailed script. Try incorporating testimonials and self-promotion. You may even want to host your demo.

If you need funding, shoot a couple of scenes of your planned work. Present this "demo" to potential investors.

Work with the earliest generation tape available. Remember, it will be about fourth-generation by the time your client views it.

Distribute your demo on VHS. If someone requests a different format, make it available.

Package your demo creatively. Use full-sleeve insert covers and colored spine labels and tape shells.

Mail demos in bubble-lined envelopes. Be sure to include a cover letter and supplementary materials.

Follow up your demo presentations with a personal sales effort. Dispel the high-cost myth. Explain how your videos will benefit your client.

Rely on these key ideas and you'll create a demo impressive enough to convince even Uncle Bob.

PART VI
Tell the Story Sooner

Internet and Disc Distribution

Though the traditional media distribution routes can deliver the highest quality audio and video to your audience, the Internet and discs can deliver your videos *now*.

You can mail your videos to your viewers on lightweight (low postage) silver platters that will play in either CD-ROM or DVD-ROM computer drives or DVD-video players. CD-R and CD-RW drives have fallen in price; many new computers come with them installed as standard equipment. Prices for rewriteable DVD drives are following the same path. Discs are more than just carriers for your videos—videotape in another form; their menuing and random-access capabilities open new creative opportunities for you as a video author.

The time is ripe for making your own discs. Though the quality of the videos you burn onto a CD today will be relatively poor, and the length will remain short, the quality of the videos you burn to DVDs tomorrow will be high and their length could be measured in hours. The basic tools and techniques used for the latter, however, will remain similar to those for the former. Start today burning CDs, and you'll be ready to use DVD-RW drives as soon as they become affordable.

When it comes to reducing the cost of distribution, the Internet beats discs. There are a number of streaming video services that will host your videos for free. Again, the video you stream through the Internet today will be short, small in frame dimensions, low in quality and jerky in motion. However, as high bandwidth connections to the Internet become increasingly common, the quality, frame size, motion and length of the videos you will be able to stream will increase. Receiving full screen, VHS quality video will soon become commonplace. There is no time like the present to learn the tools and techniques for delivering your video over the 'Net.

The final section of the *Handbook* will get you started working in both these media. It introduces the hardware and software necessary for authoring discs. It holds helpful tips for shooting, editing and encoding video for 'Net delivery—both in true streaming formats and in "download and play" formats such as MPEG.

At Videomaker we have worked and hoped for the day when television would become a democratic means of expression, a medium no less approachable for every person than pen and paper. In recent years, advances in video production tools, especially the computer, have put this medium into millions of hands. Now, the Internet promises to be the unregulated pipeline that can distribute videos from its creators to its viewers. We hope you use this final section of the *Handbook,* and all the sections that came before it, as a guide into this new era of visual communication.

54
Internet Video: From Camcorder to Computer to the World

Larry Lemm

The Internet is becoming as important as television for video distribution. Learn the basics of this blossoming field, and get your video productions out on the 'Net with ease and style.

Those predictions from a few years back are beginning to come to fruition. Internet video is better looking, easier to use and so important that the big television networks are stepping up to the Internet—lest they be left behind like the radio networks they replaced.

With high-speed Internet access such as cable modems and Digital Subscriber Line connections reaching more and more homes across the country, the idea of being able to get high-quality video entertainment is no longer a "wait until" scenario. Recent studies have also indicated that teenagers would rather surf the Internet than vegetate in front of the TV. It seems that now is the last chance to catch the rising tide of Internet video, wait any longer and you'll be racing around trying to catch up to the pack. So here is a brief

rundown of the different Internet video technologies, how they work and what they can do for you.

What You Need to Get Started

You only need to have a few things to get started with Internet video. First, you'll need a camcorder. It doesn't have to be one of the new digital camcorders, although a digital camcorder will make the whole process easier and of generally higher-quality. Second, you'll need a computer equipped with either a FireWire port (if you own a digital camcorder) or a digitizing card (if you own an analog camcorder).

Keep in mind that if you are going to be making less-than-full-frame video, and most Internet video needs to be less than full frame, you can often make do with an inexpensive digitizing card. You'll also need a software package that will take your raw video and allow you to edit it together, and another that will convert the edited clips into one of the widely used Internet Video formats (if your editing software doesn't already do that).

Lastly, you'll need an Internet connection and a Web site to host your newly created Internet video clips. While this may sound like a lot of hardware and software to purchase, it really isn't that complicated. Finally, once you go through the process once or twice, it will seem like old hat—and in general should be easier to pick up than the first time you were confronted with a linear editing suite.

The Types of Internet Video

There are two basic types of Internet video. First, there are *download-then-watch videos*. These are files that a viewer will click and download in their entirety, then watch later. An example of this type of Internet video file would be a MPEG clip, a Video-for-Windows file (.avi) or a standard QuickTime video.

Second, there are *streaming videos*. Streaming video simply means that the viewer watches one part of an Internet video clip while the next part downloads. It takes away most of the waiting associated with the download-then-watch video clips, but at a price: streaming video is usually more compressed than download-then-watch video, resulting in lower video quality. Examples of streaming video are RealVideo clips, Microsoft MediaPlayer Technology 7 (.wmv) files and the newest version of QuickTime, which offers streaming QuickTime clips.

Now that we at least have a rough idea of what types of video are on it, and what it takes to get video out onto it, let's look at some of the decision you'll have to make when taking your videos to the Internet.

Camcorders: Analog or Digital

Often times, your decision of what type of camcorder to use for your Internet videos

will already have been made for you. It will be the camcorder that you already own. However, if you are looking to purchase a new camcorder to make Internet video with, by all means, get a digital camcorder equipped with a FireWire (or i.LINK) port. A digital camcorder with one of these ports allows you to transfer your digital video right onto your computer without having to redigitize the video. This will give you higher-quality Internet video right from the beginning.

If you already have a good analog camcorder, don't feel like you have to throw it out and run out and buy a digital camcorder. The optics and manual settings of a high-quality older analog camcorder are often better for shooting video than those of a new stripped-down digital camcorder. While digitizing your video might be a little more of a process, your analog camcorder can be used for years to come.

FireWire Ports and Digitizing Cards

After shooting the video with your camcorder, whether it is an analog or digital camcorder, your next step to making it in to an Internet video is to get the raw video clips onto your computer so that you can edit them together into a cohesive video production.

If you are using a digital camcorder, your easiest bet is to get a new computer that already has a FireWire or i.LINK port built-in and configured on it. Some examples of this are Apple G4s and iMac DVs, Sony VAIO Digital Studios and certain models of HP and Compaq computers.

If you already have a digital camcorder and a newer computer, you can install aftermarket FireWire cards that plug into the expansion ports of your computer. If installing expansion cards isn't your cup of tea, make sure you take your computer to a competent nerd to handle the installation, because installing video editing equipment is not for the technically faint of heart.

If you are using an analog camcorder, you'll need to get a computer equipped with a digitizing card. Many companies sell turnkey editing computers with digitizing cards pre-installed. However, if you already have a computer that can handle digital video, you can have a digitizing card installed. Once again, if doing difficult computer upgrades isn't for you, have someone else do your upgrades for you. If you are currently using an analog camcorder, but plan on getting a digital camcorder soon, you might be best served by getting a dual-input card that has both analog inputs and a FireWire port. An example of this is the Fast DV Master, or the Canopus DVRex-M1.

Computer-based Editing Software

Almost every computer with a built-in FireWire or i.LINK port or built-in digitizer will come with some sort of video editing software pre-installed. You can use this software to assemble your video clips, or you can install your choice of editing software for editing the videos.

The editing software that is pre-installed will often be a "light" version that doesn't offer the effects and transitions that the "full" version of the software allows. Don't worry about that too much though, for the most part fancy effects and transitions are things that you want to leave out of Internet video productions because of the way that computers compress video.

The only way you can send video through the Internet is to heavily compress it, and the way that this compression works favors some types of production techniques over others. There are two types of video compression.

Interframe compression compares consecutive frames of video, looking for frames where most of the pixels are not moving. Perhaps the best known Interframe compression scheme is MPEG compression. It is also known as a *lossy* compression scheme, because information is lost during the compression. For example, if you had a clip of video of a runner running across the screen, most of the frame would remain unchanged, only the part where the runner is actually in the frame would change. With interframe compression, the video would "recycle" the parts of the frames that are static background, while refreshing the parts of the frames where the runner actually moves across the screen.

Intraframe compression works on the premise that when you know that a pixel is going to be one color, you can guess that the pixels surrounding the pixel with known color are likely to be the same color. That way, the software only has to keep absolute tabs on a certain number of pixels, while letting using the law of averages to guess what color other pixels should be. It works well with solid colors, but intraframe compression still has a hard time guessing when complex patterns are involved. Intraframe compression is also lossy,

Streaming Software Galore

There are three main providers of streaming video software. The first, and perhaps best known of the bunch is Microsoft. They became aggressive in the streaming market in 1997, and bought stake in almost every streaming video company in existence. They bought some companies outright. Because of this heavy investment in the industry, nobody was surprised when they introduced their NetShow streaming software. It was the first main competitor the RealVideo, and now, known as Windows MediaPlayer, combines many multimedia functions in one easy-to use interface.

RealVideo was started by a company called Progressive Networks. They eventually changed their name to RealNetworks, but their most famous software, RealVideo, lives on today. RealVideo was the number one choice of streaming video for most of 1997 and 1998. It is almost the de facto standard of the streaming realm, but is losing ground to the irresistible force known as Microsoft. RealVideo and NetShow work in similar manners, and are both easy to use and very popular on the Internet.

The next streaming software company is Apple. They make QuickTime, and the latest version of QuickTime offers streaming video, in addition to the traditional download-then-watch scheme employed by the software. QuickTime is generally accepted as having a better compression scheme than MPEG video, and the streaming version of QuickTime is very comparable to the quality of RealVideo or NetShow.

Tips for Shooting and Editing Internet Video

Use a tripod and limit camera moves. When you move the camcorder, whether intentionally or accidentally, the whole picture goes into motion, making compression more difficult and raising the amount of information that needs to be transmitted.

Stick to simple, static shots. Use solid-colored backdrops and wardrobes. Patterns are difficult for computer video to process, and will often come out looking like a pixelated mess.

Avoid complicated transitions when editing. Your new editing software might have a lot of fancy transitions, but that swirling transition that looks so good when you run it from your computer will look like a jagged, halting pixel hash when you compress it and run it from a Web site. Stick to straight cuts unless absolutely necessary for storytelling.

Short is sweet. The shorter your video is, the quicker it will download. Try to get to your point quickly, and if your video runs more than a few minutes, break it into smaller files so that a viewer can download the beginning and see if he or she wants to download the rest. If you feel that you need to keep a longer video together, use video streaming instead of a download and watch video format.

55
Top Tools for Streaming Internet Video

Joe McCleskey

See video clips at www.videomaker.com/handbook.

In the Fall of 1998, more than a million Americans tuned to their computers to watch President Bill Clinton's videotaped testimony for the White House sex scandal. Why did these Americans tune to their computers instead of their televisions? The answer lies in streaming video.

Streaming video is an exciting Internet technology that allows nearly anyone the opportunity to broadcast video to a worldwide audience.

While most people don't have the high-speed Internet connection needed to serve streamed video, many companies will serve your video for you. You may not get the huge audience that The President's testimony drew, but with streaming video, it's possible for just about any home videographer to set up a virtual home broadcasting studio to make videos available to the world without having to deal with broadcasters, public television stations or the FCC. In short, it's a wonderful opportunity for video to become more democratized than it ever has been in the past.

In this chapter, we'll talk to people to see which streaming video products they prefer, and why. You can use their experiences to help you decide what type of streaming software will best fit your needs. Along the way we'll look at the streaming video marketplace—companies that currently make the technology available, and some of the features of each. In this way, we hope to assist you in using this exciting new technology to reach a wider audience than you ever thought possible

Technical Issues

Before we get started, let's cover a few basics of streaming video technology as it exists today. Right now, if you have Internet access, you can easily download a variety of plug-ins for your Web browser that allow you to play streamed video on your computer. Most of them are absolutely free, and require no technical expertise beyond the ability to download

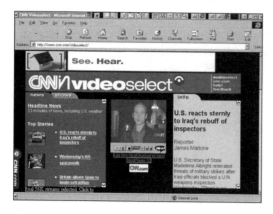

Figure 55.1 News sites like CNN.com make frequent use of streaming video.

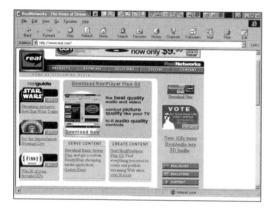

Figure 55.2 Broadcast.com offers a wide range of content in both streaming video and streaming audio formats.

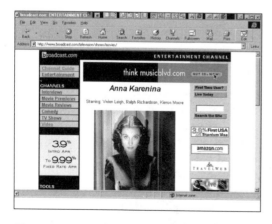

Figure 55.3 RealNetworks' RealGuide is a good place to start looking for streamed content.

and install a program on your computer. Once you install the software, you're ready to sample a few streaming video sites. Good initial stopping points include www.cnn.com (see Figure 55.1), www.broadcast.com (see Figure 55.2) and www.real.com/realguide/ (see Figure 55.3).

As you surf around the Web, however, you'll notice that different sites use different types of streaming video technology. Some make use of RealNetworks' RealMedia software, while others use Apple QuickTime or Microsoft Windows MediaTechnologies software (see Figure 55.4). Many sites offer their video in two formats, or even all three.

After you've viewed several streaming videos on the Web, you'll note that the technology is far from mature. Small screen sizes and low frame rates still plague the software for most users. Many of the people interviewed for this article expressed frustration with the existing data transfer speeds of the Internet. Still, the future promise of streaming video is great, because we can expect the speed of home user's Internet connections to increase dramatically in the next few years. When the speed increases, we will have larger, smoother-playing videos with better-sounding audio available on our Web sites.

However, until that time comes, we'll have to content ourselves with the limited data rates that are currently available.

At any rate, we're not here to discuss viewing streaming video; we're here to discuss the various products available or serving streaming video to your audience. Currently, Microsoft'sWindows Media Player, Apple's QuickTime and RealNetworks' RealPlayer split the marketplace. For most videographers, setting up and maintaining a highly technical Web server capable of streaming video would be an expensive and time-consuming task. Luckily, there are companies that specialize in providing hosting services for streamed video. If you

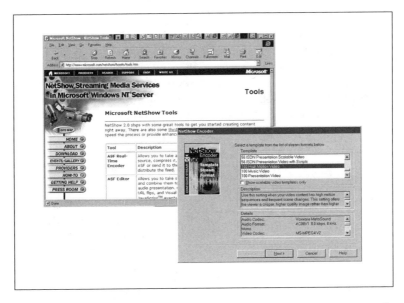

Figure 55.4 *Microsoft's NetShow streaming video server is integrated into the Windows NT 4.0 operating system.*

have a Web site, your ISP may already support one of the streaming formats.

Server? We Don't Need No Stinkin' Server

As we just mentioned, streaming video works best if you use a computer that is especially set up to serve video. They call this computer a *streaming server*. It is usually a separate computer from the Web server computer, because transmitting the video can require a serious amount of computer horsepower, and a streaming server can take advantage of more methods of transmitting information over the Internet (called *protocols*) than a Web server. However, if you don't plan to have a large number of viewers, you can stream video from a regular Web server.

Such programs as VivoActive VideoNow ($149) and Geo Interactive Emblaze VideoPro ($295), in fact, do not even use a streaming server. Emblaze VideoPro doesn't even use a player, it automatically loads one using Java when a viewer starts a video.

VideoNow works on any Web server using the HTTP (Hyper Text Transport Protocol) protocol that the Web uses. This means that you can simply add your VideoNow files into your Web site. Though this method has some drawbacks in terms of playback quality and number of users that it can serve, it's the easiest way for anybody with a Web site to stream audio and video to the world.

Low cost and ease of use were foremost in Kimberlee Grant's mind when she decided to use VideoNow to add video content to her up-and-coming Web 'zine. She's working on a way to stream video and animation content on her Web site as a means of learning the best way to incorporate the streaming video medium into Web content. She adds, "I looked into other streaming video solutions, but I didn't care for the prices I encountered, nor for the complicated server requirements. I just wanted to stream video from my site, and VideoNow did that for me, at a level of quality that's watchable."

RealNetworks, the makers of RealMedia, acquired Vivo Software a few years ago. The two products are distinct, with VideoNow being an entry-level serverless product, and RealMedia offering a professional-level multiple-viewer product, so the two products have a path for existing side by side.

Streaming MPEGs

One of the first companies to offer a streaming media solution for the Web was Xing Technology. Xing Technology's early success in the field of MPEG video and audio encoding led to the development of StreamWorks (see Figure 55.5). Unlike most streaming applications, which require you to convert your video from a QuickTime, MPEG or Video for Windows file into a proprietary streaming format (such as Microsoft's Advanced Streaming Format, or RealNetwork's RealMedia Format), StreamWorks streams MPEGs. StreamWorks supports MPEG-1 video, MPEG-1 Layer 3 audio (.MP3s) and MPEG-2 audio.

To get an idea of how the StreamWorks software operates in practice, we spoke with some of the techies at Internetwork Broadcasting (www.internetwork.com), a company that helps corporations, organizations and individuals place their streaming audio and video programming onto the Internet. Current clients of Internetwork Broadcasting include the College Radio Network (an all-audio offering) and Junior Dub's Irie Reggae page, which regularly streams live reggae events to the world. On the whole, Internetwork's employees reported that Xing's StreamWorks Live software is easy to use, and provides a high-quality stream—especially in the audio realm.

Xing uses a pricing structure based on the number of people you want to be able to serve at one time. Currently, Xing's StreamWorks Live costs $3,000 per 50-seat license (or 50 simultaneous users). For each additional 50 seats, Xing charges an additional $1,500. This pricing structure is reflective of the company's current marketing goals: corporate and educational intranets. The production of streaming video files for Web or intranet delivery via StreamWorks is very simple and inexpensive. To use the StreamWorks encoder, you need to convert your existing QuickTime or Video for Windows files into an MPEG-compressed file. This can

be done using Xing's MPEG Encoder ($249), or any of the other MPEG encoders available on the market. If you plan to stream video, and don't want to have to deal with converting your video into a non-MPEG format, finding a Streaming Video Provider that uses StreamWorks could be your best bet.

Battle Between Two Giants

Since the introduction of streaming video, many things have changed in the market. The most obvious change is in the corporate playing field. While two years ago there were 10-15 companies with streaming software, that number dwindled through corporate buyouts. Microsoft bought Vosaic, and RealNetworks, at the time known as Progressive Networks, bought VivoActive. VDOLive, the first company to offer streaming of live events shifted focus to videoconferencing, while other companies failed to gain enough users to gather a foothold in the market. Today, there are two serious contenders left in this young market: RealNetworks and Microsoft.

Recently, RealNetworks' grouped together their highly successful RealVideo and RealAudio streaming technologies under the single name of RealMedia. Now, a series of corporate partnerships and power-plays has helped place RealNetworks in the forefront among streaming video software companies.

Figure 55.5 *Xing's Streamworks allows streaming of MPEG-encoded video and audio.*

Currently, about 85 percent of all sites that stream video use RealNetworks software.

RealNetworks offers a variety of products—from the free RealPlayer browser plug-in, to a full line of RealSystem G2 server software packages. To get started right away learning how to stream video or audio, RealNetworks offers their Basic Server and RealProducer Encoder software as free downloads from their Web site. The Basic Server can handle up to 25 streams of RealAudio and/or RealVideo.

Upgrading to the Basic Server Plus ($695) increases the number of available streams to 40, and includes several software enhancements. The Basic Server Plus includes RealProducer Plus, a content creation and publishing application. The Basic Server Plus also offers optional support for RealFlash animations for an additional $295.

RealNetworks also offers a number of pre-configured software solutions for a variety of specific streaming media applications: the Classroom Solution, the Intranet Server, the Hosting Solution and the Commerce Solution. (Prices for these applications vary widely, from $1,895 for the Classroom Solution to $21,595 for the full-blown Internet Solution, which includes support for up to 400 simultaneous streams.)

On the content-creation side, RealNetworks offers RealProducer Plus G2 software ($150). This software incorporates scalable data rates and encoding for a wide variety of file types (including real-time sources as well as .avi, .wav, .mov, .qt, .au and .mpg files). This software also includes the ability to create animation, video or audio content fully embedded in the HTML code. This allows for the triggering of HTML commands as the video or audio content streams—including the display of new pictures and text information at specific times.

To get an idea of how RealMedia's streaming video production and server software titles compare to their competitors, we spoke with Jo Sager of JamTV, a hip music/streaming video site (www.jamtv.com). JamTV offers a wide variety of streaming audio and video content, as well as music and entertainment news, tour dates, live webcasts and other information for the music enthusiast. According to Sager, one of the most appealing aspects of RealNetworks' software is accessibility. In her words, "It is fairly user friendly, and can be learned quickly and easily by just about anybody." Jo also had the following advice for those who are thinking of going the streaming-video route, "Experiment with all the tools and gadgets out there to see what works best for you. Also, don't forget to test, test, test."

Of course, any great change in the Web-based digital media creation and delivery marketplace must include the world's biggest software production company—Microsoft. Their creation and promotion of Advanced Streaming Format (.asf), a single standard streaming-media file format, has become a significant part of the recent Microsoft anti-trust lawsuits. The fact that Microsoft offers the tools to create .asf files from existing audio or video content as a part of its free Windows Media Player software has gotten plenty of media-creation software companies up in arms. To further the debate, Microsoft also offers NetShow Services as a free upgrade for owners of Windows NT Server, thus enabling anyone with a video capture card and a capable server to create and stream NetShow content for free.

NetShow Tools includes additional tools for encoding .asf files and creating scripts for embedded HTML commands. Also available for purchase from Microsoft is the Netshow Theater Server, which allows broadcasting of high-quality MPEG and MPEG-2 content over a high-bandwidth network. The potential applications include education, video-on-demand for hotels and airlines, and corporate intranets. NetShow Theater Server sells for $2,499 plus an additional charge per client accessing NetShow (5 licenses sell for $499; 20 licenses, $1,779).

The Battle Rages

At the time of this writing, the battles between Microsoft and RealNetworks for supremacy in the streaming video market show no signs of abating. The winner in this all-out war is you. Right now, if you have access to the video-capture and server technology (see Figure 55.6) you need, it's possible for you to create your own streaming video files for little or no initial investment. So get out there and start building your own streaming video programs today.

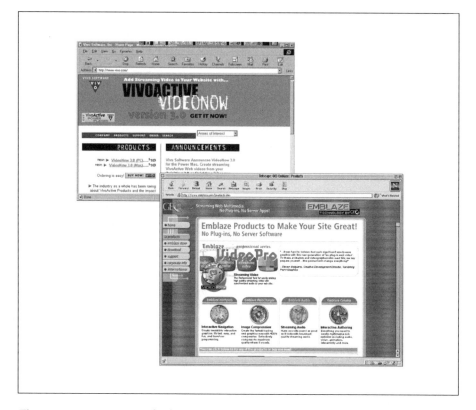

Figure 55.6 *No matter which software you use, streaming video is a great way to reach an audience of millions.*

56
How do I Shoot Video for Streaming?

Larry Lemm

See video clips at www.videomaker.com/handbook.

This is a sneak peek at the script for our upcoming video on streaming. Look for a streamed version soon.

[Roll Title: *Shooting for Streaming*]
Setting: Talent in Production Studio

TALENT A: Hi, I'm _____.

TALENT B: And I'm _____, today we will show you how to shoot video to be streamed through the Internet.

TALENT A: Streamed video is heavily compressed. The tricks of the compression are what determine what style of shooting streams well and what doesn't stream well.

TALENT B: One type of video compression separates the changing from the static parts of the frames in a particular clip of video.

[SHOW VIDEO OF A PERSON MOVING ACROSS STATIC BACKGROUND. USE TOOL TO ISOLATE MOVING PART, THEN SHOW THE ISOLATED MOVING

PERSON AGAINST A BLACK BACKGROUND]

TALENT A: Then it recycles all the parts of the clip that are not sometimes moving, and encodes the moving parts of the images.

TALENT B: When the compressed video is transmitted, parts of the information get lost in the Internet traffic, but the decompression software compensates for this by using its best guess as to where things should be in the frame.

TALENT A: The official term for compression of this sort is Lossy Video Compression.

[Superimpose Title over bottom of screen: Lossy: (adj) *Compression scheme that loses information, then interpolates what's lost from the remaining data.*]

TALENT A: (continues) Lossy compression is needed, because the amount of information that digital video requires

313

is far greater than the amount of information the Internet can transfer to an individual computer over a standard phone line.

TALENT B: We measure the amount of information transferred in bits per second, and we call this speed bandwidth.

[Superimpose Title over bottom of screen: Bandwidth: (n) *The amount of information that can be transferred over a network connection in a given period of time.*]

TALENT B: (continues) Bandwidth is important because most connections to the Internet do not have enough bandwidth to send video without compression. For example, uncompressed video digitized at only the resolution and frame rate of standard television would require the bandwidth of 471 standard modems operating at 56.6 kilobits per second.

TALENT A: Because the bandwidth is limited and the compression is lossy, video will be choppy and unclear if there is too much movement. There are, however, some tricks to shooting video to be streamed.

TALENT B: The first trick is when in doubt use talking heads! A talking head is a tight shot of just a person's head talking.

[ZOOM CAMERA IN TO TALKING HEAD SHOT]

TALENT B: (continues) If we were to stream this over the Internet

[SCREEN SHOT OF CONTINUATION OF VIDEO, STREAMED DOWN TO NET-CASTABLE LEVELS, PLAYED ON A PC/TV]

TALENT B: (continues) the quality of the video would still be high. However, if we were to add more moving information to the shot

[MOVE SHOT OUTSIDE—WIDE ANGLE IN FRONT OF MOVING TRAFFIC, BUT

STILL TAKE SCREEN SHOT OF STREAMED VIDEO COMPRESSED TO NETCASTABLE LEVELS, PLAYED ON A PC/TV]

TALENT B: (continues) it would become jumpy and hard to watch.

[END SCREEN SHOTS FROM PC/TV]

TALENT A: The video and the sound get choppy because the compression has to refresh much more information.

TALENT B: You should remember that if the camera is moving, everything in the shot is moving too. To keep your streamed video from slowing down, always try to shoot from a tripod.

[SHOW CAMERA BEING MOUNTED ONTO TRIPOD]
[BACKDROP CHANGES TO SOLID HIGH-CONTRAST COLOR]

TALENT A: The colors you choose can affect how well your video will compress. Bright solid colors are best. Dark colors can confuse the compression software, by blending together with subtle shadows.

TALENT B: The contrast of colors is crucial too. Always try to make sure your subject contrasts distinctly from the background.

TALENT A: Make sure the background is a solid color because the compression will make hash out of patterns.

[SHOW PLAID JACKET SHOT FROM SHORT STREAMED STREAMING VIDEO]

TALENT B: Patterns, either in the foreground or background must be constantly refreshed even when there is only the slightest movement. So make sure your subjects get rid of the plaid shirts.

TALENT A: You want to have bright, even lighting. Shadows might make a scene more visually compelling, but [the lighting changes to a Film

Noir'esque shadow] if you have shadows in your shot, the compression will have trouble with them. It is a good idea to shoot video to be streamed in the daytime.

TALENT B: Panning and zooming are bad for streaming too. When you are panning or zooming, the whole background is in motion.

TALENT A: That asks more from the compression than it can provide. Here is an example of a panning shot being streamed.

[SCREEN SHOT OF A PANNING SHOT, STREAMED, PLAYED ON A PC/TV HAVE IT END BEFORE TALENT SPEAKS AGAIN]

TALENT B: You also have to be careful not to have too many rapid-paced MTV-style cuts in your video. Each time you change the scene, the compression has to load a whole new image, and it can lead to choppy video.

[SCREEN SHOT OF THIS PART OF THIS VIDEOSTREAMED, PLAYED ON PC/TV]

TALENT A:[Voice-over] [Camera cuts between separate shots] [Back to first shot]

See, it is having [Cut to second shot] trouble dealing [first shot] with all these cuts [second shot] back and forth.[first shot]

[END SCREEN SHOT]

TALENT B: It might seem that the tricks you need to use for streaming limit the creativity of the videographer.

TALENT A: But this is far from truth. A true mark of creativity is being able to make a compelling video within the framework and limitations of the bandwidth and compression.

TALENT B: These are temporary limitations anyway. Only the pioneers of video streaming will have to contend with them. When more Internet bandwidth is available, you will be able to shoot video in any style you wish, and still have it look smooth and not be jumpy.

TALENT A: Until that point, you have to work within the framework of what video the software can compress, and what video it can't compress. There are also tricks to properly compressing your video to be streamed.

57
Why and How Would I Get a Streaming Server?

Larry Lemm

To provide video over the Internet, you need to have a server. Hosting your own streaming server is a good idea only if you already have a Web server with at least a T1 line running to it. If you have a Web server with no more than phone or ISDN lines you will not be able to stream video effectively.

Many people who create video content would rather spend their days shooting and editing video than maintaining an ordinary Web server, much less one that streams video. For them, a remote server hosted by someone else is a good idea.

If you do have your own Web server, you can provide a hyperlink to the video hosting company's Web site. Alternately, you could keep all your Web content—including your video content—on a remote video streaming server and avoid the need for maintaining a Web server of your own. Another advantage of enlisting a remote host: a large video serving company could also draw more people to your video than you could draw to your own web page alone.

There are many considerations to keep in mind when choosing a streaming server. The first and foremost is how many "streams" you are allotted. A single stream would allow only one viewer to watch one video clip at a time. The more streams that are available, the more viewers can be watching at one time. For many Web sites, 10 streams would be plenty. If you had 20 minutes of streamed video on demand, 10 streams would allow up to 720 people to watch that video in a 24-hour period.

Some servers charge you by the number of streams you want available to potential viewers, while others charge you by the number of streams they actually served to viewers of your video.

If you are going through the trouble of having someone else host your video, you should choose one of the multiple protocol streaming software packages. Along with choosing a server, you will need to decide what software you want to use to serve your video. Each package has its own features and drawbacks. Some programs require clients to download browser plug-ins to watch the video; others do not; still others cram a plug-in into your browser with the power of Java.

The streamers that don't require a plug-in transmit video through the Web's HTTP protocol. These are convenient for the user because they are easiest to setup and use. However, they do not take advantage of alternative protocols that are designed to deliver video more quickly and accurately. Non-HTTP software packages use "non-reliable" protocols to send everything but the control information and are thus inherently faster.

When shopping for a streaming server company, consider the amount of bandwidth that it makes available for your content. The more bandwidth you have, the faster and better your video will stream to viewers. Some servers may use a process known as multicasting. Multicasting sends only one stream of video to a collection of servers. Those servers then re-transmit the video to viewers in their area. This process saves bandwidth on the backbone of the Internet, and allows for more video to be streamed without clogging the 'Net with video. The downside is that multicasting eliminates the on-demand aspect, you can watch a multicast only while it is being transmitted; you can't initiate one on your own.

Choosing the right server is the most important decision you make after deciding that streaming video is for you. If the idea of setting up a high-bandwidth video server makes your head spin with unknown acronyms and arcane nerdspeak, then choose a remote video hosting company to stream your video for you.

58
Slide into "Thin Streaming"

Larry Lemm

Someday soon Internet video-on-demand will democratize video distribution, allowing everyone the opportunity to distribute video to a worldwide audience without having a multi-billion dollar television-broadcasting studio.

But what is there to do in the meantime? How can I put my productions on the Internet today?

The answer lies in the slideshow. I remember the first time I watched video-on-demand via the Internet. "Amazing," I thought to myself as I clicked on the button that started the video. Having already downloaded and installed the utility that allowed this technological miracle, I was ready to experience "click-and-watch" video. What a letdown.

The video was tiny, roughly the size of a saltine cracker on my 17-inch monitor. The picture was barely discernible, with large digital-artifacts appearing where the software's compression utility hadn't quite done its math correctly. The most obvious problem was the lack of motion. The video sputtered along at two or three frames per second. I had almost given up hope for Internet video, when I remembered how I was connected to the Internet: through an old, slow, copper phone wire.

Then I imagined the super-fast connection the phone companies are promising over the next couple of years. If they can transmit almost-video over my ancient phone line now, full-motion, full-screen video-on-demand will be a reality when the super-fast Internet connection comes. Until then, I can distribute my work on the Web in the form of a slideshow.

Internet slideshows are a fast and inexpensive way to get your ideas out to a mass audience, without having to resort to slow, chunky and tiny video. By using the basic story-telling concepts of a storyboard, you can easily turn any video into a multimedia web-based slideshow. Slideshows also play great on a television, opening your Web slideshow to a wider potential audience of Web TV surfers.

Making Slides to Show

There are two basic ways to create digital slides. First, you can use a still camera. Your still camera can be either a standard film-based camera, in which case you would have to use a scanner to digitize the photos, or a digital camera that saves stills on a disk in an Internet-ready format. Or, you can use a video camera, and take stills from video (see Figure 58.1).

The first method requires little or no special equipment, save a scanner if you are using a film-based still camera. If you have a FireWire based digital camcorder it is easy to create stills that are transferable to your computer through a simple connector. If you want to use existing analog videotape taken from a standard camcorder, making digital stills requires a special piece of equipment called a digitizer. This can be either a special video-capture board that you install into your computer, or an external device like Play Inc's Snappy (see Figure 58.2), that plugs into the ports in the back.

Either way, you can use these tools to capture "still frames" from your video. These devices will usually save the still frames in any of the standard digital photo formats including .gif, .jpg, .bmp, and .tga. After you have selected the stills you want to use, you'll need to put them in an order that tells a story. For example, if you are making a how-to slideshow about organic gardening your first slide could be of your untilled garden. The next could be a slide of tilling the garden, followed by slides of rebuilding the soil with compost, planting the garden, chemical-free pest control techniques, harvesting and so forth. After you have selected your slides, you can create an audio track that explains each step, and match the changing slides to your audio track.

Tools to Slide the Show

Now that you have slides to show, you are ready to put them on the Internet. For those

Figure 58.1 Your camcorder is all you need to gather images for your Internet slideshow.

Figure 58.2 Video stills can be captured using an inexpensive device like Play, Inc.'s Snappy.

of you who aren't experienced in HTML Web programming, I'd recommend using an existing video streaming program like the Real Video Producer (see Figure 58.3) or a special slideshow program such as InMedia's Slides and Sounds.

To make a slideshow in Slides and Sounds (see Figure 58.4), simply place your selected slides in order in the slideshow creation menu. Captions and sound effects can be added to each slide by

Figure 58.3 *RealVideo allows you to make streaming slideshows by showing a video frame every four seconds.*

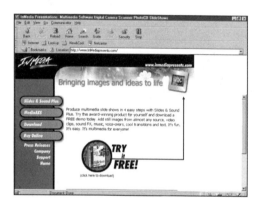

Figure 58.4 *InMedia's Slides and Sounds is designed to make creating Internet slideshows easy.*

simply clicking the "add sound" and "add caption" buttons. Transitions between slides can then be selected from the F/X menu. Blank slides can also be created to add simple titles at the start or finish of your slideshow. It's that easy. To finish the slideshow, save it as a file that can be e-mailed to family and friends, or select the "save as HTML" option from the file menu.

This will create a Web ready HTML file that will play on most browsers (in Internet Explorer, you might have to go to the "view" menu, "options" section,

"security level" button, and select the medium security setting to permit a small file to be temporarily used to play the slideshow).

Getting Your Slideshow on the Web

Now that your slideshow is ready for the web, you'll need to publish it to your Web page. There are numerous Web hosting companies, and each offers its own package of options for your site. Users of InMedia's Slides and Sound who don't want to host their own Web site can have a slideshow served by InMedia.

If you plan to use a video-streaming package to create your slideshow, make sure that the hosting company you choose supports the streaming package you plan on using. There are some sites on the Web that will even give you a free web site to display your slideshow. Geocities (www.geocities.com) and Tripod (www.tripod.dcom) will host a small site for free as long as it is a not-for-profit endeavor. (See Figure 58.5.)

Until the majority of the wired world has a high-bandwidth Internet connection that streams video full-size and full motion, slideshows are the low-bandwidth alternative of choice for camcorder enthusiasts.

Making a Slideshow in HTML

It is easy to make a series of Web pages act like a slideshow is you already know the basics of HTML. To perform this feat of coding magic, simply use this tag above of your HTML header (make it the first thing listed in an HTML source).

<META HTTP-EQUIV="Refresh" content="5;

URL=http://www.yoururl.com/nextslide.htm">

In this example, the refresh call is the "slideshow" command, the content number (5 in the example) is the number of seconds the page will wait before cycling to the next page in the slideshow, which is the URL listed in the tag. To use this bit of HTML magic, replace the sample URL with the URL of your next slide, and replace the 5 in the content call with the number of seconds you want your slideshow to be displayed.

Create your web pages for the slideshow with the images and text you want displayed, as you would create any web page. If you use an HTML generator such as Pagemill, you can create the pages normally, then add the Meta-Refresh tag above any other coding. An example of this style of slideshow is available at www.adventureliving.com/home/slidesh ow/index.html.

Using RealVideo to Make a Slideshow

Video streaming software such as VivoActive or RealVideo can also be used to make an Internet slideshow. The downside of this method is that viewers will have to download a special player plug-in to watch your slideshow. On the upside,

the streaming packages allow for a continuous soundtrack of narration or music to be added to your slideshow.

The RealVideo encoder, for example, has a pre-defined slideshow setting that will take a video clip, and stream it with high-quality sound, and a single frame from the video is shown every four seconds. This is the easiest way to make a slideshow from a video. Another way to use RealVideo to make a slideshow is to make an audio-only RealMedia file. Then you create a series of web pages holding the images you want shown in your slideshow. The next step is to make a text file that will list the web pages you want synchronized to the audio with a time marker next to each, and it will begin loading that page at that point in the audio file. As the audio file plays, the web pages are automatically displayed.

A stellar example of a RealVideo slideshow is at www.starwars.com/dew-back/index.html. Here you will see George Lucas explain some moviemaking magic while his Web site employs some Web-Jedi tricks.

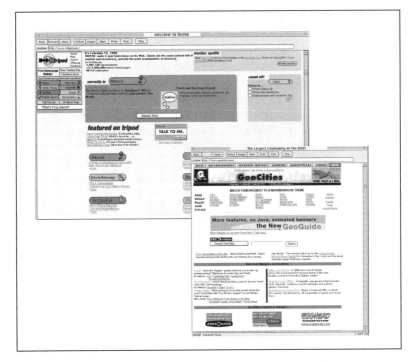

Figure 58.5 An easy way to get your show on the Web (if you're a non-profit outfit) is to utilize a free host like Geocities or Tripod.

59
Put MPEGs on your Home Page

Joe McCleskey

See video clips at www.videomaker.com/handbook.

Got a Web page? Got a camcorder? Here's how to put your video clips on the Web. MPEG (emm-peg): *Moving Picture Experts Group* 1)A working group of digital video experts who meet regularly under the auspices of the International Standards Organization (ISO) and the International Electro-technical Commission (IEC) to develop standards for compressed digital video. 2) A compressed digital video clip, often found on the World Wide Web or in multimedia CD-ROM products.

If you've spent any time at all on the World Wide Web, you're probably familiar with the MPEG acronym by now. That's because MPEGs are one of the main types of digital video files available on Web pages, some of the others being Microsoft's Windows Media (formerly known as Video for Windows), Real-Networks' RealMedia and Apple's QuickTime. MPEG compression is the older brother to MPEG-2 compression which is used to encode video for DVDs as well as DSS. MPEG compression has been designed to compress digital video

to manageable size while retaining picture quality.

"Okay," you say, "that's all very well and good, but how do I get these MPEGs onto my home page?"

Glad you asked. What follows is a concise guide to putting your own short video clips onto the World Wide Web. We'll cover the shooting and digitizing of these clips in somewhat less detail; what we're really after here is the simplest way to 1) compress your digital video files using the MPEG-1 CODEC, and 2) post these files to a Web page using the store-and-forward method.

When you've finished with the article, you'll be able to put the entire Web audience just a few mouse clicks away from viewing your short video clips.

Store and Forward

Currently, the most popular method for distributing digital video on the Internet is known as the "store and forward" method. The concept is pretty simple: a videographer makes a digitized video clip available on his or her Web page, where the Internet public at large may download it onto their own computer and play at their leisure. It's nearly identical to the shareware concept of computer software distribution, with one condition: the shared software is a video clip instead of a software package or application.

There are some drawbacks to this method. The biggest problem is time; video files tend to be rather large, so it takes some time to download them on a typical (56.6 kilobaud) modem. A thirty-second MPEG clip, for example, can easily occupy 1MB of hard drive space—which, in turn, will take approximately five minutes to download on a 56.6Kb modem.

Fortunately, there are ways to make the process a little bit more palatable. Common practice is to place a single frame of the video clip onto your Web page near the location of the MPEG itself; this gives potential viewers a look at what they're getting before they commit some of their valuable online time.

Limitations

Before you get ready to start digitizing your favorite video clips, you'll need to consider the method that your audience will most likely use to view your MPEGs. Most multimedia systems offer only a small, low-res window for viewing digital video.

Brand-new computer systems can handle bigger window sizes and better resolutions, but you can't count on everyone in your potential audience having a brand-new computer. For this reason, we suggest offering your MPEGs in a 320x240 format or smaller, and 15 frames per second or less. This tiny window will place limits on the kind of video clips you can use. The most obvious limit is size; if you want to show a stunning wide-angle shot

of a mountain range, for example, it won't look like much once you've reduced it to fit, stutteringly, into a tiny little box on a computer screen.

These and other factors will unfortunately make most of your existing video footage difficult to watch in a small window. The solution: either sort through the footage you have for the appropriate shots, or start from scratch and shoot a video project with the above-mentioned concepts in mind.

Another problem with MPEGs is audio. Many software MPEG encoders won't handle MPEG audio, and just as many software MPEG playback applications won't play MPEG audio even if you take the time and effort to include it in your video clip. For this reason, MPEG video artists might find themselves operating in a visual-only medium. (The easiest way to get around this? Go with the standard .avi or .mov formats instead of MPEG.)

Shooting Video for Multimedia

Once you decide on what footage to use, you'll have to get the video into the computer with a video digitizer. Digitizers are available in a wide range of performance levels and prices.

Note that a 60-field-per-second, full-screen video digitizer is not necessary to produce a suitable digital video clip. Many of the older low-cost digitizers were designed with the small multimedia video presentation in mind, so if you've got one of these, you're set to go. If you're in the market for a new video digitizer, however, and you're willing to spend a little extra, it's a good idea to get the best model you can afford. A better digitizer will not only give you a better overall image (even at these small sizes); it'll be there for you when you're ready to upgrade your DTV workstation.

One more thing to look for if you're in the market for a video digitizer: check to see if the model you're interested in comes with MPEG encoding software bundled. Many of the newer video digitizers come complete with software for nonlinear editing, 3D animation, photo enhancement and other applications. For our purposes, a software-only MPEG encoder is a direct hit.

Software MPEG Encoding

Now that you've digitized your shot-for-multimedia footage, it's time to use MPEG to bring your video files down to a more manageable size.

You can accomplish this by two means: through a hardware MPEG encoder, such as might be integrated with your capture card, or by using a software MPEG solution. An hardware encoding solution is usually faster (real-time or better), but generally costs more, while a software solution might leave you waiting, but with more money in your wallet.

Once you've successfully encoded them to MPEG, your video clips should occupy a much smaller space, requiring less online time for your audience to download. Now it's time to post it to a Web page. In order to do so, you'll either have to pay someone who knows how (i.e. an Internet service provider or consultant), or roll up your sleeves and get friendly with HTML, the language of the Web.

Post It

Don't panic: the basics of HTML (hypertext markup language) are quite easy to learn. For a primer on Web production fundamentals, take a look at "A Beginner's Guide to HTML" (www.ncsa. uiuc.edu/General/Internet/WWW/HTML Primer.html). An alternative would be to use one of the WYSIWYG ("What You See Is What You Get") Web page editors, such as Microsoft's FrontPage or Adobe's PageMill, which handle all the "coding" for you. There are also shareware and even freeware editors available for download, so don't feel compelled to learn HTML.

Besides knowledge of the basics of HTML (or at least a method of creating it), you're going to need access to the right kind of Internet service. Specifically, you'll need the type that allows users to create and post their own Web pages. The major online service providers (AOL, CompuServe, etc.) offer limited Web pages to their clients, but the large number of individual pages this creates makes it necessary to severely limit them in size. For this reason, your best option for Web publishing is likely to be one of the many free web-hosting service providers, such as iDrive (http://www.idrive.com), Driveway (http://www.driveway.com) or K-Turn (http://www.kturn.com).

Here's how it usually works: you create your HTML files, then "post" them to your assigned site, either through FTP or through the HTML editing program you're using (e.g., Microsoft FrontPage allows you to manage your site through an Explorer-like interface, with folders and file icons).

Students who are lucky enough to attend schools that offer them Web space can take advantage of this opportunity, but bear in mind that you'll probably be on your own for the actual posting. Consult your campus information center to find out how your school handles student Web pages.

For those who already have some experience producing Web pages, here's a tip: posting MPEGs (or any other kind of video file, for that matter) is easier than it may seem. Just link the MPEG file (video.mpg, for example) the same way you would create an internal link to another Web page. The HTML might look like this: Click a <href="video.mpg"> here </a> to download my latest MPEG.

When you do this, a person viewing your page has only to click on the word "here" to download your video clip. (Note: in the above example, you'd have to make sure that the file "video.mpg" was in the same directory as the page listing the link. Otherwise, you'd have to put the directory information ahead of the file name, e.g. "mpegs\video.mpg".)

Confused? Don't worry; a visit to the above-mentioned HTML tutorial should help to sort things out a bit. And if all else fails, you can always bribe one of your computer-nerd friends with a six pack of Jolt Cola; this always works for me.

What For?

Now that you know how to post MPEGs on the Web, what will you do with this knowledge? Here's a short list of applications you might consider: Illustrate a process. Provide talking-head narration. Make a preview of a longer work. Create a weekly 30-second Internet TV show. Advertise a product. Smile and wave to your Web audience. Give the Web a "virtual" tour of your backyard. Introduce your pets. Expose a scandal. Sell your car. Create a video personal ad.

Or anything else you might think of.

60

Burn Your Own:
A Guide to Creating Your Own CDs and DVDs

Loren Alldrin

See video clips at www.videomaker.com/handbook.

Been to a video store lately? If so, you've probably noticed that you can rent movies on something other than good-old VHS tape. More and more releases are now available on DVD, a shiny disc that holds moving images and sounds in a digital format. DVD delivers excellent audio and video quality plus the potential for inter-activity: viewer-selectable camera angles, alternate edits, multiple languages and much more. Intrigued? As a videographer, you should be!

Unfortunately, due to the high cost of hardware putting your own videos onto DVD has been an unaffordable dream for most. But this is rapidly changing—hardware prices are dropping to the point where burning your own DVD (or CD-ROM or Video CD) isn't outrageously expensive. What it still is, though, is confusing. Many videographers have tried to burn their own video discs but found themselves lost in a maze of acronyms, technical jargon and fast-moving standards.

Sound familiar? If so, you've come to the right place. We'll do our best to explain the basic concepts of recording to DVD-Video, CD-R video and Video CD discs.

The Big Picture

When you boil it all down, there are two main issues at work when it comes to recording video onto a disc. First, you have the physical disc format itself. There are just two to consider—CD and DVD. Both use tiny pits in a reflective medium to represent digital data, but the newer DVD technology packs much more data on the disc. You can think of both as simple storage devices that store any type of data without bias.

The bigger issue is what type of data a given disc holds and how that data is structured. DVD-Video, for example, is simply a DVD disc with MPEG-2 video, audio and a file structure that makes sense to a DVD player. A Video CD's medium is just like that of an audio CD, but the video and audio files are saved in the specific format a Video CD player expects.

Keep in mind that CD and DVD can store any type of information, not just the latest DVD Video movie or Microsoft Office 2004. This means you can use a CD or DVD to hold digital video and audio that's not formatted for a specific hardware device (such as a DVD player). Since this is sort of like a cross between the strictly data and strictly video applications, you'd use a computer to read and play back these files. It's just good to remember that DVD Video, for example, represents just one way to store digital video and audio on a DVD.

If you're reading between the lines, you've probably already figured out the good news—computers that have a CD-R drive are physically capable of writing a Video CD, and those with a DVD-R drive can write a DVD Video disc. When it comes to burning 1s and 0s into that shiny silver disc, you're already equipped.

But as we mentioned above, the physical format of the disc is only half of the equation. The other half is getting your video and audio files in the correct format—and in the right place—for playback in a stand-alone player. This is where specific software comes into play to encode and author.

Blend, Pour, Bake

Whether you're making just one copy for yourself or creating a master to send off for duplication, the DVD and CD-ROM creation process is essentially the same: digitize, encode, author and burn. (See Figure 60.1.)

The encoding step converts digital video and audio files from one format to another, usually reducing their size dramatically along the way. Encoding for a DVD Video, for example, involves compressing video files with the MPEG-2 standard. The software then encodes audio files into any of several formats (surround, 5.1 surround, stereo, etc.) recognized by DVD players. The encoding process is similar for a Video CD, but software uses lower-quality MPEG-1 compression for the video and audio.

At the encoding stage, DVD makers have numerous parameters at their disposal for controlling image and sound quality, as well as the amount of video that will fit on the disc. Good DVD encoding fits the required amount of video on the disc at the highest-possible quality, with no wasted space. For a major motion picture, this might equate to roughly two hours of video. Drop the quality down to the VHS level, and you could fit 10 or

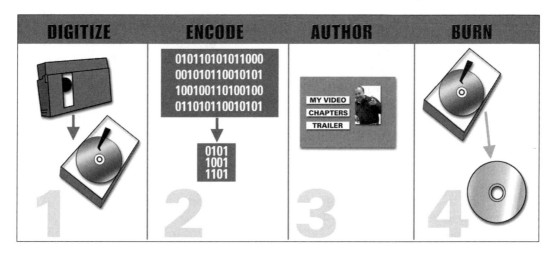

Figure 60.1 *1) Transfer your video from tape to hard drive. 2) Apply the appropriate compression/decompression (codec) scheme to your video. 3) Design the operating interface for your video. 4) Copy your finished video to a disc.*

more hours of video on the smallest-capacity DVD. Be forewarned: depending on quality settings, the software you use and the speed of your computer, encoding MPEG-2 can be a lengthy process.

Authoring software then takes the encoded video and audio files and formats them according to their intended purpose. For DVD, the encoding software records various special codes required for correct playback, as well as such optional goodies as chapter points, navigational aids, menus and overlayed graphics. Because capabilities of the Video CD aren't as advanced, Video CD authoring software is simpler and less costly than DVD authoring software. Some software packages combine encoding and authoring into one step.

The final step is to actually burn the audio, video and other special files to a blank DVD-R or CD-R disc (about $30 and $1 respectively). Everything is just a series of numbers at this point, and the recorder will lay these bits onto the disc as fast as the burner hardware (or the operator) will allow. Since DVD and CD-R burners range in speed from 2x to 12x and beyond (CD-R), the process doesn't happen in real time. If your hardware is very fast, for example, you may be able to burn a full 74-minute Video CD in about 5 minutes.

Burning For You

Is everything rosy in the world of home-cooked DVDs? Not entirely. DVD-R drives record permanently on a disc, offer the best compatibility with home DVD players, but still cost several thousands of dollars. Most professional DVD authors use DVD-R drives.

Rewritable DVD drives have dipped below the $1,000 mark, and may be available for less than $500 by the time you read this. Unfortunately, the rewriteable DVD market is in the middle of a bitter format war. Three competing standards (DVD-RW, DVD+RW and DVD-RAM) are available and each has its advantages and disadvantages. Compatibility with exist-

ing home DVD Video drives and computer DVD-ROM drives is one of the hottest topics, so be sure to check for late-breaking news before you purchase a drive.

Folks wanting to burn their own Video CDs have it easy. They may already have everything they need, provided their CD-R bundle included the correct software.

ABCs of Acronyms

One of the first obstacles to get over when moving from a video cassette to a silver disc is the mountain of acronyms that seems to pile up with every new technology. Here are some of the key acronyms and terms that you need to know:

CD (Compact Disc): a digital "bit bucket" that can store any type of data, be it audio, video or computer software. One CD holds roughly 650 megabytes (MB) of data.

CD-R (CD Recordable): a CD you can record your own data on. You can't erase or re-record a CD-R.

CD-ROM (CD Read Only Memory): a CD pre-recorded with computer data that can't be changed or erased.

CD-RW (CD Rewritable): a CD you can write and re-write to thousands of times.

DVD (Digital Versatile Disc): like a CD, holds digital data of any kind. Today's DVDs hold roughly 5, 9 or 13 gigabytes (GB) of data on one, two or three data layers. A 13 GB DVD holds the equivalent to about 20 normal CDs.

DVD-R (DVD Recordable): a DVD that you can record your own data onto only one time. DVD-ROM (DVD Read Only Memory): a DVD pre-recorded with computer data that can't be changed or erased.

DVD-RW, DVD-RAM, DVD+RW: competing rewriteable DVD standards embroiled in a heated format scuffle.

DVD-Video: a DVD disc that holds MPEG-2 video and any of several different types of audio. When you rent a DVD, this is what you get.

MPEG-1 (Motion Picture Experts Group, first standard): a highly efficient way of compressing digital video for lower resolutions (i.e. 320 horizontal lines).

MPEG-2 (MPEG, second standard): offers better quality and higher resolution than MPEG-1. DVD players use MPEG-2 with a horizontal resolution of 720 lines.

VCD (Video CD): just like a regular CD, but holds MPEG-1 video and audio for playback in reasonably fast CD-ROM and DVD-ROM drives and current-generation DVD video players; as well as Video CD players not readily available in the U.S.

For a look at the physical differences between a rewritable CD and at recordable CD, take a look at Figure 60.2.

Keeping up with digital video is all about learning new technology and new techniques and the latest trend away from tape and toward DVD is no exception. Someday soon, a silver disc may be the final destination for all your video projects.

When that day comes, you'll be ready to burn.

Web Links

The Internet is one of the best places to stay abreast of rapid-fire changes in the DVD and CD-ROM markets. Visit these informative Web sites for the latest news and product information.

General Info

DVD FAQ (Frequently Asked Questions)
www.videodiscovery.com/vdyweb/dvd/dvdfaq.html

DVD for Not-so-Dummies
www.nimbuscd.com/dvd.html

MPEG Home Page
www.cselt.stet.it/mpeg

MPEG Pointers and Resources
www.mpeg.org

Canopus
www.canopuscorp.com

Creative Labs
www.soundblaster.com

Digigami
www.digigami.com

Fast Multimedia
www.fastmultimedia.com

Vitec Multimedia
www.vitecmm.com

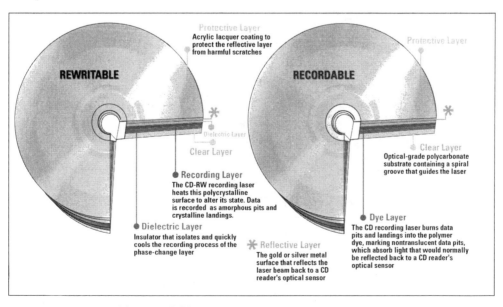

Figure 60.2 *Rewritable/recordable Comparison*

61
Eight Steps to Streaming

Stephen Muratore

Dear Videomaker,

Thanks for the recent articles telling how we might stream video through the Internet. They make me feel that streaming is something I could and should try. Still, I don't know where to begin. I edit video on a Windows computer, but I don't have an Internet service provider, or a Web site. And I don't want to spend a lot of money. For me, this is supposed to be a hobby.

Jenny Small
Internet

Dear Jenny,

Lots of folks find themselves in the same boat. There are a number of ways to get Internet service, a Web site and a streaming server inexpensively, or even free. Here's one of them. Follow these steps and you will have a Web site streaming one of your videos in 2-3 hours: all for free. For starters, we'll assume you've got a short video in one of the standard digi-

tal forms, say .avi or .mov, on your hard drive, and that your computer has a CD-ROM drive and a modem.

Step 1: Get a free Internet Service Provider. To make it simple, go to http://www.netzero.com/ and register. If you don't have an Internet service provider you'll have to call NetZero at 1-888-279-8132 to order their sign-up software on a CD-ROM. This is the only step in the list that isn't completely free. The disc is free, but you'll have to pay $3.50 for shipping and you'll have to click on a banner ad every 10 minutes to prevent NetZero from dropping your connection, but hey, it's free.

Step 2: Get a free Web server. If your Internet service provider won't host your Web site at no extra charge, you should consider getting a new one. In the meantime, use a free Web hosting service like Geocities.

Surf on over to http://geocities. yahoo.com/home/. Click "Get a free home page." Click "I'm a new user Sign me up!"

Then follow the instructions to the button that says; "Build a Page." There, you've got a Web site with a single page in it. You could actually upload your streaming video files right to this Web site, and they would play fairly well, but to truly stream your videos, Geocities would charge you about $5 per month and I said this would be free, so:

Step 3: Get a free streaming server. Surf to www.popcast.com. and click "Join Today. Its Free!" You now have 10MB of space on a free streaming server.

Step 4: Get free encoding software. Go to Microsoft's Media Technologies Web site (http://www.microsoft.com/widows/windowsmedia/en/download/default.asp). Click the "Download" button next to "Windows Media Tools-Intel." Download and install this bundle of tools. Be sure to install the Windows Media Encoder because Popcast serves only files encoded in Windows Media, ".asf" format.

Step 5: Encode a video clip. Open the Windows Media Encoder and select a video file on your hard drive. For starters, pick one 30 seconds in length or shorter. Select your choices among the various options offered by the encoder. In the end

it will create a new file, with an .asf extension, containing your video.

Step 6: Upload your streaming video file to your streaming server. Go back to Popcast.com, sign in and click where it says, "Want to upload another file? Click here." Then give it the location of the encoded video file and fill out the rest of the form describing your video. Then hit the upload button. Now you've got a streaming video file on a streaming server. Popcast gives you the URL of this file.

Step 7: Link this file to your Web home page. Copy the URL of your streaming video file; go back to Geocities and log in. Click "Edit pages." Follow the instructions on editing your Web pages, or use html editing software of your own, to create a link to your streaming video. You'll need to put the "Powered by Popcast" logo somewhere near the video link on your home page. You agreed to that when you signed up for your free streaming server.

Step 8: Sit forward and enjoy. Take pride in the fact that the little video you now see streaming through the Internet to the world is your very own. And hey, it was free.

Contributing Authors

Art Aiello is a video hobbyist and marketing communications specialist.

Loren Alldrin, once Contributing Editor to Videomaker, is a freelance video and music producer.

John Bishop is an instructor at University of California at Los Angeles.

Robert Borgatti is a writer and video producer at a community college.

Mark Bosko is a freelance writer and an independent video and film producer.

Larry Burke-Weiner is a professional gaffer.

Tim Cowan is a freelance writer specializing in video and computer subjects.

Armand Ensanian is a professional videomaker, photographer, and former columnist for *Video Review*.

Stephen Jacobs is an English and data processing instructor at National Technical Institute for the Deaf at RIT.

Robert J. Kerr is a consultant, teacher and writer in the video industry.

Larry Lemm is a freelance writer.

Geoff Best is a professional videographer and freelance writer.

Michael Loehr is a foreign documentarian.

Janis Lonnquist is a writer and producer with clients including Intel and America's Funding Source.

Joe McCleskey is a multimedia producer and freelance writer.

Norm Medoff is a university professor, author, and video workshop instructor.

Stephen Muratore is the Editor in Chief of *Videomaker* Magazine, having been with Videomaker since 1991.

Robert Nulph is a producer, director, and professor.

Stray Wynn Ponder is a writer and producer of television commercials and industrial training videos.

Michael Rabiger teaches filmmaking at Columbia College, Chicago. He is the author of *Directing the Documentary*.

William Ronat is the owner of a video production company.

Bill Rood is an engineer at KTXL Channel 40 in Sacramento, CA.

Jim Stinson has a masters' degree in film production, teaches high school video, and writes about computers.

John K. Waters is a freelance writer.

David Welton is a community college instructor.

Bernard Wilkie designed special effects for the BBC for over 25 years.

Matt York is the Publisher-Editor of *Videomaker* Magazine.

Jargon:
A Glossary of Videography Terms

8mm Compact videocassette format, popularized by camcorders, employing 8-millimeter-wide videotape. [See *Hi8.*]

A/B roll editing Two video sources played simultaneously, to be mixed or cut between.

action axis Imaginary line drawn between two subjects or along a line of motion as an aid in maintaining continuity of screen direction. Sometimes referred to as the "180-degree rule."

ad-lib Unrehearsed, spontaneous act of speaking, performing, or otherwise improvising on-camera activity without preparation.

AFM See *audio frequency modulation.*

AGC See *automatic gain control.*

ambient sound (ambience) Natural background audio representative of a given recording environment. On-camera dialog might be primary sound; traffic noise and refrigerator hum would be ambient.

analog An electrical signal is referred to as either analog or digital. Analog signals are those signals directly generated from a stimulus such as a sound hitting a microphone's element, which the microphone converts to an audio signal that is analogous to the original sound wave.

animation Visual special effect whereby progressive still images displayed in rapid succession creates the illusion of movement.

aperture See *iris.*

artifacts Unwanted visual distortions that appear in a video image, such as cross-color artifacts, cross-luminance artifacts, jitter, blocking, ghosts, etc.

artificial light human-made illumination not limited to "indoor" variety fluorescent bulbs, jack-o'-lanterns, a car's headlights, all qualify. Typically, has lower color temperature than natural light, and thus more reddish qualities.

aspect ratio Proportional width and height of picture on screen. Current standard for conventional receiver or monitor is four by three 43; 169 for HDTV.

assemble edit Recording video and/or audio clips in sequence immediately following previous material; does not break control track. Consecutive edits form complete program. [See *edit, insert edit.*]

ATV (amateur television) Specialized domain of ham radio, transmits standard TV signals on UHF radio bands.

audio dub Result of recording over pre-recorded videotape soundtrack, or a portion thereof, without affecting pre-recorded images.

audio frequency modulation (AFM) Method of recording hi-fi audio on videotape along with video signals.

audio mixer Device with user-adjustable controls used to blend multiple sound inputs into desired composite output. [See *mix.*]

automatic exposure Circuitry that monitors light levels and adjusts camcorder iris accordingly, compensating for changing light conditions.

automatic gain control (AGC) Camcorder circuitry that adjusts incoming audio levels automatically.

available light Amount of illumination normally present in a particular environment natural light, artificial, or a combination.

back light Illumination from behind, creates a sense of depth by separating foreground subject from background area. Applied erroneously, causes severe silhouetting. [See *fill light, key light, three-point lighting.*]

barndoors Accessory for video lights, two- or four-leaf folding flaps that control light distribution.

batch capture The ability of certain computer-based editing systems to automatically capture whole lists or "batches" of clips from source videotapes.

Betamax More commonly known as "Beta," half-inch videotape format developed by Sony, eclipsed by VHS in home video market popularity. [See *ED Beta.*]

bidirectional Microphone pickup pattern whereby sound is absorbed equally from two sides only. [See *omnidirectional, unidirectional.*]

black box Generic for wide variety of video image manipulation devices with perceived mysterious or "magical" capabilities, including proc amps, enhancers, SEGs, and TBCs.

bleeding Video image imperfection characterized by blurring of color borders; colors spill over defined boundaries, "run" into neighboring areas.

BNC (bayonet fitting connector) Durable "professional" cable connector, attaches to VCRs for transfer of high-frequency composite video in/out signals. Connects with a push and a twist.

boom Extension arm used to suspend a microphone or camera over sound or subject(s) being recorded. Objective is to keep production gear out of camera's view or to provide a unique angle to shoot video from.

booming Camera move above or below subject with aid of a balanced "boom arm," creating sense of floating into or out of a scene. Can combine effects of panning, tilting, and pedding in one fluid movement.

C See *chrominance.*

cable/community access Channel(s) of a local cable television system dedicated to community-based programming. Access centers provide free or low-cost training and use of video production equipment and facilities.

cameo lighting Foreground subjects illuminated by highly directional light, appearing before a completely black background.

Cannon See *XLR.*

capture card A piece of computer hardware that captures digital video and audio to a hard drive, typically through a FireWire (IEEE 1394) port.

cardioid The most common type of unidirectional microphone; pickup pattern resembles a heart-shaped figure.

CCD (Charge Coupled Device) Light-sensitive computer chip in video cameras that converts images into electrical flows. Less prone to image irregularities—burn-in, lag, streaking—than are older image sensors. [See *pickup.*]

character generator A device that electronically builds text which can be combined with a video signal. The text is created with a keyboard and program that has a selection of font and backgrounds.

chroma Characteristics of color a videotape absorbs with recorded signal, divided into two categories AM (amplitude modulation) indicates color intensity; PM (phase modulation) indicates color purity.

chromakey Method of electronically inserting the image from one video source into the image of another through areas designated as its "key color." It is frequently used on news programs to display weather graphics behind talent.

chrominance Portion of video signal that carries color information (hue and saturation, but not brightness); frequently abbreviated as "C," as in "Y/C" for luminance/chrominance. [See *luminance.*]

clapstick Identification slate with hinged, striped top that smacks together for on-camera scene initiation. Originally used to synchronize movie sound with picture. [See *lip-sync.*]

closeup Tightly framed camera shot in which principal subject is viewed at close range, appearing relatively large and dominant on screen. Extent of view may be designated "medium closeup" or "extreme closeup." [See *long shot, medium shot.*]

color bars Standard test signal containing samples of primary and secondary colors, used as reference in aligning color video equipment. Generated electronically by a "color bar generator," often viewed on broadcast television in off-air hours. [See *test pattern.*]

color corrector Electronic device that dissects the colors of a video signal, allowing them to be adjusted individually.

color temperature The tint of "white" light, measured in "degrees Kelvin." Light from an incandescent bulb typically measures 2,800K, while that from a quartz lamp measures 3,200K and that from noontime sunlight measures 5,600K. [See *artificial, natural light.*]

comet tailing Smear of light resulting from inability of camera's pickup to process bright objects—especially in darker settings. Object or camera in motion creates appearance of flying fireball. [See *lag.*]

component video Signal transmission system, resembling S-video concept, employed with professional videotape formats. Separates luminance and two chrominance channels to avoid quality loss from NTSC or PAL encoding.

composite video Single video signal combining luminance and chrominance signals through an encoding process, including image's separate RGB (red, green, blue) elements and sync information.

compositing Superimposing multiple layers of video. Each layer may move independently.

composition Visual makeup of a video picture, including such variables as balance, framing, field of view, texture—all aesthetic considerations. Combined qualities form image that's pleasing to view, and effectively communicates.

compression Reducing the digital data in a video frame, typically from nearly one megabyte to 300 kilobytes or less, by throwing away information the eye can't see. JPEG, Motion-JPEG, MPEG, DV, Indeo, Fractal and Wavelet are all compression schemes.

condenser Microphone with built-in amplifier, the type installed on camcorders. Also called capacitor or electret condenser, requires battery or external power source. [See *electret condenser.*]

continuity [1 visual] Logical succession of recorded or edited events, necessitating consistent placement of props, use of wardrobe, positioning of characters, and progression of time [2] directtional] Consistency in camera-subject relationships, to avoid confusing a viewer's perspective.

contrast Difference between a picture's brightest and darkest areas. When high, image contains sharp blacks and whites; when low, image limited to variations in gray tones.

control track A portion of the videotape containing information to synchronize playback and linear videotape editing operations.

Control-L A two-way communication system used to coordinate tape transport commands for linear editing. Primarily found in Mini DV, Digital8, Hi8 and 8mm camcorders and VCRs. [See *Control-S, synchro edit.*]

Control-S A one-way communication system that treats a VCR or camcorder as a slave unit, with edit commands emanating from an external edit controller or compatible deck. Primarily found on 8mm VCRs and camcorders. [See *Control-L, synchro edit.*]

cookie (cucalorus) Lighting accessory consisting of random pattern of cutouts that cast patterened shadows when light passes through. Used to imitate shadows of natural lighting.

crawl Text or graphics—usually special announcements or credits—that move across screen horizontally, typically from bottom right to left. Produced with a character generator.

cross-fade Simultaneous fade-in of one audio source or lighting effect as another fades out; may overlap temporarily. Also called a dissolve.

cucalorus (cookie) Lighting accessory consisting of random pattern of cutouts that forms shadows when light passes through it. Used to imitate shadows of natural lighting.

cue [1] Signal to begin, end, or otherwise influence on-camera activity while recording. [2] Presetting specific starting points of audio or video material so it's available for immediate and precise playback when required.

cut [1] Instantaneous change from one shot to another. [2] Director's command to immediately inate on-camera action and recording.

cutaway Shot of other than principal action (but peripherally related), frequently used as transitional footage or to avoid a jump cut.

cuts-only editing Editing limited to immediate shifts from one scene to another, without smoother image transition capabilities such as dissolving or wiping. [See *cut, edit.*]

D1, D2, D3, D5, Digital-S, DVCPRO, DVCAM, Digital Betacam Entirely digital "professional" videotape recording formats capable of multigeneration duplication without picture degradation.

decibel (dB) Measure of audio signal strength based on a logarithmic scale. Also the unit of measure for sound pressure level (loudness).

depth of field Range in front of a camera's lens in which objects appear in focus. Varies with subject-to-camera distance, focal length of camera lens and camera's aperture setting.

desktop video (DTV) Fusion of personal computers and home video components for elaborate videomaking capabilities rivaling those of broadcast facilities.

diffused light Illuminates relatively large area indistinctly; produces soft light quality with soft shadows. [See *directional light.*]

diffuser Gauzy or translucent material that alters the quality of light passing through it to soften shadows and produce less intense lighting.

diffusion filter Mounted at front of camcorder lens, gives videotaped images a foggy, fuzzy, dreamy look. [See *filter.*]

digital The digital signal is composed of bits (ones and zeros). Digital information can be manipulated by computers.

digital audio Sounds that have been converted to digital information.

digital video effects (DVE) Electronic picture modification yielding specialty image patterns and maneuvers tumbling, strobing, page turning, mosaic, posterization, solarization, etc.

digitization The process of converting a continuous analog video or audio signal to digital data for computer storage and manipulation.

digitizer Device that imports and converts analog video images into digital information for hard drive-based editing.

directional light Light that illuminates in a relatively small area with distinct light beam; usually created with spotlight, yields harsh, defined shadows. [See *diffused light.*]

dissolve Image transition effect of one picture gradually disappearing as another appears. Analogous to audio and lighting cross-fade. [See *cross-fade.*]

distribution amp (distribution amplifier) Divides single video or audio signals, while boosting their strength, for delivery to multiple audio/video acceptors. Allows simultaneous recording on multiple VCR's from the same source, especially useful for tape duplication.

dolly Camera movement toward or away from a subject. Effect may appear same as zooming, which reduces and magnifies the image, but dollying in or out maintains perspective while changing picture size.

dollying Camera movement toward or away from a subject. Effect may appear same as zooming, which reduces and magnifies the image, but dollying in or out maintains perspective while changing picture size.

dropout Videotape signal voids, viewed as fleeting white specks or streaks. Usually result of minute "bare spots" on a tape's magnetic particle coating, or tape debris covering particles and blocking signals.

DTV Desktop video.

dub [1] Process or result of duplicating a videotape in its entirety. [2] Editing technique whereby new audio or video replaces portion(s) of existing recording.

DV effects **(DVE)** Electronic analog-to-digital picture modification yielding

specialty image patterns and maneuvers tumbling, strobing, page turning, mosaic, posterization, solarization, etc.

dynamic Microphone type, also called "moving coil." Works much like a loudspeaker in reverse, employing a simple diaphragm, magnet and wire coil to convert sound waves into an electrical signal.

ED Beta (extended Beta) Improved version of the original half-inch Betamax video format, yielding sharper pictures with 500-line resolution. [See *Betamax.*]

edit Process or result of selectively recording video and/or audio on finished videotape. Typically involves reviewing raw footage and transferring desired segments from master tape(s) onto new tape in a predeined sequence. [See *assemble edit, in-camera editing, insert edit.*]

edit control protocols protocols Types of signals designed to communicate between editing components including computers, tape decks and camcorders. Allows components to transmit instructions for various operations such as play, stop, fast forward, rewind, etc.

edit controller Electronic programmer used in conjunction with VCRs/camcorders to facilitate automated linear videotape editing with speed, precision and convenience.

edit decision list **(EDL)** Handwritten or computer-generated compilation of all edits (marked by their time code in points and out points) to be executed in a video production.

edited master Original recorded videotape footage; "edited master" implies original copy of tape in its edited form. Duplications constitute generational differences.

EDL **(EDL)** Handwritten or computer-generated compilation of all post-production edits to be executed in a video work.

EFP **(Electronic field production)** Film-style production approach using a single camera to record on location.

Typically shot for post-production application, non-live feed.

electret condenser Microphone type incorporating a precharged element, eliminating need for bulky power sources. [See *condenser.*]

encoder Device that translates a video signal into a different format—RGB to composite, DV to MPEG, etc.

ENG (Electronic news gathering) Use of portable video cameras, lighting and sound equipment to record news events in the field quickly, conveniently, and efficiently.

enhancer (Image enhancer) Video signal processor that compensates for picture detail losses and distortion occurring in recording and playback. Exaggerates transitions between light and dark areas by enhancing high frequency region of video spectrum.

EP (Extended play) Slowest tape speed of a VHS VCR, accommodating six-hour recordings. [See *LP, SP.*]

equalization Emphasizing specific audio or video frequencies and eliminating others as signal control measure, usually to produce particular sonic qualities. Achieved with equalizer.

essential area Boundaries within which contents of a television picture are sure to be seen, regardless of masking differences in receiver displays. Also called the "critical area," and "action safe area," it encompasses the inner 80 percent of the screen.

establishing shot Opening picture of a program or scene. Usually a wide and/or distant perspective, orients viewer to overall setting and surroundings.

extra Accessory talent not essential to a production, assuming some peripheral on-camera role. In movie work, performers with fewer than five lines are called "under fives."

f-stop Numbers corresponding to variable size of camera's iris opening, and thus amount of light passing through lens. The higher the number, the smaller the iris diameter thus less light enters.

F/X Special effects. Visual tricks and illusions—electronic or on camera—employed in film and video to define, distort or defy reality.

fade Gradual diminishing or heightening of visual and/or audio intensity. "Fade out" or "fade to black," "fade in" or "up from black" are common terms.

feed Act or result of transmitting a video signal from one point to another.

feedback [1video] Infinite loop of visual patterns from signal output being fed back as input; achieved by aiming live camera at receiving monitor. [2audio] Echo effect at low levels, howl or piercing squeal at extremes, from audio signal being fed back to itself; achieved by aiming live microphone at receiving speaker.

field Half a scanning cycle. Two fields comprise a complete video frame. Composed of either all odd lines or all even lines.

field of view Extent of a shot that's visible through a particular lens; its vista.

fill light Supplementary illumination, usually from a floodlight positioned midway between camera and subject, which lightens or eliminates shadows created by key light. [See *three-point lighting.*]

film-style Out-of-sequence shooting approach, to be edited in appropriate order at post-production stage. Advantageous for concentrating on and completing recording at one location at a time, continuity and convenience assured.

filter Transparent or semi-transparent material, typically glass, mounted at front of camcorder lens to change light passing through. Manipulates colors and image patterns, often for special effect purposes.

FireWire (IEEE 1394 or i.LINK) A high-speed bus that was developed by Apple Computer. It is used, among other things, to connect digital camcorders to computers.

flare flashes and evident in picture, caused by excessive light beaming into a camera's lens and reflecting off its internal glass elements.

flat lighting Illumination characterized by even, diffused light without shadows, highlights or contrast. May impede viewer's sense of depth, dimension.

floodlight Radiates a diffused, scattered blanket of light with soft, indistinct shadows. Best used to spread illumination on broad areas, whereas spotlights focus on individual subjects.

fluid head Tripod mount type containing viscous fluid which lubricates moving parts, dampens friction. Design facilitates smooth camera moves, alleviates jerkiness. [See *friction head.*]

flying erase head Accessory video head mounted on spinning headwheel, incorporated in newer camcorders and VCRs to eliminate glitches and rainbow noise between scenes recorded or edited.

focal length Distance from a camera's lens to a focused image with the lens focused on infinity. Short focal lengths offer a broad field of view (wide-angle); long focal lengths offer a narrow field of view (telephoto). Zoom lenses have a variable focal length.

follow focus Controlling lens focus so that an image maintains sharpness and clarity despite camera and/or subject movement.

format Videotape and video equipment design differences—physical and technical—dictating compatibility and quality. In most basic sense, refers to standardized tape widths, videocassette sizes. [See *Betamax, D1/D2, 8mm, three-quarter-inch, VHS.*]

frame 1)One complete image. In NTSC video a frame is composed of two fields. One 30th of a second. 2) The viewable area or composition of an image.

frame rate Number of video frames per second. NTSC is 30 frames per second. On the Internet, frame rate is dependent upon the bandwidth available and the multimedia format from which the video file is produced.

framing Act of composing a shot in the camcorder's viewfinder for desired content, angle, and field of view—overall composition.

freeze frame Single frame paused and displayed for an extended period during video playback; suspended motion perceived as still snapshot.

frequency Number of vibrations produced by a signal or sound, usually expressed as cycles per second, or hertz (Hz).

frequency response Measure of the range of frequencies a medium can respond to and reproduce. Good video response maintains picture detail; good audio response accommodates the broadest range, most exacting sound.

friction head Tripod mount type with strong spring that counterbalances camera weight, relying on friction to hold its position. More appropriate for still photography than movement-oriented videomaking. [See *fluid head.*]

full-motion video A standard for video playback on a computer; refers to smooth-flowing, full-color video at 30 frames per second, regardless of the screen resolution.

gaffer Production crew technician responsible for placement and rigging of all lighting instruments.

gain Video amplification, signal strength. "Riding gain" means varying controls to achieve desired contrast levels.

gel Colored material placed in front of a light source to alter its hue or intensity. Useful for effect, or for correcting mismatches in lighting, as in scenes lit by both daylight and artificial light.

generation Relationship between a master video recording and a given copy of that master. A copy of a copy of the original master constitutes a second-generation duplication.

generation loss Degradation in picture and sound quality resulting from an analog duplication of original master video recording. Copying a copy and all successive duplication compounds generation loss. Digital transfers are free of generation loss.

genlock (generator locking device) Synchronizes two video sources, allowing part or all of their signals to be displayed together. Necessary for overlaying computer graphics with video, for example.

ghosting Undesirable faint double screen image caused by signal reflection or improperly balanced video circuitry. "Ringing" appears as repeated image edges.

glitch Momentary picture disturbance.

grain Blanketed signal noise viewed as fuzziness, unsmooth images— attributable to lumination inadequacies.

grip Production crew stagehand responsible for handling equipment, props, and scenery before, during, and after production.

hard disk Common digital storage component in a computer. For video use, hard disks need 1) an access time of less than 10 milliseconds; 2) a sustained throughput (data transfer rate) of 3 Megabytes per second; and 3) a maximum housekeeping of 33 milliseconds (one video frame).

HDTV (high- television) "In the works" television system standard affording greater resolution for sharper pictures and wide-screen viewing via specially-designed TV equipment.

head Electromagnetic components within camcorders and VCRs that record, receive and erase video and audio signals on magnetic tape.

headroom Space between the top of a subject's head and a monitor's upper screen edge. A composition consideration.

hi-fi (high fidelity) Generalized defining audio quality approaching the limits of human hearing, pertinent to high-quality sound reproduction systems.

Hi8 (high-band 8mm) Improved version of 8mm videotape format characterized by higher luminance resolution for a sharper picture. Compact "conceptual equivalent" of Super-VHS. [See *8mm.*]

hiss Primary background signal interference in audio recording, result of circuit noise from a playback recorder's amplifiers or from a tape's residual magnetism.

horizontal resolution Specification denoting amount of discernable detail across a screen's width. Measured in lines, the higher the number the better the picture quality.

image enhancer Video signal processor that compensates for picture detail losses and distortion occurring in recording and playback. Exaggerates transitions between light and dark areas by enhancing high frequency region of video spectrum.

image sensor A video camera's image sensing element, either CCD (charge coupled device) or MOS (metal oxide semiconductor); converts light to electrical energy. [See *CCD.*]

in-camera editing Assembling finished program "on the fly" as you videotape simply by activating and pausing camcorder's record function.

incident light That which emanates directly from a light source, measured from the object it strikes to the source. [See *reflected light.*]

indexing Ability of some VCRs to electronically mark specific points on videotape for future access, either during the recording process (VISS VHS index search system) or as scenes are played back (VASS VHS address search system).

insert edit Recording video and/or audio on tape over a portion of existing footage without disturbing what precedes and follows. Must replace recording of same length. [See *edit, assemble edit.*]

Interlace The concept of splitting a complete frame of video into two fields of odd and even lines. Under the interlaced method, every other line is scanned during the first pass, then the remaining lines are scanned during the second pass in order to complete the frame.

interlaced video Process of scanning frames in two passes, each painting every other line on the screen, with scan lines alternately displayed in even and odd fields. NTSC video is interlaced; most computers produce a noninterlaced video signal. [See *noninterlaced video.*]

iris Camcorder's lens opening or aperture, regulates amount of light entering camera. Diameter is measured in f-stops. [See *f-stop.*]

jack Any female socket or receptacle, usually on the backside of video and audio equipment; accepts plug for circuit connection.

jitter Video image aberration seen as slight, fast vertical or horizontal shifting of a picture or portion of one.

jog/shuttle Manual control on some VCRs, facilitates viewing and editing precision and convenience. Jog ring moves tape short distances to show a frame at a time; shuttle dial transports tape forward or reverse more rapidly for faster scanning.

jump cut Unnatural, abrupt switch to/from shots identical in subject but slightly different in screen location. Awkward progression makes subject appear to jump from one screen location to another. It can be remedied with cutaway or shot from a different angle.

Kelvin Temperature scale used to define the color of a light source; abbreviated as "K." [See *color temperature.*]

key light Principal illumination source on a subject or scene, normally positioned slightly off center and angled to provide shadow detail. [See *back light, fill light, three-point lighting.*]

keystoning Perspective distortion from a flat object being shot by a camera at other than a perpendicular angle. Nearer portion of object appears larger than farther part.

lag Camera pickup's retention of an image after the camera has been moved, most common under low light levels. Comet tailing is a form of lag.

lavaliere Small, easily concealed, unobtrusive and aesthetically pleasing microphone, typically attached to clothing for interview settings.

linear editing Tape-based editing. Called linear because scenes are laid in a line along the tape.

lip sync Proper synchronization of video with audio—lip movement with audible speech. [See *synchronous sound.*]

long shot Camera view of a subject or scene, usually from a distance, showing a broad perspective.

LP (long play) Middle tape speed of a VHS VCR, accommodating four-hour recordings. [See *EP, SP.*]

LTC (longitudinal time code) Frame identification numbers encoded as an audio signal and recorded lengthwise on the edge of a tape. [See *time code, VITC.*]

luminance Black-and-white portion of video signal, carries brightness information representing picture contrast, light and dark qualities; frequently abbreviated as "Y."

lux Amount of lumens in a square meter. Means of measuring a camcorder's low-light sensitivity—minimum amount of illumination required to record an "acceptable" image. Lower the lux reading the greater the sensitivity.

macro Lens capable of extreme closeup focusing, useful for intimate views of small subjects.

master Original recorded videotape footage; "edited master" implies original tape in its edited form.

matched dissolve Dissolve from one image to another that's similar in appearance or shot size.

medium shot Defines any camera perspective between long shot and closeup, whereby subjects are viewed from medium distance. [See *closeup, long shot.*]

memory effect Power-loss phenomenon alleged of NiCad—camcorder batteries, attributed to precisely repetitive partial discharge followed by complete recharge, or long- overcharge. Considered misnomer for "voltage depression" and "cell imbalance."

MIDI (musical instrument digital interface) System of communication between digital electronic instruments allowing synchronization and distribution of musical information.

mike "Mic," short for "microphone."

mix [1audio] Combining two or more sound sources, with various channels controlled to achieve desired balance of single audio signal output. Executed with audio mixer. [2video] Combining video signals from two or more sources.

model release Agreement to be signed by anyone appearing in a video work, protecting videomaker from right of privacy lawsuit. Specifies event, date, compensation provisions, and rights being waived.

monitor [1video] Television set without receiving circuitry, wired to camcorder or VCR for display of live or recorded video signals. Most standard TVs have dual-function capability as monitor and receiver. [See receiver.] [2audio] Synonymous with speaker.

monopod One-legged camera support. [See *tripod.*]

montage A sequence of shots assembled in juxtaposition to each other to communicate a particular idea or mood. The implied relationship between seemingly unrelated material communicates messages.

mosaic Electronic special effect whereby individual pixels comprising an image are blown up into larger blocks—a kind of checkerboard effect. [See *DVE.*]

MPEG **(MPEG-1)** A video compression standard set by the Motion Picture Experts Group. Involves changing only those elements of a video image that actually change from frame to frame and leaving everything else the same.

MPEG-2 The highest quality digital video compression currently available. MPEG-2 is less blocky than MPEG-1 and is used in DVDs and DSS satellite TV systems.

natural light Planetary illumination—from sun, moon, stars—be it indoors or out. Has higher color temperature than artificial light, and thus more bluish qualities. [See *artificial light, color temperature.*]

neutral-density filter **(ND)** Mounted at front of camcorder lens, reduces light intensity without affecting its color qualities. [See *filter.*]

NiCad **(nickel cadmium)** Abbreviation coined and popularized by SAFT America for lightweight camcorder battery type designed to maintain power longer than traditional lead-acid batteries.

noise Undesirable video or audio signal interference; typically seen as snow, heard as hiss.

noninterlaced video Process of scanning complete frames in one pass, painting every line on the screen, yielding higher picture quality than that of interlaced video. Most computers produce a noninterlaced video signal; NTSC is interlaced. AKA progressive scan.

nonlinear editing Digital "random access" editing that uses a hard drive instead of tape to store images. Random access allows easy arrangement of scenes in any order. Also eliminates the need for rewinding and allows for multiple dubs without generation loss.

nonsynchronous sound Audio without precisely matching visuals. Usually recorded separately, includes wild sound, sound effects, or music incorporated in post-production. [See *synchronous sound.*]

NTSC U.S. television broadcasting specifications. NTSC refers to all video systems conforming to this 525-line 59.94-field-per-second signal standard. [See *PAL.*]

Off-line Describing older non-linear editing equipment not capable of master-quality output. These systems were most useful for creating edit decision lists (EDLs), which were later used in high-end edit suites to actually create the edit master tapes.

omnidirectional Microphone pickup pattern whereby sound is absorbed eqally from all directions. [See *bidirectional, unidirectional.*]

outtake Footage not to be included in final production.

over-the-shoulder shot View of primary camera subject framed by another subject's shoulder and back of head in foreground. Common in interview situations, perspective affords sense of depth.

PAL (phase alternate line) 625-line 50-field-per-second television signal standard used in Europe and South America. Incompatible with NTSC. [See *NTSC.*]

pan Horizontal camera pivot, right to left or left to right, from a stationary position.

pedding Vertical camera movement, rising or lowering, with camera levelness maintained. Approaching closer to either floor or ceiling, the up/down equivalent of dollying.

phone plug Sturdy male connector compatible with audio accessories, particularly for insertion of microphone and headphone cables. Not to be confused with phono plug.

phono plug Also called "RCA" or "RCA phono," popular cable connector for home audio as well as video components. Standard connection for direct audio/video inputs/outputs. Not to be confused with phone plug.

pickup [1] A video camera's image sensing element, either CCD (charge coupled device) or MOS (metal oxide semiconductor); converts light to electrical energy. [See *CCD.*] [2] A microphone's sound reception.

pickup pattern Defines a microphone's response to sounds arriving from various directions or angles. [See *bidirectional, omnidirectional, unidirectional.*]

PiP (picture in picture, p-in-p, pix in pix) Image from a second video source inset on a screen's main picture, the big and small pictures usually being interchangeable.

playback Videotaped material viewed and heard as recorded, facilitated by camcorder or VCR.

playback VCR Playback source of raw video footage (master or workprint) in basic player/recorder editing setup. [See *recording VCR.*]

point of view shot (POV) Shot perspective whereby the camera assumes subject's view, and thus viewers see what the subject sees as if through his/her/its eyes.

polarizing filter Mounted at front of camcorder lens, thwarts undesirable glare and reflections. [See *filter.*]

post production (post) Any video production activity following initial recording. Typically involves editing, addition of background music, voiceover, sound effects, titles, and/or various electronic visual effects. Results in completed production.

posterization Electronic special effect transforming a normal video image into a collage of flattened single-colored areas, without graduations of color and brightness. [See *DVE.*]

POV See *point of view.*

pre-roll [1] Slight backing-up function of camcorders and VCRs when preparing for linear tape-to-tape editing; ensures smooth, uninterrupted transitions between scenes.

proc amp (processing amplifier) Video image processor that boosts video signal's luminance, chroma, and sync components to correct such problems as low light, weak color, or wrong tint.

Progressive scan A method of displaying the horizontal video lines in computer displays and digital TV broadcasts. Each horizontal line is displayed in sequence (1, 2, 3, etc.) until the screen is filled; as opposed to interlaced (first fields of odd-numbered lines, then fields of even-numbered lines).

props Short for "properties," objects used either in decorating a set (set props) or by talent (hand props).

PZM (pressure zone microphone) Small, sensitive condenser mike, usually attached to a 5-inch-square metal backing plate. Senses air pressure changes in tiny gap between mike element and plate. [See *condenser.*]

QuickTime Computer system software that defines a format for video and audio data, so different applications can open and play synchronized sound and movie files.

rack focus   Shifting focus during a shot in progress, typically between background and foreground subjects, creating respective clarity and blurriness.

raw footage   Pre-edited recordings, usually direct from the camcorder.

RCA plug   A popular cable connector for home audio as well as video components. Standard connection for direct audio/video inputs/outputs.

reaction shot   Cutaway view showing someone's or something's response to primary action/subject. [See *cutaway.*]

real time   Occurring immediately, without delay for rendering. If a transition occurs in real time, there is no waiting, the computer creates the effect or transition on-the-fly, showing it on demand.

real-time counter   Tallying device that accounts for videotape playing/recording by measure of hours, minutes, and seconds.

receiver   Television set that includes a tuner as well as an audio amplifier and speaker. Accommodates broadcast RF signals, whereas a monitor accepts composite video signals only. [See *monitor.*]

recording VCR   Recipient of raw video feed (master or workprint) and recorder of edited videotape in basic player/recorder editing setup. [See *playback VCR.*]

reflected light   That which bounces off the illuminated subject. Light redirected by a reflector. [See *incident light.*]

reflector   Lighting accessory helpful for bouncing light onto a subject as well as filling in shadows. Often made of lightweight reflective metal or poster board covered with metallic material.

remote   [1] Videomaking performed "on location," outside controlled studio environment. [2] Equipment allowing from-a-distance control, usually without physical connections.

render   The processing a computer undertakes when creating an applied effect, transition or composite.

rendering time   The time it takes an NLE computer to composite source elements and commands in it's edit decision list into a single video file so the sequence, including titles and transition effects, can be played in full motion (30 frames per second).

resolution   Amount of picture detail reproduced by a video system, influenced by a camera's pickup, lens, internal optics, recording medium, and playback monitor. The more detail, the sharper and better defined the picture. [See *horizontal resolution.*]

Rewritable Consumer (RC)   Time code sent trhoug Control-L interface permitting extremely accurate edits. Each frame is assigned a unique address expressed in hoursminutessecondsframes.

RF (radio frequency)   Combination of audio and video signals coded as a channel number, necessary for television broadcasts as well as some closed-circuit distribution.

RF converter   Device that converts audio and video signals into a combined RF signal suitable for reception by a standard TV.

RGB (red, green, blue)   Video signal transmission system that differentiates and processes all color information in separate red, green, and blue components—primary colors of light—for optimum image quality. Also defines type of color monitor.

ringing   Undesirable faint double screen image caused by signal reflection or improperly balanced video circuitry. "Ringing" appears as repeated image edges.

roll   Text or graphics—usually credits—that move up or down the screen, typically from bottom to top. Typically produced with a character generator or computer. [See *crawl.*]

rough cut   Raw, tentative edit of footage in the approximate sequence, length, and content of finished program. Gives preliminary indication of eventual actual work.

rule of thirds   Composition theory based on the screen being divided into thirds vertically and horizontally and the placement of important elements along those lines.

S-video   Also known as Y/C video, signal type employed with Hi8 and S-VHS video formats. Transmits chrominance and luminance portions separately via multiple wires, thereby avoiding the NTSC encoding process and its inevitable picture quality degradation.

S/N Ratio   See *signal-to-noise ratio.*

safe title area   The area that will produce legible titles on most TV screens; 80% of the screen, measured from the center.

scan converter   Device that changes scan rate of a video signal, possibly converting it from noninterlaced to interlaced mode. Allows computer graphics to be displayed on a standard video screen, for example.

scan line   Result of television's swift scanning process which sweeps out a series of horizontal lines from left to right, then down a bit and left to right again. Complete NTSC picture consists of 525 scan lines per frame.

scan rate   Number of times a screen is "redrawn" per second. Computer displays operate at different scan rates than standard video.

scene   In the language of moving images, a sequence of related shots usually constituting action in one particular location. [See *shot.*]

scrim   Lighting accessory made of wire mesh, lessens intensity of light source without softening it. Half scrims and graduated scrims reduce illumination in more specific areas.

script   Text specifying content of a production or performance, used as a guide. May include character and setting profiles, production directives (audio, lighting, scenery, camera moves), as well as dialogue to be recited by talent. [See *storyboard.*]

SECAM   (sequential color and memory) 625-line 25-frame-per-second television signal standard used in France and the Soviet Republic. Incompatible with NTSC; PAL and SECAM are partially compatible. [See *NTSC, PAL.*]

SEG   See *special effects generator.*

selective focus   Adjusting camera focus to emphasize desired subject(s) in a shot. Selected area maintains clarity, image sharpness while remainder of image blurs. Useful for directing viewer's attention.

sepia   Brassy "antique" look characteristic of old photographs. For video images, tone achieved with a special lens filter or electronically with a special effects filter.

shooting ratio   Amount of raw footage recorded relative to the amount used in edited, finished program.

shot   All pictorial material recorded by a camera. More strictly speaking, shots are intentional, isolated camera views which collectively comprise a scene. [See *scene.*]

shotgun   Highly directional microphone with long "barrel" designed to pick up sound from extreme subject-to-mike distances.

signal-to-noise ratio   (S/N) Relationship between signal strength and a medium's inherent noise. Video S/N indicates how grainy or snowy a picture will be, plus color accuracy; audio S/N specifies amount of background tape hiss present with low- or no-volume recordings. Higher the S/N the cleaner the playback.

Skylight   (1A) or haze (UV) filter Mounted at front of camcorder lens, virtually clear glass absorbs ultraviolet light. Also excellent as constant lens protector. [See *filter.*]

SMPTE   Time code standard for film, video, and audio named for the Society of Motion Picture and Television Engineers, which sanctions standards for recording systems in North America. [See *time code.*]

snoot   Open-ended cylindrical funnel mounted on a light source to project a narrow, concentrated circle of illumination.

snow   Electronic picture interference; resembles scattered snow on the television screen. Synonymous with chroma and luma noise.

solarization Electronic special effect distorting a video image's original colors, emphasizing some and de-emphasizing others for a "paint brush" effect. [See *DVE.*]

sound bite Any recorded video or audio-only segment salvageable for use in edited program—usually a highlight phrase or event. Common component of broadcast news.

sound effects Contrived audio, usually prerecorded, incorporated with a video soundtrack to resemble the real thing. Blowing on a microphone, for example, might simulate wind to accompany hurricane images.

soundtrack The audio portion of a video recording, often multifaceted with voiceover, background music, sound effects, etc.

SP (standard play) Fastest tape speed of a VHS VCR, accommodating two-hour recordings. [See *EP, LP.*]

special effects (FX) Tricks and illusions—electronic or on camera—employed in film and video to define, distort, or defy reality.

special effects F/X. Tricks and illusions—electronic or on camera—employed in film and video to define, distort, or defy reality.

special effects generator (SEG) Video signal processor with vast, but varying, image manipulation capabilities involving patterns and placement as well as color and texture mixing, multiplying, shrinking, strobing, wiping, dissolving, flipping, colorizing, etc.

spotlight Radiates a well-defined directional beam of light, casting hard, distinct shadows. Best used to focus illumination on individual subjects, whereas floodlights blanket broader areas.

stabilizer Video signal processor used primarily for tape dubbing to eliminate picture jump and jitter, maintain stability.

star Filter Mounted at front of camcorder lens, gives videotaped light sources a starburst effect. Generally available in four-, six-, and eight-point patterns. [See *filter.*]

stereo Sound emanating from two isolated sources, intended to simulate pattern of natural human hearing.

stock shot Common footage—city traffic, a rainbow—conveniently accessed as needed. Similar to a "photo file" in the photography profession.

storyboard Series of cartoon-like sketches illustrating key visual stages (shots, scenes) of planned production, accompanied by corresponding audio information. [See *script.*]

Streaming Playing sound or video in real time as it is downloaded over the Internet as opposed to storing it in a local file first. Avoids download delay.

strobe Digital variation of fixed-speed slow motion, with image action broken down into a series of still frames updated and replaced by new ones at rapid speed. [See *DVE.*]

Super-VHS (S-VHS, S-VHS-C) Improved version of VHS and VHS-C videotape formats, characterized by separate carriers of chrominance and luminance information, yielding a sharper picture. [See *VHS, VHS-C.*]

superimposition (super) Non-inherent titles or graphics appearing over an existing video picture, partially or completely hiding areas they cover.

sweetening Post-production process of adding music and sound effects to or otherwise enhancing, purifying, massaging a final audio track.

swish Extremely rapid camera movement from left to right or right to left, appearing as image blur. Two such pans in the same direction—one moving from, the other moving to a stationary shot—edited together can effectively convey passage of time.

switcher Simplified SEG, permits video signal mixing from two or more sources—cameras, time base correctors, character generators—for dissolves, wipes, and other clean transition effects.

sync (synchronization) Horizontal and vertical timing signals or electronic pulses—component of composite signal, supplied separately in RGB systems. Aligns video origination (live camera, videotape) and reproduction (monitor or receiver) sources.

synchronous sound Audio recorded with images. When the mouth moves, the words come out.

talent Generic for the people or creatures assuming primary on-screen roles in a videotaping.

tally light Automatic indicators on camera front and within viewfinder that signal recording in progress—seen by both camera subject(s) and operator.

TBC Electronic device that corrects timing inconsistencies in a videotape recorder's playback, stabilizing the image for optimum quality. Also synchronizes video sources, allowing image mixing. [See *sync.*]

telecine converter Imaging device used in conjunction with a movie projector and camcorder to transfer film images to videotape.

telephoto Camera lens with long focal length, narrow horizontal field of view. Opposite of wide-angle, captures magnified, closeup images from considerable distance.

teleprompter (prompter) Mechanical device that projects and advances text on mirror directly in front of camera's lens, allowing talent to read their lines while appearing to maintain eye contact with viewers.

test pattern Any of various combinations of converging lines, alignment marks, and gray scales appearing on screen to aid in video equipment adjustment for picture alignment, registration, and contrast. Often viewed on broadcast television in off-air hours. [See *color bars.*]

Thin Streaming Low bandwidth delivery of streaming video, audio or animated content.

three-point lighting Basic lighting approach employing key, back, and fill lights to illuminate subject with sense of depth and texture.

three-quarter-inch (U-matic) Most popular professional/industrial video format employing larger cassettes and three-quarter-inch tape, as opposed to the half-inch width of VHS and Beta "consumer" formats. Related equipment is generally larger and sturdier, format's recording considered superior to any consumer video format except Mini DV.

three-shot Camera view including three subjects, generally applicable to interview situations.

tilt Camera move in a vertical direction, down or up, from a stationary position. Follows movement, contrasts differences in size between two subjects, or gives viewer point-of-view sense of a subject's height.

time base corrector (TBC) Electronic device that corrects timing inconsistencies in a videotape recorder's playback, stabilizing the image for optimum quality. Also synchronizes video sources, allowing image mixing. [See *sync.*]

time code Synchronization system, like a clock recorded on your videotape, assigning a corresponding hours, minutes, seconds, and frame-number designation to each frame. Expedites indexing convenience and editing precision. [See *SMPTE.*]

time-lapse recording Periodically videotaping a minimal number of frames over long durations of actual time. Upon playback, slow processes such as a flower blooming may be viewed in rapid motion.

timeline editing A computer-based method of editing in which video and audio clips are represented on a computer screen by bars proportional to the length of the clip.

titling Process or result of incorporating on-screen text as credits, captions, or any other alphanumeric communication to video viewers. [See *character generator.*]

tracking [1] Lateral camera movement aligned with moving subject; background appears to move. Camera should maintain regulated distance from subject. [2] Positioning of video and/or audio heads over a videotape's recorded signals. [See *head.*]

tripod Three-legged camera mount offering stability and camera placement/movement consistency. Most are lightweight, used for remote recording. [See *monopod.*]

tuner Television and VCR component that receives RF signals from an antenna or other RF sources and decodes into separate audio and video signals.

two-shot Camera view including two subjects, generally applicable to interview situations.

U-matic See *three-quarter-inch.*

umbrella Lighting accessory available in various sizes usually made of textured gold or silver fabric. Facilitates soft, shadowless illumination by reflecting light onto a scene.

unidirectional Highly selective microphone pickup pattern, rejects sound coming from behind while absorbing that from in front. [See *bidirectional, omnidirectional.*]

VCR Videocassette recorder. Multifunction machine intended primarily for recording and playback of videotape stored in cassettes.

vectorscope Electronic testing device that measures a video signal's chrominance performance, plotting qualities in a compass-like graphic display.

vertical interval time code (VITC) Synchronization signals recorded as an invisible component of the video signal, accessed for editing precision. [See *time code.*]

VHS (video home system) Predominant half-inch videotape format developed by Matsushita and licensed by JVC. [See *Super-VHS.*]

VHS-C (VHS compact) Scaled-down version of VHS using miniature cassettes compatible with full-size VHS equipment through use of adapter. [See *Super-VHS.*]

video card The PC card that controls the computer's monitor (usually with a form of VGA). Not to be confused with capture, overlay or compression cards, which handle NTSC video.

video prompter A mechanical device that projects and advances text on a mirror directly in front of a camera lens, allowing talent to read lines while appearing to maintain eye contact with viewers.

videocassette recorder (VCR) Multifunction machine intended primarily for recording and playback of videotape stored in cassettes.

vignette Visual special effect whereby viewers see images through a perceived keyhole, heart shape, diamond—whatever. In low-budget form, achieved by aiming camera through cutout of desired vignette.

vignetting Undesirable darkening at the corners of a picture, as if viewer's peering through a telescope, due to improper matching of lens to camera—pickup's scope exceeds lens size.

VITC (VITC) Synchronization signals recorded as an invisible component of the video signal, accessed for editing precision. [See *time code.*]

voiceover (VO) Narration accompanying picture, heard above background sound or music. Talking typically applied to edited visual during post-production.

waveform monitor Specialized oscilloscope testing device providing a graphic display of a video signal's strength. Plus, like a sophisticated light meter, aids in precise setting of picture's maximum brightness level for optimum contrast.

whip pan (swish pan) Extremely rapid camera movement from left to right or right to left, appearing as image blur. Two such pans in the same direction—one moving from, the other moving to a stationary shot—edited together can effectively convey passage of time.

white balance Electronic adjustment of camera to retain truest colors of recorded image. Activated in camcorder prior to recording, proper setting established by aiming at white object.

wide-angle Camera lens with short focal length and broad horizontal field of view. Opposite of telephoto, supports viewer perspective and tends to reinforce perception of depth.

wild sound Nonsynchronous audio recorded independent of picture ie. rain on roof, five o'clock whistle—often captured with separate audio recorder. [See *nonsynchronous sound.*]

windscreen Sponge-like microphone shield, thwarts undesirable noise from wind and rapid mike movement.

wipe Transition from one shot to another wherein the new shot is revealed by a moving line or pattern. In it's simplest form, it simulates a window shade being drawn.

wireless microphone Consisting of radio transmitter and receiver, utilizes low-power radio signal for cable-free operation.

workprint Copy of a master videotape used for edit planning and rough cut without excessively wearing or otherwise jeopardizing safekeeping of original material. Also called "working master."

wow and flutter Sound distortions consisting of a slow rise and fall of pitch, caused by speed variations in audio/video playback system.

XLR (Ground-Left-Right) Three-pin plug for three-conductor "balanced" audio cable, employed with high-quality microphones, mixers and other audio equipment.

Y Symbol for luminance, or brightness, portion of a video signal; the complete color video signal consists of R, G, B and Y.

Y/C Also known as S-video. Video signal type employed with Hi8 and S-VHS video formats. Transmits chrominance and luminance portions separately via multiple wires, thereby avoiding the NTSC encoding process and its inevitable picture quality degradation.

zoom Variance of focal length, bringing subject into and out of closeup range. Lens capability permits change from wide-angle to telephoto, or vice versa, in one continuous move. "Zoom in" and "zoom out" are common s.

zoom ratio The ratio of the shortest focal length to the longest in a zoom lens.

Index